METHODS FOR THE STUDY OF MARINE BENTHOS

Methods for the Study of Marine Benthos

EDITED BY

N. A. HOLME

AND

A. D. McINTYRE

SECOND EDITION

IBP HAND BOOK 16

BLACKWELL SCIENTIFIC PUBLICATIONS
OXFORD LONDON EDINBURGH
BOSTON PALO ALTO MELBOURNE

Editorial offices:
Osney Mead, Oxford, OX2 0EL
8 John Street, London, WC1N 2ES
9 Forrest Road, Edinburgh, EH1 2QH
52 Beacon Street, Boston
Massachusetts 02108, USA
706 Cowper Street, Palo Alto,
California 94301, USA
99 Barry Street, Carlton
Victoria 3053, Australia

First published 1971
Second edition 1984

Printed by
Galliard (Printers) Ltd
Great Yarmouth, Norfolk

DISTRIBUTORS
USA and Canada
Blackwell Scientific Publications Inc.,
P.O. Box 50009, Palo Alto,
California 94303

Australia
Blackwell Scientific Book Distributors,
31 Advantage Road, Highett,
Victoria 3190

British Library
Cataloguing in Publication Data
Methods for study of marine benthos.
—2nd ed.
1. Benthos—Research 2. Sampling
I. Holme, N.A. II. McIntyre, A.D.
574.92 QH91.8.B4

ISBN 0-632-00894-6

Contents

List of Contributors

JOHN B. BUCHANAN, *Dove Marine Laboratory, Cullercoats, North Shields, Tyne and Wear NE30 4PZ, England*

DENNIS J. CRISP, *Marine Science Laboratories, Menai Bridge, Gwynned LL59 5EH, North Wales*

ANASTASIUS ELEFTHERIOU, *Marine Laboratory, Victoria Road, Aberdeen AB9 8DB, Scotland*

J. MALCOLM ELLIOTT, *Freshwater Biological Association, The Ferry House, Far Sawrey, Ambleside, Cumbria LA22 0LP, England*

DEREK V. ELLIS, *Department of Biology, University of Victoria, British Columbia, Canada, V8W 2Y2*

JOHN C. GAMBLE, *Marine Laboratory, Victoria Road, Aberdeen AB9 8DB, Scotland*

MICHAEL HICKMAN, *Department of Botany, University of Alberta, Edmonton, Alberta, Canada, T69 2E9*

NORMAN A. HOLME, *Marine Biological Laboratory, Citadel Hill, Plymouth PL1 2PB, England*

ALASDAIR D. McINTYRE, *Marine Laboratory,Victoria Road, Aberdeen AB9 8DB, Scotland*

FRANK E. ROUND, *Department of Botany, University of Bristol, Woodland Road, Bristol BS8 1UG, England*

RICHARD M. WARWICK, *Institute for Marine Environmental Research, Prospect Place, The Hoe, Plymouth PL1 3DH, England*

PAUL F. WILLERTON, *Department of Marine Science, Plymouth Polytechnic, Drake Circus, Plymouth PL4 8AA, England*

Preface to the Second Edition

During the 12 years since the first edition of this Handbook was prepared for the International Biological Programme (IBP) there have been not only considerable advances in techniques for sampling, sorting and interpretation of data from programmes of benthic studies, but also a radical change in outlook, which is reflected in the objectives to which such programmes are directed. Except in a few instances (notably the exploration of hydrothermal vents in the deep sea), sampling programmes are now directed less towards exploration of new habitats, and more towards a fuller understanding of the interrelations between organisms, both within the benthic community and with the overlying waters. With this has been linked a need to study fluctuations in benthic communities under altered environmental conditions, leading to projects described as 'ecologial impact studies' or 'environmental monitoring programmes'.

This second edition attempts to take into account such changing needs and there has been some rearrangement of text to cover the considerable amount of material published in the past 12 years. In some instances our task has been lightened by the appearance of a work summarizing or reviewing advances in a particular field, but elsewhere we have shortened or eliminated sections which no longer appeared immediately relevant in order to make way for new material.

The objective of the Handbook remains broadly the same as outlined in the Preface to the first edition—that is, to be a general introduction to the subject and to indicate the range of possibilities, rather than to enter into precise details, which are obtainable in the literature cited.

We are aware that among some workers—notably those studying the Baltic Sea—significant advances have been made in the intercalibration of sampling, sorting and identification techniques. Nevertheless, we have refrained from laying down more than general guidelines as we are aware of the very wide range of conditions under which work is carried out worldwide in relation to sea state, depth of water, type of sediment, research ships and gear available, and conditions for processing samples and data, both ashore and afloat.

In this new edition some of the chapters (3, 8 and 9) have been revised by their original authors, others have involved the collaboration of additional contributors (1, 2, 6 and 7) and some (4 and 5) have been written afresh.

We would like to acknowledge the continuing support and advice of colleagues in our different laboratories in the preparation of this new edition. In particular the editors would like to thank A. Varley and Library staff at the MBA, G.W. Battin for redrawing some of the diagrams, R.L. Barrett for assistance with the editing and M. Rapson and S. Marriott for preparation of the typescript. The illustration for the cover was prepared by A. Rice and A. Eleftheriou. Acknowledgement of sources for diagrams and photographs is given at appropriate points in the text.

Norman A. Holme, *Plymouth*
Alasdair D. McIntyre, *Aberdeen*

Preface to the First Edition

This Handbook has been written to meet the needs of three kinds of worker: the newcomer to the field, the isolated worker without access to large libraries or the advice of colleagues, and those in related disciplines who may for some reason require to collect or study biological samples from the sea bed. Because of the very varied nature of the sea bed in different regions and because of the differing requirements of individual workers, we have from the first avoided laying down definite rules and procedures to be adopted. However, there is a right and a wrong way to set about things, and the Handbook had to contain more than a set of platitudes. The best general advice that can be given is for a preliminary survey to be made to see the range of possibilities for study. Before a large scale survey at sea is made, for example, some preliminary sampling, perhaps by dredge, should be made to show the types and size-range of the animals and plants to be studied, the nature of the deposits, and the topography of the sea bed. Only then will it be possible to decide what techniques to adopt in the main survey.

This is an International Handbook, but at the same time it is an English-language edition written by workers in one part of the world. We appreciate that we may not have been able to give adequate coverage to some techniques which may have been successfully adopted in other countries, and similarly the list of gear suppliers in the Appendix may seem to be somewhat restricted geographically. Such limitations may be remedied in a later edition, should this appear, and meanwhile it may be possible to have French and Spanish translations of this Handbook produced through the assistance of FAO.

Acknowledgements to those who have kindly given permission for reproduction of figures are given in the text. Our thanks are due to the staff of the Drawing Offices at Plymouth and Aberdeen for redrawing some of the figures, and to Mr A. Eleftheriou of Aberdeen who provided the cover drawing. Helpful discussions with colleagues in our different laboratories, and with other workers at the Arcachon meetings, have contributed much to this Handbook, and special mention must be made of the assistance given by Library staff at our Laboratories and at the National Institute of Oceanography, Wormley, Surrey, both in the preparation of this Handbook and of the Bibliography. We would also like to thank Mrs W.D.S. Kennedy and Mrs H. Readings for assistance in the editing of the Handbook, and Mr A. Varley for compiling the index.

Norman A. Holme, *Marine Biological Laboratory, Plymouth, England*

Alasdair D. McIntyre, *Marine Laboratory, Victoria Road, Torry, Aberdeen, Scotland*

July 1970

Chapter 1. Introduction: Design of Sampling Programmes

A. D. McINTYRE, J. M. ELLIOTT
AND D. V. ELLIS

Introduction

The object of this Handbook is to indicate and evaluate the equipment and techniques which are at present in general use for studying marine benthos, and to provide a guide to relevant publications. It is hoped that this volume will serve as an aid to those approaching the subject for the first time as well as to established workers, and that it will help to produce a degree of uniformity in the collection and treatment of material, and in the presentation of results which will make data from different laboratories more readily comparable.

Perhaps because of the inherent difficulties in benthos sampling, including the need for heavy gear and ship facilities, and the labour-intensive nature of sample processing, knowledge of the benthos has been slow to build up, in spite of the lead given by C.G.J. Petersen and his Danish colleagues in the early part of this century. While there have been no dramatic advances in techniques for collecting the infauna (except perhaps through recently developed suction devices) new impetus for the study of the epifauna of both sediments and rocky bottoms has been provided through observational techniques from SCUBA diving, use of manned submersibles, and underwater photography and television.

This Handbook deals primarily with the sampling and study of sediments and their biota, from the intertidal zone to the deep sea. The benthos comprises a variety of different types of organism. For zoobenthos, the division into macro- and meiofauna is a convenient way of separating size groups which, for the most part, require different sampling and processing techniques, the division usually being made between those retained on and those passing a 0·5 mm mesh. Macrofauna comprises the burrowing fauna (infauna) of sediments, and the surface living fauna (epifauna) of both rocky and sediment grounds, together with associated bottom-living fish and crustaceans. Meiofauna covers mainly the smaller metazoans, while the smallest forms—protozoans and organisms of bacterial size—are classified as microbenthos, and require special techniques not covered in this Handbook. The remaining benthic organisms—the phytobenthos—are primary producers, comprising macroalgae and unicellular bottom-living plants.

Various parts of the field covered by this Handbook have been dealt with in previous reviews. Those of Thorson (1957), Holme (1964), Hopkins (1964) and Elliott & Tullett (1978, 1983) as well as the publication of the International Commission for Scientific Exploration of the Mediterranean Sea (C.I.E.S.M.M., 1965) deal mainly with gear and techniques for macrofauna investigations, while Schlieper (1968) covers a wide range of methods for marine studies. A review by McIntyre (1969) refers briefly to problems of sampling the meiofauna, and the *Manual for the Study of Meiofauna* (Hulings & Gray, 1971) presents detailed procedures for each taxon. Recommendations for methods of studying benthos and other biota in the Baltic Sea are given in Dybern *et al.* (1976).

Information on the benthos is required in a wide range of contexts. The investigator making a qualitative study of a particular community or species, or carrying out a general survey, will usually be able to identify the optimum equipment and procedures for his purpose, and will often be particularly concerned with covering large areas of bottom and obtaining a representative collection of the species present. For quantitative work involving studies of production or food web dynamics, natural fluctuations, or the effects of pollution, more sophisticated programmes requiring a high degree of planning, must be designed.

An adequate consideration of the design of sampling programmes requires a book on its own (e.g. Cochran, 1963; Green, 1979); this chapter attempts only to comment on some of the relevant considerations, and to draw attention to the appropriate literature.

Planning the study

Ideally, the details of the operation, such as the optimum specifications of the ship, type of sampling gear, the nature of the processing and the sampling strategy should be determined by the questions that the study sets out to answer. However, it is more usual for the work to be carried out within the constraints imposed by the facilities which are available.

Preliminary considerations

The type of vessel available will often influence survey procedures to a large extent. A large oceanographic vessel must be operated formally, and established maritime routines will need to be observed, the timing and pattern of sampling being in accord with the ship's working arrangements and capabilities. It is thus important to consult the captain on the details of the programme or even to involve him at the planning stage, and to keep him informed of progress during the cruise. Good relationships and full communication between scientists and ship's staff make for a successful outcome. A smaller vessel can often be operated less formally but, even so, a plan of the operations should be prepared and discussed with the captain before starting out.

Open-sea or deep-water work imposes procedures substantially different from those required for near-shore surveys, chiefly with regard to the distance or time interval between sampling stations and the speed of sampling when on station. In deep-water surveys there can be long periods between operations because of the greater spacing between stations that may be required in open waters, and because of the time required to lower and raise instruments. It is, in general, desirable to use as large a sampler as possible for deep-water work so as to reduce the number of hauls and the time at each station. On larger ocean-going vessels a significant amount of the sorting and working-up may be done while still at sea, but for near-shore surveys, usually involving smaller vessels, it is probably more efficient to use the vessel only for collection of samples, these being processed ashore following rough sorting or sieving and preservation on board ship.

Selection of samplers

In planning the sampling it is important to appreciate that several different types of gear are likely to be required in order to sample adequately the different habitats and types of organisms likely to be encountered. It is also important to be aware of the limitations of the sampling and processing techniques employed and of the extent to which they may miss or underestimate some of the species. For macrofauna, the main difficulties usually arise at the sampling stage. The larger animals, and in particular the more active ones, are not easy to sample quantitatively. Those living on the surface may move away to avoid the sampler, and some of the burrowers may be buried too deeply to be taken by most grabs. Some occur at such low densities, or are so widely distributed (singly or in aggregations) that adequate sampling is difficult.

With meiofauna more problems arise at the processing than at the sampling stage, when the object is to separate animals of different sizes from one another and from the sediment. The arbitrary divisions into macro-, meio- and microbenthos make sense largely because different techniques are required for each group.

The two basic types of instruments are those towed horizontally, such as dredges and trawls, and those operated vertically, such as grabs, corers and air-lifts. Data from either type can usefully be supplemented by photography, television, or by direct observation by a SCUBA diver, from submersibles, or by hand collecting on the shore.

The final choice of a sampler also depends upon the purpose of the exercise. There are three major types of objective often associated with surveillance programmes, and the objective becomes progressively more difficult to fulfil as it demands a successive increase in both taxonomic and sampling effort (Elliott *et al.*, 1980). The simplest objective is a list of taxa or species with no measure of abundance. Some biotic indices may be derived from such data and the complexity of the index obtained depends upon the taxonomic level to which the organisms are identified. The minimum requirement is a sampler that adequately collects material from the different types of habitat on the bottom. Sampling to fulfil this first objective is usually referred to as 'qualitative'. A second objective may be to measure the relative abundance of species. Such information can be used to calculate some community or biotic indices based on rank order or diversity. For this purpose, the sampler must operate in a standard manner on all the types of substratum that are to be investigated. Although a qualitative sampler, e.g. a dredge, is often adequate, quantitative samplers are preferable because their performance is less biased by the operator. Sampling to fulfil this second objective should also be called qualitative, but the rather ambiguous term 'semi-quantitative' is often used. A third objective may be to estimate the number or biomass per unit area—these estimates can be used to compare spatial or temporal differences in populations using parametric tests, e.g. to detect small changes in water quality and rates of growth, reproduction and mortality. Sampling to fulfil this third objective is always called 'quantitative'. Only quantitative samplers—grabs, corers, air-lift samplers—are fully adequate for this purpose and many replicate sampling units need to be taken on each type of habitat. This means that the effort required is considerably greater than that needed to fulfil the first and second objectives.

Pilot surveys

For qualitative work, the dredges described in Chapter 6 can be used, and rigorous planning of the sampling programme will not usually be required.

It is important, however, to understand species/area or species/numbers relationships before starting a qualitative survey, and these relationships are, therefore, discussed on p. 6.

For quantitative surveys, a standardized procedure is desirable, and decisions on such matters as the layout of the stations, the type of sampler to be used, routine observations to be made at each station, the treatment of samples at sea and in the laboratory, and the presentation of results must be made at an early stage. Such decisions can be reached more easily after preliminary observations have been made in the selected area. Some information on depths and deposits can be obtained from charts, and published accounts of the regional fauna may be available, but a small-scale pilot survey should be carried out if possible. The pilot survey is essentially a reconnaissance of the ecosystem to be investigated which demonstrates the salient features of the physical environment and the biological communities. It offers an opportunity to test gear and techniques, to train personnel and to estimate costs. On beaches it could include visual examination, and in offshore areas, exploration by SCUBA divers, underwater photography and television. Preliminary qualitative sampling by dredge will indicate the types of animals and plants present and the patchiness of organisms and sediment. Some limited measurements of physical and chemical factors such as temperature and salinity may give further help with decisions on the layout of stations. It is particularly important to carry out such pilot surveys if little is known of the local flora, fauna and conditions, since time and thought expended at this stage will prove worthwhile in the avoidance of mistakes later.

Layout of stations

If there are no obvious environmental gradients, the sampling may be laid out in the form of a grid over the study area. By taking several samples at random in each grid square the whole area will be covered and a measure of environmental heterogeneity obtained. If there is a distinct gradient of, for example, depth or salinity, then stratified sampling may be best, working on transects across the gradients with stations arranged to give equal coverage of the different zones. The final layout adopted will depend on the purpose of the sampling and on the methods of analysis to be used (see p. 21) but the aim should be for each station to be reasonably representative of a wider area, whenever possible, if valid replicate hauls are to be made.

When surveying a very large area, whether qualitatively or quantitatively, the economical use of ship's time is important. Some time must elapse between stations while a sample is being sieved or rough-sorted, preserved and labelled, and repairs may have to be made to gear; it is convenient if the ship can be steaming to the next station during this period. In such circumstances, the spacing of stations at intervals of about $\frac{3}{4}$–1 hour's steaming time apart may allow the maximum amount of ground to be covered in a day.

An accurate position fix is important for finding and maintaining stations, and this aspect is considered in Chapter 2. However, degree of accuracy should be related to the purpose in hand, and it may often be advisable, in view of biological variability and environmental heterogeneity, to regard a station as an area within which sampling is carried out rather than a point position.

Number of sampling units required for a sample

This is a complex decision which involves both the size of the sampler and the number of stations, as well as the major objectives of the survey. The term sampling unit is used here to describe one member of a set forming a sample. Therefore each action of a sampler removes one sampling unit from the bottom and several sampling units make up the sample.

It is important to realize that, for both qualitative and quantitative sampling, the number of species or taxa taken in the samples usually increases as the size of the catch increases. The relationship between the number of species (S) and the number of individuals (N) often follows a power law:

$$S = aN^b \tag{1.1}$$

or:

$$\log_e S = \log_e a + b \log_e N \tag{1.2}$$

where a and b are constants with the value of the exponent b usually lying within the range 0·2–0·6. The relationship is curvilinear on arithmetic scales and linear on logarithmic scales (Fig. 1.1a,b). The number of species in a sample is similarly related to the area sampled or the number of replicate sampling units (Fig. 1.1c,d).

As the number of species is related to the size of the sample, a representative species list may be obtained by taking many small sampling units (e.g. with a grab) or a few large units (e.g. with a dredge), assuming that all samplers catch all taxa with equal efficiency. Thus, many samples taken with a grab should give a result similar to that obtained from a few hauls with a large dredge. The smaller sampler may be preferable in a heterogeneous habitat because more widely spaced points may be sampled for the same total sample size. It is also difficult to compare the species richness or diversity of samples unless similar numbers of animals are taken in each sample. For example, a single grab sample may contain 100 animals of 12 species and a

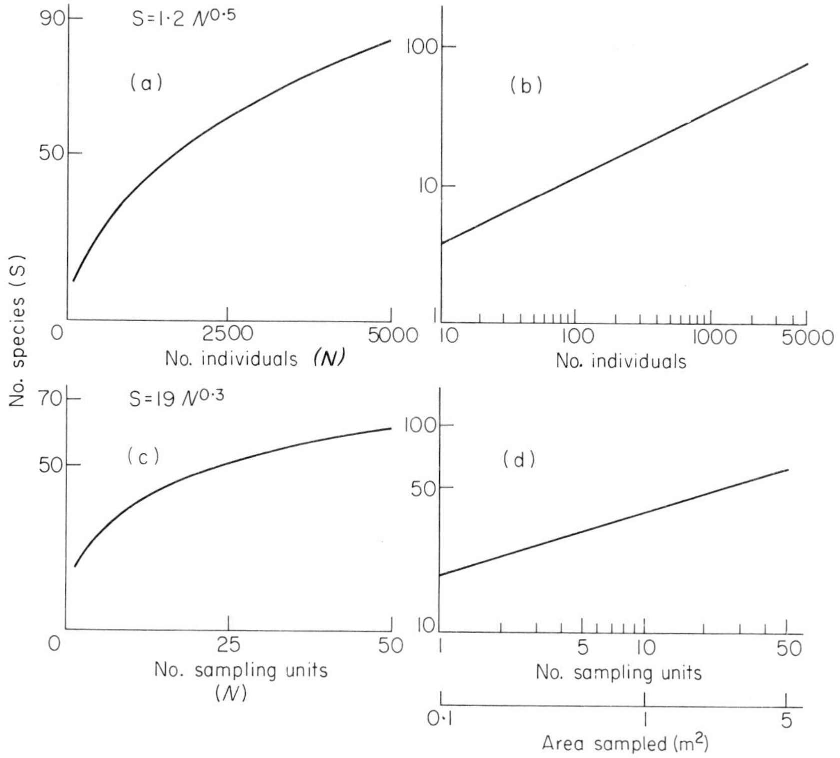

Fig. 1.1. Relationship between number of species and number of individuals on (a) arithmetic, and (b) logarithmic scales. Relationship between number of species and number of sampling units or area sampled on (c) arithmetic, and (d) logarithmic scales.

single dredge sample may contain 1000 animals of 38 species, both samples being taken from the same habitat (cf. Fig. 1.1a,b).

To determine how many sampling units are sufficient to collect a high proportion of the species, the recruitment of species to the sample total from each successive sampling unit may be examined by plotting on a cumulative basis the number of species against the number of sampling units (Holme, 1953). To eliminate bias, Ursin (1960) suggested that the number of species in one sampling unit should be obtained by using the mean of all the sampling units in the sample; the mean number in two sampling units is then obtained by taking the units in pairs and so on. The construction of a species/area cumulative curve will indicate how many sampling units are required to sample a given proportion of the species for the ground in question. Williams (1964) provides a theoretical discussion of the species/area problem.

Other workers have realized that this relationship makes it difficult to compare samples, and have suggested that comparisons should be based on the number of taxa per standard number of individuals. For example, Odum (1967) used the number of taxa per 1000 individuals to measure species richness and found that it was related to the total organic matter in several aquatic systems. Sanders (1968) used a similar approach with 'species diversity curves' to compare the diversity of marine benthos. Hurlbert (1971) considered the general problem of the comparison of species richness, i.e. the number of species or taxa present in a collection containing a specified number of individuals. He noted that the species richness usually increases with the number of individuals in a collection and therefore proposed that collections should be reduced to a common size before comparisons of species richness are made between them. This was effected by an equation that was derived from the multivariate hypergeometric distribution and describes the joint distribution of random variables representing the number of individuals of different taxa in the samples. Gage (1975) used this expression to construct 'species-richness' curves that were then used to compare the relative performance of marine samplers. Although the basic reasons for its development were similar to those used to develop equation (1.1), the assumptions implicit in Hurlbert's equation are not always valid and the simpler power law is adequate for most situations.

For macrofauna sampling, an instrument covering at least $0{\cdot}1\,m^2$ is generally used, and most of the instruments available for fully quantitative work cover either 0·1 or $0{\cdot}2\,m^2$. No absolute advantage has been demonstrated for one or other of these sampler sizes. If time is short at the collecting stage, if comparatively large animals are to be sampled, or if the work is in deep water, the larger size may be more appropriate, while the smaller has the advantage of reducing statistical error and providing a more representative coverage of the habitat since a larger number of samples can probably be taken for the same processing effort.

Having selected the gear, as discussed in Chapters 6 and 7, the next important issue is how many sampling units should be collected and how they should be grouped, given that the object of quantitative sampling is probably to make valid statements about the number of species present and about the distribution and density (number of individuals and/or biomass per unit area) of the fauna. If logistic constraints impose a limit on the total number of sampling units that can be collected and examined, then it is possible that the optimum number per station may be only one (Cuff & Coleman, 1979; Green, 1980; Cuff, 1980), but replication is appropriate if the significance of differences between stations is to be tested or if the spatial distribution of the organisms is to be assessed (Green, 1979).

In this connection it is important to recognize the difference between accuracy and precision. Quantitative estimates are accurate if they are close to the real population parameters and inaccurate if they are too high or too low. Estimates are precise if they have a low variability and imprecise if they have a wide scatter round the mean. Precision, therefore, refers to the reproducibility of the result for the same population. The precision of samplers is sometimes compared by using the coefficient of variation (CV), which is the standard deviation (s) of the sample expressed as a percentage of the sample mean ($\bar{x}$) thus:

$$\mathrm{CV} = 100s/\bar{x} \tag{1.3}$$

Unfortunately, it is not often realized that this coefficient frequently decreases with increasing mean density and cannot, therefore, be used to compare the precision of different samplers (Elliott & Drake, 1981). The precision of a sample mean should be expressed by the standard error (SE) or preferably by 95 % confidence limits (CL). Lack of precision is easily recognized by the resulting high variance and may be improved by increasing the number of samples, but low accuracy is much more difficult to recognize and correct. This is one reason for using a variety of sampling approaches.

Another question is concerned with how many samples are required to give an acceptable estimate of population numbers and/or biomass, and this depends partly on the variability of the observations. Unfortunately, in sampling with a grab, the variance of counts tends to be large. Variability arises from experimental error due to the functioning of the grab, and also from factors associated with the spatial distribution of the organisms. The first cause (which includes avoidance reactions of the animals) is discussed in Chapter 6 and the resulting variation can be reduced by selecting the most suitable grab and operating it carefully. Variation of sample counts due to the spatial distribution of fauna can be a more complex problem.

The use of several terms to describe the spatial distribution of plants and animals has caused some confusion. Some workers use the term 'dispersion', which is a mathematical term for the distribution of numbers about their mean value. Hence the term 'spatial dispersion' is used to describe the spatial distribution of individuals in a population. Unfortunately, dispersion is sometimes confused with 'dispersal', a term that should be used to describe the movement of individuals in space. There are similar problems with some of the terms used to describe the three basic types of spatial distribution:

1 Random distribution (no other terms).

2 Regular distribution (=under-dispersion, or uniform distribution, or even distribution).

3 Clumped distribution (=over-dispersion, or contagious distribution, or aggregated distribution).

The simplest terms are used here, i.e. spatial distribution, random, regular and clumped.

The following relationships between population variance and mean form the basis of most statistical tests for spatial distribution: variance equals the mean for a random distribution, is significantly less than the mean for a regular distribution, and is significantly greater than the mean for a clumped distribution. Methods used to test for these relationships are described in detail in some textbooks (e.g. Elliott, 1977). There is usually little problem in interpretation if the sample variance is significantly less or greater than the sample mean, but if the variance is not significantly different from the mean, it cannot be concluded that the spatial distribution is random. It is often not realized that no statistical test can prove randomness. If the random hypothesis is accepted, then there is little point in examining environmental factors that may affect spatial distribution because it has already been decided that the distribution is due to pure chance! When a significant departure from random cannot be detected, it is probably wiser to conclude that the distribution is effectively random but that undetectable clumping may be present. This conclusion is frequently applicable to low density populations, especially when there are few sampling units in the sample. The practical advantage of a non-significant departure from random is that a Poisson series can be used as a model for the samples. As will be seen later, there are relatively simple methods of calculating confidence limits and comparing samples from a Poisson series.

Various indices have been used to compare different degrees of clumping and most of these are reviewed and discussed by Elliott (1977). One problem with several of these indices is that the absolute degree of clumping is chiefly affected by mean density, i.e. spatial distribution is density-dependent. Therefore a relative index that excludes the effects of density is often preferable. One such index is provided by the power-law relationship between the arithmetic mean ($\bar{x}$) and variance (s^2) of a series of samples (Taylor, 1961; Taylor *et al.*, 1978):

$$s^2 = a\bar{x}^b \tag{1.4a}$$

which can be fitted in the form of the linear regression model:

$$\log_e s^2 = \log_e a + b \log_e \bar{x} \tag{1.4b}$$

where a and b are constants (for a critical evaluation of the transformation and fitting technique, see Taylor *et al.*, 1978). The constant b is a useful index of relative clumping, and can also be used in the estimation of sample size (see p. 15) and to derive a suitable transformation for the data (see p. 12).

A variety of spatial distributions have been observed in marine benthos. Holme (1950) found a regular distribution in the intertidal bivalve *Tellina tenuis* which may have been related to the foraging activity of the mollusc's siphons on the sand surface. Such a distribution is not common, and other workers have mainly described clumped distributions (Buchanan, 1967; Kosler, 1968), or were unable to detect departure from random (Clark & Milne, 1955). For the epifauna, Fager (1968) found that in a community of nine species living on sand in shallow water, seven were clumped and two effectively random. However, Wilson *et al.* (1977) demonstrated a tendency for individuals of a surface living brittle-star to spread out, tending to produce an even distribution.

Patterns of spatial distribution are often complex, and their elucidation depends upon a number of factors, including the area scale of sampling, the size of sample, time of the year, and even the size of sieve used in processing. The importance of area scale is illustrated by the distribution of echinoderms off the N.E. coast of England (Buchanan, 1967). Over a large area (about 200 square miles) the species were found to be clumped into communities bounded by the line of 20% silt in the sediment. Within the smaller areas of each community another type of clumping, described as a faunal density trend, associated with various environmental parameters, was encountered. On an even smaller scale, when 150 samples were taken within 3 square miles, some species (such as *Echinocardium cordatum*) showed a third type of clumping which was unstable and probably associated with larval settlement. Finally, when sampling was reduced to a single station—a circle of 100 m radius—*E. cordatum* was clumped only at the spawning season, and at other times the distribution was effectively random.

The size of the sampling unit is also important, since if counts drop to <1 per unit, it is often difficult to detect significant departures from a random distribution. Angel & Angel (1967) used divers to collect 256 contiguous core samples (each 7·62 cm diameter) and by variously grouping the sampling units together into blocks of different sizes, they were able to examine the effect of sample size on the analysis of spatial distribution. They found that in three of the 11 species studied the distribution patterns varied according to the block size, and concluded that the microdistribution of these species could not be determined using conventional remote sampling methods. Again, the results may depend on the size of sieve used. Jackson (1968) showed that while second-year individuals of a mollusc population were effectively random, the total population was clumped because of ovoviviparous habit. Similar results are described by Gilbert (1968), who found juveniles of *Tellina agilis* clumped, but adult distribution not significantly different from random in the field.

Once the spatial distribution is known, it is possible to calculate the standard error ($SE = \sqrt{(s^2/n)}$), or preferably confidence limits for the mean,

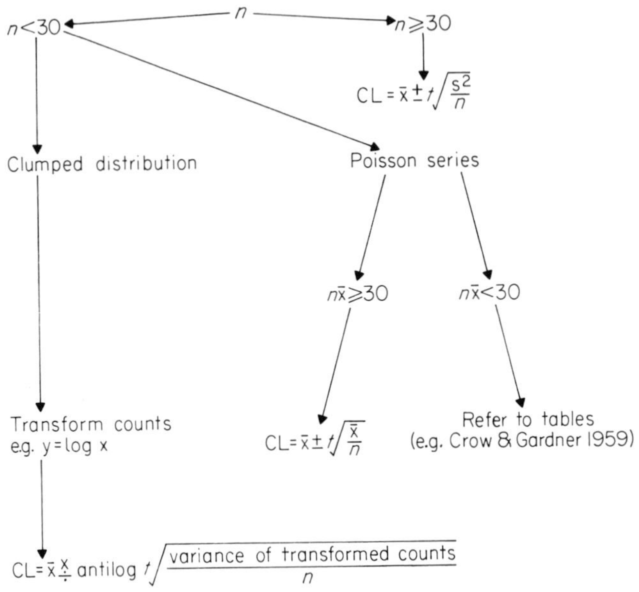

Fig. 1.2. Summary of major decisions to be made in the calculation of confidence limits for the sample mean. The value of t is found in tables of Student's t-distribution with $n-1$ degrees of freedom (column headed $2Q = 0{\cdot}05$ for 95% CL). These calculations are described in more detail in Elliott (1977).

defining the upper and lower values of a range within which the true population mean lies (Fig. 1.2). When the number of sampling units (n) in the sample is large ($n \geq 30$), the central-limit theorem is applicable, the non-normality of the data can be ignored and confidence limits can be calculated in the standard way (Fig. 1.2). If the sample is small ($n < 30$) and agreement with a Poisson series is not disproved, then confidence limits can be easily calculated or obtained from tables (Fig. 1.2). The most difficult situation is encountered with small samples from a clumped distribution, when the counts must be normalized by a suitable transformation before confidence limits can be calculated or sample means are compared. A logarithmic transformation ($y = \log x$, or $\log(x+1)$ when zero counts occur) is frequently used (e.g. Fig. 1.2). A more exact transformation may sometimes be required and can be derived from the power-law relationship (equation 1.4), using the exponent, i.e. $y = x^{1-(b/2)}$. This and more complex transformations are reviewed and discussed by Elliott (1977).

Although transformations are not required for small samples from a Poisson series when confidence limits are calculated or two means are compared, they are often required for other statistical methods, e.g. analysis of variance. The usual transformation is to replace each count by its square root ($y = x^{0 \cdot 5}$ or $\sqrt{x}$) or by $\sqrt{(x + 0 \cdot 5)}$ if some counts are < 10. Downing (1979) criticized this use of different transformations and tried to develop a common transformation. He concluded that the value of the exponent b in the power-law (equation 1.4a) was frequently close to 1·5, and therefore proposed that each count should be replaced by its fourth root, i.e. $y = x^{1-(b/2)} = x^{0 \cdot 25} = \sqrt[4]{x}$. The major weaknesses in this argument are presented in the critical comments of L.R. Taylor (1980) and W.D. Taylor (1980). Some of the examples presented by Downing show that a square root transformation for data close to random ($b \approx 1$) and a log transformation for data from a clumped distribution ($b \approx 2$) yield better results than the common transformation $x^{0 \cdot 25}$. The latter transformation is, therefore, only necessary when data from a Poisson series and a clumped distribution have to be compared.

Having selected the appropriate transformation, confidence limits can be set to the means and the number of samples required to give a satisfactory population estimate can be determined. As an example of what may be expected from a grab survey, data collected off the east coast of Scotland on a ground of medium sand have been used. The small echinoderm, *Echinocyamus pusillus*, was studied because it lives near the sand surface, is comparatively inactive, and, being conspicuous in the samples because of its colour, can be adequately collected and sorted. Over a 3-month period, three separate surveys were made, each usually consisting of 12 stations with 3 hauls with a 0·1 m^2 grab at each station. The area sampled could be regarded as homogeneous with respect to the distribution of *Echinocyamus*, and the mean of all hauls on a survey gives the estimate of population density per unit area. The precision of this estimate was provided by its standard error, and, after working on a logarithmic scale (the standard deviation had been found to increase with the mean), it was convenient to express this as a percentage of the mean on the original scale. It was found that to achieve a precision in the mean, on the untransformed scale, of 10 % with 3 hauls at each station, 47 stations must be worked—a total of 141 hauls. If wider coefficients were acceptable, then 11 stations (33 hauls), 5 stations (15 hauls) and 3 stations (9 hauls) were required for 20 %, 30 % and 40 % precision, respectively. Thus, if any great precision is required for the mean, a large number of samples must be taken; if this is necessary for each different ground or for each seasonal survey then the effort involved may be much too great.

As well as enquiring how precisely a mean estimates a population density, we may wish to know how sensitive the mean is in detecting differences in densities between surveys at different times or locations. A rough guide is that a difference between two means is usually significant when the confidence limits for the means do not overlap, and is usually not significant when the confidence limits overlap and each mean lies within the confidence limits of the other mean. The uncertain situation requiring a statistical test is when the confidence limits overlap but one or both means do not lie within the confidence limits of the other mean. It is obvious that as confidence limits decrease, the probability of detecting differences between means increases. Returning to the *Echinocyamus* surveys described above, it was found that the number of stations (with 3 hauls at each) which would need to be visited in order to detect a 50% difference between two means, were as shown below (calculated for three probabilities of detection):

Probability of detection	0·25	0·50	0·75
Number of 3-haul stations	10	21	37
Number of hauls	30	63	111

In other words, to attain a 3 out of 4 probability of detecting a 50% difference, 111 hauls would be required. Again, a discouragingly large volume of field and laboratory work.

If groups of animals are dealt with rather than single species, the results may be less variable. In a survey of benthos off the Scottish west coast, the fauna was divided into molluscs, polychaetes and crustaceans. The 95% confidence limits of the mean counts ($\bar{x}$) for different numbers of grab hauls are given in Table 1.1. This shows, for example, that for a true mean of 100 polychaetes, the limits from 5 hauls are 78–128, while for 20 hauls the narrower limits are 88–113, an increase in precision which may well not be considered worthwhile at the price of multiplying the sampling effort by four.

Table 1.1 The 95% confidence limits for mean counts ($\bar{x}$) from a benthos survey.

	5 hauls	10 hauls	15 hauls	20 hauls
Polychaeta	$\bar{x} \stackrel{\times}{\div} 1{\cdot}28$	$\bar{x} \stackrel{\times}{\div} 1{\cdot}19$	$\bar{x} \stackrel{\times}{\div} 1{\cdot}15$	$\bar{x} \stackrel{\times}{\div} 1{\cdot}13$
Crustacea	$\bar{x} \stackrel{\times}{\div} 1{\cdot}49$	$\bar{x} \stackrel{\times}{\div} 1{\cdot}32$	$\bar{x} \stackrel{\times}{\div} 1{\cdot}26$	$\bar{x} \stackrel{\times}{\div} 1{\cdot}22$
Mollusca	$\bar{x} \stackrel{\times}{\div} 1{\cdot}56$	$\bar{x} \stackrel{\times}{\div} 1{\cdot}37$	$\bar{x} \stackrel{\times}{\div} 1{\cdot}29$	$\bar{x} \stackrel{\times}{\div} 1{\cdot}25$

Taking this a stage further, Longhurst (1959) has pointed out that since there is usually less variation between hauls for the gross faunal indices of total biomass or total number of individuals than there is between numbers of individuals of each species, such faunal indices may be a useful quick overall measure of the density on particular grounds. As a result of surveys of different sediment types off West Africa, he considers that a standard 5-haul station covering $0{\cdot}5\,m^2$ will adequately estimate the faunal indices.

Thus it may appear that, in numerical studies of benthic populations, a high degree of precision would require a very large number of samples, and that moderate increases in sampling effort might not repay the extra time involved in processing. However, the acceptable level of precision will depend on the objective and if, for each study, this level can be carefully defined, it will be possible to calculate the minimum number of samples that must be taken or, after collection, the number that must be analysed. That number may turn out to be not too demanding in effort (Saila *et al.*, 1976; Walker *et al.*, 1979; Cuff & Coleman, 1979). If for any reason such calculations cannot be made, then it is suggested that, for statistical study, a standard 5-haul station with a $0{\cdot}1\,m^2$ grab should be regarded as a minimum requirement, and that even if data are not subjected to detailed analyses, the presentation of means should always be accompanied by an estimate of their precision, preferably as 95 % confidence limits. For example, if a log transformation is found necessary, then a plot of the means with 95 % CL all on a log scale should be regarded as the minimum presentation, giving at a glance confidence limits for each mean and an indication of the significance of the differences between them. The advantage of using transformed values to compare sample means is that this avoids the complications which arise on converting back to the original scale (Bagenal, 1955; Elliott, 1977).

It has been shown that the three principal factors which affect the number of sampling units required for quantitative sampling are the desired precision, the mean catch and the degree of clumping of the fauna being sampled. The effects of these factors are illustrated in Table 1.2. The required precision (D) is the 95 % confidence limits expressed as a fraction of the mean (or % as in Table 1.2). Sample mean ($\bar{x}$) refers to the mean number of animals per sample, e.g. mean number per $0{\cdot}1\,m^2$ if a sampler covering that area is used. It does not refer to mean number per m^2. The index of clumping (b) is obtained from the power-law relationship (equation 1.4) and the values for the number of sampling units (n) in the Table are for $a = 1$ in equation 1.4. If a is not equal to 1, then the values of n should be multiplied by the value of a to obtain the correct value. The values of a and b are obtained either from previous samples at the station or from previous knowledge of the usual spatial distribution of each species or taxon. Exact values of n may be obtained from the equation:

$$n = t^2 a\bar{x}^{b-2} D^{-2} \qquad (1.5a)$$

or,

$$\log n = 2\log t + \log a + (b-2).\log \bar{x} + \frac{1}{2\log D} \qquad (1.5b)$$

where n = number of sampling units, t = Student's t, $\bar{x}$ = mean catch, and D = index of precision.

Table 1.2 shows that, for all sample means and levels of clumping, the number of sampling units increases as the required precision decreases. When the spatial distribution of the invertebrates is close to random ($b = 1$), the required number of sampling units (n) for each level of precision is inversely

Table 1.2 Number (n) of sampling units per sample required for different levels of precision (95% CL expressed as a percentage of the sample mean) for different values of the sample arithmetic mean ($\bar{x}$) (N.B. $\bar{x}$ is the mean number per sample, not mean number m^{-2}), and for different values of b in equation 1.5; the value of t was 2 for $n > 25$ but more exact values were used for $n < 25$. The value of a in equation 1.5 is assumed to be 1 for this table but if $a \neq 1$, then multiply the appropriate value of n by a to obtain the correct estimate of n. (After Elliott & Drake, 1981.)

Precision of sample (100 D)	Sample mean ($\bar{x}$)								Index of clumping (b)
	0·5	1	5	10	20	50	100	1000	
10%	800	400	80	40	22	10	6	1	
20%	200	100	22	12	7	3	1	1	For
40%	50	26	7	4	1	1	1	1	random
60%	24	13	3	1	1	1	1	1	distribution
80%	14	8	1	1	1	1	1	1	$b = 1$
100%	10	6	1	1	1	1	1	1	
10%	566	400	178	126	90	56	40	15	
20%	140	100	45	32	24	16	12	5	Some
40%	35	26	13	10	8	6	5	1	clumping
60%	17	13	7	5	4	2	1	1	$b = 1·5$
80%	11	8	5	3	2	1	1	1	
100%	8	6	2	2	1	1	1	1	
10%	400	400	400	400	400	400	400	400	
20%	100	100	100	100	100	100	100	100	Clumped
40%	26	26	26	26	26	26	26	26	$b = 2$
60%	13	13	13	13	13	13	13	13	Data
80%	8	8	8	8	8	8	8	8	follow
100%	6	6	6	6	6	6	6	6	log normal
10%	283	400	894	1265	1789	2828	4000	12 649	
20%	70	100	224	316	447	707	1000	3162	Very
40%	19	26	56	79	112	177	250	791	clumped
60%	10	13	26	35	50	79	111	351	$b = 2·5$
80%	6	8	14	20	28	44	63	198	
100%	4	6	9	13	18	28	40	126	

proportional to the sample mean ($\bar{x}$). The relationship is similar for moderate levels of clumping ($b < 2$), but the values of n are always higher than the corresponding values for effectively random distributions. For example, if 95 % confidence limits for the estimate of population density are to be within 40 % of the mean value ($D = 0{\cdot}4$) and preliminary samples gave mean catches ($\bar{x}$) of about 5, values of $a \approx 1$ and $b \approx 1{\cdot}5$, then 13 sampling units ($n = 13$) would be required in future samples to maintain the precision. The corresponding value for an effectively random distribution ($a = 1$, $b = 1$) is only 7 sampling units (cf. values in Table 1.2).

For populations with $b = 2$, the number of sampling units is not affected by the sample mean and 26 sampling units would be required for a 40 % level of precision. Sampling problems are acute for very clumped populations ($b >$ 2), and the number of sampling units for each level of precision increases markedly as the sample mean increases, i.e. the relationship is the opposite to that shown for $b < 2$. Fortunately, the value of b usually lies between 1 and 2, and, therefore, quantitative sampling is usually feasible.

The above discussion deals with macrofauna, but the problems for meiofauna are similar. A comparison of meiofauna populations from widely separated sites and in several seasons off the Scottish coast (McIntyre, 1964) showed that the means and variances tended to increase together. In the statistical analysis, a logarithmic transformation was used for the nematodes, and a square root transformation for the copepods. Vitiello (1968) made counts of the meiofauna at a single station and his data on the more numerous groups fitted a negative binomial distribution. Studies of meiofauna distribution are perhaps more complex because of the difficulties of recognizing microhabitats, because of a wider range of sample size and because subsampling is frequently done. Again, however, replicates should always be collected, and standard errors or 95 % confidence limits indicated.

Benthic studies of pollution

At the present time of financial stringency in science, an increasing proportion of ecological research is related to applied objectives. Thus, apart from work directly on resource species and on predator-prey relationships of commercial relevance, a significant effort in the marine field is currently focussed on such topics as the provision of environmental impact statements, the study of effects of dumping or other waste disposal, and the assessment of damage to the marine ecosystem from activities such as oil or gravel exploitation. It is often difficult to use the highly mobile water column and plankton in such work, but the sea bed is recognized as a sink for most contaminants which enter the marine environment, and the more stable sediments along with their contained organisms tend to be favoured for study in this context (Gray *et al.*, 1980). A survey of benthos has all too often been a convenient standby for those faced with pollution monitoring, whether in intertidal zones or in offshore regions. There are, therefore, frequent examples of extensive repeated surveys describing the density and distribution of benthic organisms, many of which have not got beyond the stage of unpublished or 'grey literature' reports, purporting to provide comment on the biological impact of marine pollution. The relevance of such exercises should be carefully considered. The natural variability in benthic populations is very great indeed. Alterations in the detailed structure of communities resulting from changes in the relative number of species and abundance of individuals take place throughout the year and between years due to a considerable variety of causes, including physical effects of storms, natural fluctuations in spawning success, hydrographically induced variation in settlement of recruits, competition within and between species, and effects of predation. The impact of pollution can add to this variability, but is clearly only one of many factors causing change. If a monitoring survey is to detect pollution-related changes in biotic communities it must be designed with this purpose specifically in view and must be amenable to statistical analysis, otherwise the changes that will almost certainly be recorded will not be unequivocally attributable to anything but natural fluctuations.

There are several valid field approaches by which pollution effects may be recognized in the benthos. One is by sampling along some expected gradient of contaminant concentration. This can be done if the source of the contamination is known. Thus an effluent outfall, an oil production platform or the centre of a dumping ground can be taken as a point source and stations laid out at intervals from it. Given some knowledge of the hydrographic regime in the area, and given also chemical measurements in the water, the sediments, and the biota, changes in the benthos along the gradient might reasonably be correlated with contaminant concentrations. In New York Bight, for example, the decreasing concentration of certain heavy metals with distance from a dumping point conforms approximately to a log-normal distribution that can be used as a model to determine how the sampling effort can be focussed where it is most needed (Saila *et al.*, 1978). The transect approach has the advantage that, since it depends on demonstrating change along a gradient, seasonal and other such factors are not relevant. However, the general environmental characteristics of the area must be known, since a significant change along the transect in, for example, depth and the particle diameter of the sediment, could cause a change in the fauna. Indeed, the stratification of samples in relation to sediment characteristics can help to reduce variability and such stratification can be arranged after collection on the basis of the grab contents (Walker *et al.*, 1979).

If comparable stations can be found inside and outside an identifiable area of impact and if the assumption can be made that organism abundance at these 'treatment' and 'control' stations respond similarly to environmental conditions, a control-treatment pairing (CTP) design is then applicable (Skalski & McKenzie, 1982). This use of estimates of proportional abundance between pairs of stations helps to reduce problems of confounding impact effects with changes due to natural causes. The value of the CTP design is enhanced if the stations can be established and sampled before pollution occurs.

Another approach is to develop a sufficient understanding of the natural fluctuations in the community studied so that any non-natural change will be recognized. For example, a programme off the Northumberland coast of England has developed a long time series of observations on the soft-bottom benthos in a relatively restricted subtidal area and is beginning to permit the classification of the species according to how they fluctuate over several years (Buchanan *et al.*, 1978). While this type of approach gives the most complete understanding of the benthic community, it is long-term and extremely labour intensive, particularly in the early years when the basic data are being built up. Thus in the Northumberland study, collections were required every 8 weeks in the initial part of the programme, although the communities eventually became so well known that surveys could be scheduled with a frequency of only one per year (Buchanan & Warwick, 1974).

General considerations

While the overall design of the programme will depend on the objectives, in practice the details of the sampling will, as suggested above, be controlled to a considerable extent by the facilities available—gear, ships, staff and time—so that the final arrangement will be a compromise between the desirable and the practicable.

Sample-sorting is one of the most labour-demanding aspects of benthic work and the time involved is determined to a large extent by the mesh of the sieve used to screen the sediment in the initial processing. In many ecological studies, or when a comprehensive picture of the fauna is required, it may be desirable to retain all sizes of organism. But in pollution studies, for example, there may be valid arguments for collecting a limited range of sizes. Thus a relatively coarse sieve of 1 mm (or even occasionally 2 mm) mesh that could reduce both the work and the cost substantially, may be attractive if it retains a sufficient proportion of the key species. A finer sieve ($\leq 0{\cdot}5$ mm) will, for most sediments, retain significantly more small species and juvenile stages of larger ones. These smaller organisms may be more sensitive to stress or to habitat changes, thus providing an earlier warning of pollution, but at the expense of a great deal more work at the sorting stage. If the survey can be done during a part of the year when spawning is at a minimum, there may be a stronger argument for the use of a relatively fine sieve which will retain small adults, but at that time of the year will not overburden the sorter with large numbers of juvenile stages. The relative desirability of using 1 mm or 0·5 mm sieves is a perennial source of discussion for the pollution biologist (Hartley, 1982; see also pp. 194–6) and the decision must be taken by the individual operator in the light of the relevant circumstances. In describing screen dimensions it is important to specify the aperture size rather than the number of meshes per linear unit, since wire dimensions may not be standardized.

Whatever the objective, it is clear that before any substantial survey is initiated the very large amount of time required for subsequent sorting of samples should be taken into account. In one survey in New York Bight it was estimated that the collection of a single grab sample cost $25 and that the sorting and analysis of the sample to taxa cost $300 (Saila *et al.*, 1976). Estimates from other sources of the time taken to sort fauna range from an hour or so to several days or even weeks per sample in the case of deep-sea material, indicating not only the variability between samples but also the diverse interests of individual scientists. In addition, it is difficult for a single worker to undertake the sorting and also the identifying of the full range of species to be found in the course of a general survey, and any need for assistance from taxonomic specialists should be recognized at the planning stage. For some groups, the non-specialist may be able quickly to master taxonomic keys, but the danger should be noted of using in one part of the world a key based on species from another part. Further, while ecological studies may require that organisms be identified to the species or the lowest possible level, this may not be necessary in all programmes. Thus in environmental impact studies such taxonomic accuracy may be unnecessary if the significance of changes at the species level cannot be evaluated.

Ideally, a survey should be costed in a systematic manner. Thus statistical procedures referred to above can be extended to calculate the number of samples which will give some specified confidence range at minimum cost, or to indicate how to optimize the precision on a fixed budget (Saila *et al.*, 1976). Frequently, however, sufficient data are not available for sophisticated planning of this kind, and in coastal waters, where the collection of additional samples may involve little extra effort in the field, it is not uncommon to collect more than it is initially intended to work up, to allow for the loss or damage of main samples, or for the later need of additional replicates. In the deep sea, however, because samples are difficult and expensive to obtain and replicate sampling may not be practicable, and because of the lower density of deep-sea fauna, each cast of the gear should, as suggested earlier, aim to obtain the biggest appropriate sample. The difficulty of collecting deep water samples influences subsequent treatment, which should be carefully organized to make maximum use of the material. For example, in the initial sieving for macrofauna, no material should be discarded that could later be used for meiofauna extraction or sediment analysis. It should also be remembered that most animals brought up from deep sea will not survive the temperature shock of normal washing on deck. They may reach the surface alive but to keep them alive, or even to fix them from the living state, arrangements must be made to wash the sample in refrigerated water, or to transfer it immediately to an appropriate constant temperature room.

The seasonal element may be of considerable importance in shallow water work and this also should be recognized at the planning stage. If the survey consists of a single sampling series, due allowance must be made for biological (e.g. spawning, migration, etc.) and environmental factors (e.g. monsoons) which might lead to non-representative sampling.

Data analysis

In the previous sections on design of programmes, passing references have been made to methods of analysing the data collected from benthos surveys, and it is clear that the programme design and data analysis should be planned together. A wide selection of approaches is available for data analysis, ranging from general methods for describing community structure and illuminating the relationships to detailed statistical techniques for testing specific hypotheses.

An extensive investigation will often produce a very large volume of data and some reduction may be desirable to exclude those which have no biological relevance, to facilitate the use of particular analytical techniques or simply to reduce the number of calculations (Clifford & Stephenson, 1975). The most common form of data reduction is probably the elimination of some species, and several criteria have been used, e.g. frequency in the collections, level of abundance (Day *et al.*, 1971) or habitat-constancy (Boesch, 1976). Whatever criteria are adopted (and it may be useful to have more than one), they must be relevant to the ecological questions to be answered. It is not always the commonest species that are the most important.

The basic data are usually the numbers and weights of each species, but a full analysis will include the recording of such population factors as sex, size and age, while at the community level, relative abundance, species diversity and faunal homogeneity are important and knowledge of the trophic relationships could be revealing.

The classical concepts of marine benthic communities tend to have been applied on the assumption that communities could often be clearly discerned by an experienced researcher examining large collections of samples. The subjective element in this was criticized, and Thorson (1957) put forward a number of recommendations for standardization of the procedures, involving the definition of 1st, 2nd and 3rd order 'characterizing species' and 'associated animals', and including the use of suggested standard sampling units—$0{\cdot}1\ m^2$ for depths of 0–200 m; $0{\cdot}2\ m^2$ for 200–2000 m; and $1\ m^2$ for the abyssal zone. Emphasis has not always been placed primarily on the animals and Jones (1950) proposed a scheme of classification of animal associations based on environmental factors—depth, salinity and the nature of the substratum. Ellis (1969) recognized the difficulties of delimiting ecosystems, and used indices of dispersion, density, biomass and respiration rate to identify objectively which species in his samples were sufficiently important to merit autoecological study.

This more objective, numerical approach is now widely used to classify data from benthos surveys, and groupings can be studied by many forms of multivariate analysis (e.g. Green, 1979; Walker *et al.*, 1979), to indicate patterns, or by the use of indices to assess temporal and spatial relations (e.g. Hummon, 1974). Detailed discussion of these techniques is beyond the scope of this handbook but they are reviewed by Clifford & Stephenson (1975) and Gray (1981). As already discussed, the study of changes in benthic communities to detect pollution stress is difficult because of natural variability, but the use of appropriate data analyses may help. Thus it has been suggested (Gray & Mirza, 1979) that data from an unpolluted community in equilibrium fits a log-normal distribution and that stress which results in an increase in the abundance of some species is shown by a break in the straight line of the log-normal plot.

For the calculations involved in dealing with the indices referred to above, and in the application of statistical techniques such as multivariate analysis to benthos work, computers are useful and, indeed, essential if even moderately large numbers of samples are involved. Computer programs which can readily be written for sediment analysis also eliminate much of the vast amount of work normally involved in statistical analysis of grain-size data in computing modes, medians, arithmetic means, standard deviation, skewness and kurtosis (Schlee & Webster, 1967).

References

Angel, H.H. & Angel, M.V., 1967. Distribution pattern analysis in a marine benthic community. *Helgoländer wissenschaftliche Meeresuntersuchungen*, **15**, 445–454.

Bagenal, M., 1955. A note on the relations of certain parameters following a logarithmic transformation. *Journal of the Marine Biological Association of the United Kingdom*, **34**, 289–296.

Boesch, D.F., 1976. A new look at zonation of benthos along the estuarine gradient. pp. 245–66 in B.C. Couell (ed.) *Ecology of Marine Benthos*. University of South Carolina Press, Columbia, 467 pp.

Buchanan, J.B., 1967. Dispersion and demography of some infaunal echinoderm populations. *Symposia of the Zoological Society of London*, **20**, 1–11.

Buchanan, J.B. & Warwick, R.M., 1974. An estimate of benthic macrofaunal production in the offshore mud of the Northumberland coast. *Journal of the Marine Biological Association of the United Kingdom*, **54**, 197–222.

Buchanan, J.B., Sheader, M. & Kingston, P.R., 1978. Sources of variability in the benthic macrofauna off the south Northumberland coast, 1971–1976. *Journal of the Marine Biological Association of the United Kingdom*, **58**, 191–209.

C.I.E.S.M.M., 1965. Méthodes quantitatives d'étude du benthos et echelle dimensionelle des benthontes. *Colloque du Comité du Benthos* (Marseille, Novembre, 1963). Conseil International pour l'Exploration de la Mer Méditerranée (Monaco), Paris.

Clark, R.B. & Milne, A., 1955. The sublittoral fauna of two sandy bays on the Isle of Cumbrae, Firth of Clyde. *Journal of the Marine Biological Association of the United Kingdom*, **34**, 161–180.

Clifford, H.T. & Stephenson, W., 1975. *An Introduction to Numerical Classification*. Academic Press, New York, 229 pp.

Cochran, W.G., 1963. *Sampling Techniques*, 2nd ed. John Wiley & Sons, New York, 413 pp.

Crow, E.L. & Gardner, R.S., 1959. Table of confidence limits for the expectation of a Poisson variable. *Biometrica*, **46**, 441–453. (New Statistical Tables Series No. 28.)

Cuff, W., 1980. Comment on optimal survey design—Reply. *Canadian Journal of Fisheries and Aquatic Sciences*, **37**, 297.

Cuff, W. & Coleman, N., 1979. Optimal survey design: lessons from a stratified random sample of macrobenthos. *Journal of the Fisheries Research Board of Canada*, **36**, 351–361.

Day, J.H., Field, J.G. & Montgomery, M.P., 1971. The use of numerical methods to determine the distribution of the benthic fauna across the continental shelf of North Carolina. *Journal of Animal Ecology*, **40**, 93–125.

Downing, J.A., 1979. Aggregation, transformation, and the design of benthos sampling programs. *Journal of the Fisheries Research Board of Canada*, **36**, 1454–1463.

Dybern, B.I., Ackefors, H. & Elmgren, R., 1976. Recommendations on methods for marine biological studies in the Baltic Sea. *The Baltic Marine Biologists*, **1**, 98 pp.

Elliott, J.M., 1977. Some methods for the statistical analysis of samples of benthic invertebrates. *Scientific Publications, Freshwater Biological Association*, **25** (2nd ed.), 156 pp.

Elliott, J.M. & Drake, C.M., 1981. A comparative study of seven grabs used for sampling macroinvertebrates in rivers. *Freshwater Biology*, **11**, 99–120.

Elliott, J.M., Drake, C.M. & Tullett, P.A., 1980. The choice of a suitable sampler for benthic macroinvertebrates in deep rivers. *Pollution Report of the Department of the Environment*, **8**, 36–44.

Elliott, J.M. & Tullett, P.A., 1978. A bibliography of samplers for benthic invertebrates. *Occasional Publications, Freshwater Biological Association*, **4**, 61 pp.

Elliott, J.M. & Tullett, P.A., 1983. A supplement to a bibliography of samplers for benthic invertebrates. *Occasional Publications, Freshwater Biological Association*, **20**, 26 pp.

Ellis, D.V., 1969. Ecologically significant species in coastal marine sediments of southern British Columbia. *Syesis*, **2**, 171–182.

Fager, E.W., 1968. A sand-bottom epifaunal community of invertebrates in shallow water. *Limnology and Oceanography*, **13**, 448–464.

Gage, J.D., 1975. A comparison of the deep-sea epibenthic sledge and anchor-box dredge samplers with the van Veen grab and hand coring by divers. *Deep-Sea Research*, **22**, 693–702.

Gilbert, W.H., 1968. Distribution and dispersion patterns of the dwarf tellin clam, *Tellina agilis*. *Biological Bulletin, Marine Biological Laboratory Woods Hole*, **135**, 419–420.

Gray, J.S., 1981. *The Ecology of Marine Sediments*. Cambridge University Press, London, 185 pp.

Gray, J.S., Boesch, D., Heip, C., Jones, A.M., Lassig. J., Vanderhorst, R. & Wolfe, D., 1980. The role of ecology in marine pollution monitoring. Ecology Panel Report. *Rapports et Procès-verbaux des Reunions. Conseil International pour l'Exploration de la Mer*, **179**, 237–252.

Gray, J.S. & Mirza, F.B., 1979. A possible method for the detection of pollution-induced disturbance on marine benthic communities. *Marine Pollution Bulletin*, **10**, 142–146.

Green, R.H., 1979. *Sampling Design and Statistical Methods for Environmental Biologists*. John Wiley & Sons, New York, 257 pp.

Green, R.H., 1980. Comment on optimal survey design—Comment. *Canadian Journal of Fisheries and Aquatic Sciences*, **37**, 296.

Hartley, J.P., 1982. Methods for monitoring offshore macrobenthos. *Marine Pollution Bulletin*, **13**, 150–154.

Holme, N.A., 1950. Population-dispersion in *Tellina tenuis* da Costa. *Journal of the Marine Biological Association of the United Kingdom*, **29**, 267–280.

Holme, N.A., 1953. The biomass of the bottom fauna in the English Channel off Plymouth. *Journal of the Marine Biological Association of the United Kingdom*, **32**, 1–49.

Holme, N.A., 1964. Methods of sampling the benthos. *Advances in Marine Biology*, **2**, 171–260.

Hopkins, T.L., 1964. A survey of marine bottom samplers. *Progress in Oceanography*, **2**, 213–256.

Hulings, N.C. & Gray, J.S., 1971. A manual for the study of meiofauna. *Smithsonian Contributions to Zoology*, **78**, 84 pp.

Hummon, W.D., 1974. SH^1: A similarity index based on shared species diversity, used to assess temporal and spatial relations among intertidal marine Gastrotricha. *Oecologia (Berlin)*, **17**, 203–220.

Hurlbert, S.H., 1971. The nonconcept of species diversity: a critique and alternative parameters. *Ecology*, **52**, 577–586.

Jackson, J.B.C., 1968. Bivalves: spatial and size-frequency distributions of two intertidal species. *Science, New York*, **161**, 479–480.

Jones, N.S., 1950. Marine bottom communities. *Biological Reviews of the Cambridge Philosophical Society*, **25**, 283–313.

Kosler, A., 1968. Distributional patterns of the eulitoral fauna near the Isle of Hiddensee (Baltic Sea, Rugia). *Marine Biology*, **1**, 266–268.

Longhurst, A.R., 1959. The sampling problem in benthic ecology. *Proceedings of the New Zealand Ecological Society*, **6**, 8–12.

McIntyre, A.D., 1964. Meiobenthos of sub-littoral muds. *Journal of the Marine Biological Association of the United Kingdom*, **44**, 665–674.

McIntyre, A.D., 1969. Ecology of marine meiobenthos. *Biological Reviews of the Cambridge Philosophical Society*, **44**, 245–290.

Odum, H.T., 1967. Biological circuits and the marine systems of Texas. p. 99–157 in T.A. Olson & F.J. Burgess (eds.) *Pollution and Marine Ecology*. Interscience Publishers–John Wiley & Sons, New York, 364 pp.

Saila, S.B., Anderson, E.L. & Walker, H.A., 1978. Sampling design for some trace elemental distributions in New York Bight sediments. pp. 167–177 in K.L. Dickson, J. Cairns Jr. & R.J. Livingston (eds) *Biological Data in Water Pollution Assessment: Quantitative and Statistical Analyses*. American Society for Testing and Materials, Philadelphia, 184 pp.

Saila, S.B., Pikanowski, R.A.& Vaughan, D.S., 1976. Optimum allocation strategies for sampling benthos in New York Bight. *Estuarine and Coastal Marine Science*, **4**, 119–128.

Sanders, H.L., 1968. Marine benthic diversity: a comparative study. *American Naturalist*, **102**, 243–282.

Schlee, J. & Webster, J., 1967. A computer program for grain-size data. *Sedimentology*, **8**, 45–53.

Schlieper, C. (ed.), 1968. *Methoden der Meeresbiologischen Forschung*. Vels Gustav Fisher Verlag, Jena, 322 pp.

Skalski, J.R. & McKenzie, D.M., 1982. A design for aquatic monitoring programmes. *Journal of Environmental Management*, **14**, 237–251.

Taylor, L.R., 1961. Aggregation, variance and the mean. *Nature, London*, **189**, 732–735.

Taylor, L.R., 1980. New light on the variance/mean view of aggregation and transformation: Comment. *Canadian Journal of Fisheries and Aquatic Sciences*, **37**, 1330–1332.

Taylor, L.R., Woiwood, I.P. & Perry, J.N., 1978. The density-dependence of spatial behaviour and the rarity of randomness. *Journal of Animal Ecology*, **47**, 383–406.

Taylor, W.D., 1980. Comment on 'Aggregation, transformations, and the design of benthos sampling programs.' *Canadian Journal of Fisheries and Aquatic Sciences*, **37**, 1328–1329.

Thorson, G., 1957. Sampling the benthos. *Memoirs of the Geological Society of America*, **67(1)**, 61–73.

Ursin, E., 1960. A quantitative investigation of the echinoderm fauna of the central North Sea. *Meddelser fra Danmarks Fiskeri-og Havundersøgelser*, N.S. **2(24)**, 204 pp.

Vitiello, P., 1968. Variations de la densité du microbenthos sur une aire restreinte. *Recueil de Travaux de la Station Marine d'Endoume*, **43(59)**, 261–270.

Walker, H.A., Saila, S.B. & Anderson, E.L., 1979. Exploring data structure of New York Bight benthic data using post-collection stratification of samples, and linear discriminant analysis for species composition comparisons. *Estuarine and Coastal Marine Science*, **9**, 101–120.

Williams, C.B., 1964. *Patterns in the Balance of Nature*. Academic Press, London, 324 pp.

Wilson, J.B., Holme, N.A. & Barrett, R.L., 1977. Population dispersal in the brittle-star *Ophiocomina nigra* (Abildgaard) (Echinodermata: Ophiuroidea). *Journal of the Marine Biological Association of the United Kingdom*, **57**, 405–439.

Chapter 2. Position Fixing of Ship and Gear

N. A. HOLME AND P. F. WILLERTON

Navigation

It is important to determine the positions at which samples or measurements are taken, as this will enable observations to be repeated at a later date or, alternatively, identify a unique position for publication and the subsequent use of other researchers. While the accuracy required is a function of the particular purposes of the survey, the accuracy obtainable depends on many factors, largely outside the control of the user, and that is the subject of this chapter.

Navigation, like all of the true sciences, is concerned with measurement and in its day to day practice is subject to mistakes and errors, as are any other scientific disciplines. Mistakes are avoided by good professional practice and by due attention to the possibility of error, whether from quantifiable or less predictable causes.

Required accuracy

This will largely depend upon the precise nature of the sampling programme being undertaken and the variations in submarine topography and fauna. In some places it will not be possible, even from an anchored vessel, to obtain replicate samples, whilst in others, perhaps farther offshore, the sediments and their fauna may not vary significantly over a wide area, so requiring a lesser degree of position-fixing accuracy. At this point it may be worth while examining four basic terms which should be considered during the planning stage. These are absolute position, relative position, repetition rate and repeatability.

Absolute position

This is a description of a point on the Earth's surface by a Cartesian co-ordinate system which would enable another observer to find the same point. It is usually expressed in terms of latitude and longitude or some alternative grid system. This concept appears a simple one, yet it is a trap for the unwary in that the untrained observer tends to assume that a map or chart is a statement of truth and that it is a faithful representation of the surface of our planet. The assumption is made that we know the shape of the Earth and can, therefore, describe it in mathematical terms which allow these spherical parameters to be drafted on to a plane (map) surface. This assumption is not true as firstly we still do not know the Earth's precise shape and secondly different areas of the world are mapped using approximations to the spheroid relevant to that area; consequently co-ordinates obtained from adjacent maps or from satellite navigation systems may be incompatible without correction. This has led to many confusing and expensive situations in recent years as the requirement for accuracy has increased offshore.

Relative position

This is the description of a location relative to some other point or object. Classically, such a position is defined in terms of bearing and distance from a lighthouse or other conspicuous feature. Alternatively, in offshore areas, underwater acoustic beacons may be used for positioning within their pattern. The advantage of relative positioning is that for many purposes it gives more adequate location information, provided reference to absolute position is not required. Caution must be exercised in converting from relative to absolute position as the co-ordinates of the reference object can frequently be erroneous.

Repetition rate

This is a measure of the frequency with which positional fixes can be obtained. This factor is particularly important when making scientific observations from a moving ship. It takes into account the time interval between receipt of successive sets of data, the time taken to present this data in the required form, and the ability of the system to detect the ship's movement over a given distance.

Repeatability

This is a measure of the extent to which the observer can use the position fixing system to bring him back to the same point on a subsequent occasion. This is an important consideration when time series observations are contemplated. The repeatability of a particular system will vary; the optical methods are generally reliable, but electro-magnetic systems are prone to random errors as distance from transmitters and base lines increases. Soundings taken by acoustic means or line frequently lack a unique identity, whereas some sonar and beacon types can be used to relocate position with great accuracy.

Obtainable accuracy

Position fixing may be thought of in terms of recording results on graph paper. If a number of lines meet at a given point we have greater confidence that we can describe the coordinates of that point accurately. The greater the number of lines from different sources passing through or close to the same point, the greater the degree of confidence.

It is in the nature of a navigator's science to be familiar with the source of position lines, to be aware of possible mistakes in their observation and plotting, and to apply appropriate corrections or reject unreliable data. In general terms the accuracy of the result will depend upon the sophistication and resolution of the system(s) used, the repetition rate, and the distance from origins. Cost may be a prime consideration, and in the planning stage the advice of someone familiar with navigation systems applicable to the area will prove useful. Too frequently the value of work is diminished by lack of repeatability, or highly expensive equipment lies redundant because it is not suited to the work in hand.

Systems now exist which can give an accuracy offshore of ± 30 m. It is possible, if the Navstar satellite system becomes available for civilian use, that this will improve to better than ± 10 m. However, for a conventionally equipped ship working out of sight of land relative accuracy may be of the order of ± 300 m and absolute accuracy ± 500 m. These last are approximations and vary with, for example system coverage, weather and sea state, atmospheric conditions, time of day, season, and sunspot activity.

Means of fixing position

Sources of positional information are as follows:

1 Optical—bearings, transits and included angles obtained by alidades, sextants and shore-based theodolites.
2 Acoustic—depths and ranges taken by measuring the echo return of sound waves underwater.
3 Linear measures—lines, chains and wires are still used occasionally for distance measurement in the horizontal and vertical planes.
4 Electro-magnetic—position lines, usually range measurements or their derivatives, found by measuring echo return time or phase difference between two signals.

A typical shipboard system primarily designed for navigational use will include compasses (Gyro and magnetic) on which bearings can be taken, as well as echo sounder and lead lines to measure water depths and a log to record speed through water. In addition, there will be a radio direction finder for obtaining bearings of shore-based radio transmitters, radar which can supply range and bearing information, and one or more of the electronic navigation aids such as Decca Navigator, Loran 'C', Omega and Transit Satellite Navigation. Sextants, which will provide a means of fixing position by astronomical and terrestrial observations of included angles, should also be carried.

Table 2.1 gives a list of navigational instruments with a subjective assessment of the accuracy found in practice, assuming correct observational techniques on a single measurement, tabulated errors applied and statutory specifications maintained. The table shows that high orders of accuracy can be obtained under ideal conditions. For short-range work, parallax error due to the distance between the scientific observation and the positioning fixing system must be allowed for.

Return to the same position

For repetitive sampling in sight of land transits (lining up of pairs of points on the land) are the most accurate means of returning to the same place, although if the marks used are not shown on the chart it may be difficult to determine absolute position. Transits are also useful for approaching a spot the position of which is determined by Decca Navigator or other means.

Table 2.1. Accuracy of position-fixing systems. (nm = nautical miles)

Instrument	Method of measurement	Accuracy	Remarks
Compasses (Magnetic and Gyro)	Bearings	$\pm 1°$	i.e. ± 32 m at 1 nm range. Use included angles to avoid compass errors
Echo sounder	Time of echo return	5% of depth	Assuming no regular calibration or density compensation
Lead line	Linear measure	± 10 cm	Assuming hard bottom, shallow water, stationary vessel and line stretched calibrated and vertical
Log	Stream or hull fitting	3% for towed logs 1% for electro-magnetic logs	Assuming calibration. Error increases in heavy seas. Only distance measured is that in the line of advance
RDF	Bearings	class 1: $\pm 2°$ class 3: $\pm 10°$	Random errors large at times
Radar	Range	$1\frac{1}{2}$% of range scale in use	i.e. ± 167 m when using 6 nm range scale
	Bearing	$\pm 2\frac{1}{2}°$	i.e. ± 485 m at 6 nm range (assuming stabilized display)
Decca Navigator	Hyperbolic lattice	$< \pm 3$ nm	Very accurate (± 30 m) on base line but up to 3 nm error at limits of range (300 nm)
Loran 'C'	Hyperbolic lattice	$< \pm 10$ nm	± 50 m obtainable when stable ground wave received but decreasing with range and propagation conditions
Omega	Hyperbolic lattice	$< \pm 2$ nm	Better performance by daylight paths, still experimental.
Transit satellite	Digital display (lat and long) from doppler ranges	$\pm 0{\cdot}25$ nm	Assuming two frequency systems. Low repetition rate (could be several hours in low latitude). Most systems indicate an updated Dead Reckoning position between satellite passes
Sextant	Astro	± 3 nm	
	Terrestrial	$\pm 1'$ of arc	Very accurate at short range, beware of parallax

Offshore, out of sight of land, the best means is to mark the position physically, i.e. lay one or more buoys or acoustic beacons. Both solutions are, however, highly susceptible to interference, natural or malicious. Alternatively 'raw' (uncorrected) position lines from electronic navigation systems can be used, but propagation conditions can alter with time of day and season. Also change of instrument or its repair can alter its 'index error' thus making some difference to the results. Determination of absolute position remains a problem using any of these methods.

Hydrographic survey equipment

In addition to the navigational instruments listed above there is an increasing number of devices which may be described as hydrographic survey equipment. These have now reached a high degree of sophistication and when working in specially sited electronic position fixing patterns, arrays of acoustic beacons or inertial navigation devices backed by satellite passes, before being statistically screened by on-board computers, make relative positioning offshore to within one metre possible. Such systems are used extensively for geophysical studies where very close lines of seismic shots have to be run, rigs positioned, pipelines laid, etc.

Reference works

Reference should be made to the most recent works, which include descriptions of modern navigation methods. For more traditional navigation methods see the *Admiralty Manual of Navigation* (Admiralty, Vol. I, 1970; Vol. II, 1973), the *Admiralty Manual of Hydrographic Surveying* (Admiralty, 1965), the *American Practical Navigator* (U.S. Navy Hydrographic Office, Vol. I, 1977; Vol. II, 1981), or equivalent works in other languages.

Further titles include *Navigation for Watchkeepers* (Fifield, 1980); *Shiphandling for Masters, Mates and Pilots* (Willerton, 1980); *Admiralty Sailing Directions* (NP1-NP73); *Admiralty Tidal Stream Atlases* (NP209, etc); *Admiralty List of Radio Signals* (Vol. 5 contains notes and data on position fixing systems); and *Radar and Electronic Navigation* (Sonnenberg, 1978).

Depth finding

Wire sounding

Sounding with the hand lead, although now largely superseded by echo sounding, is sometimes required for checking soundings and for obtaining a sample of the bottom, which is picked up by tallow placed in a hollow at the lower end of the lead.

For deep-sea work, soundings are made with 0·7 mm galvanized piano wire, single strand, to the end of which a lead weight is attached. In depths of <2000 m a simple sounding tube with flap valve, weighted to about 12 kg, may be used to obtain a sample of the bottom under favourable circumstances. In deeper water a rather heavier weight or sounding tube is used, designed so that weights are released on striking the bottom.

Echo sounding

Depths are usually measured by the echo sounder, an instrument for measuring the time interval between the emission of a sound signal from a transducer in the ship's hull and the returning echo from the sea bed. For deep-sea sounding, frequencies of 10 or 12 kHz are used, but higher frequencies of 15, 30 or 50 kHz are used for sounding on the continental shelf and for fish detection.

The accuracy of an echo sounder is dependent on a constant speed of movement of the recording stylus. Ordinary echo sounders have a stabilizing mechanism or may depend on manual adjustment of a governor to attain the correct timing. It is, in any case, important to make frequent checks on the speed of rotation, using a stop-watch. It is also necessary to allow for the depth of transmitting and receiving transducers below water level.

On the continental slope and in the deep sea the returning echo is often weak, and ship's noise may interfere with satisfactory reception of the signal. This, and a requirement for more accurate timing of the recording stylus, led to the development of the Precision Depth Recorder for research and survey purposes, particularly in the deep sea. A Precision Depth Recorder usually has high-quality ceramic or crystal transducers, mounted in a streamlined torpedo-shaped 'fish' suspended from a cable attached to a boom projecting clear of the ship's side. The 'fish' is towed a few metres below the sea surface, giving it increased stability, freedom from aeration, and decreased interference from the ship's noise. The signals are recorded on a graphic chart recorder (Fig. 2.1), for example the Mufax, which has a helical rotating stylus traversing the chart once per second. The speed of rotation is governed by a tuning-fork within the instrument, and so is independent of fluctuations in the electricity supply. At one revolution per second the full width of the chart corresponds to about 400 fathoms or 750 m, so that in deep water the returning echo will arrive after several traverses of the chart have been made. Some confusion may, therefore, arise as to the actual depth. This is overcome by a 'gating' procedure by which, at certain intervals of time, transmission

Fig. 2.1. Precision Depth Recorder, the modified Mufax recorder on which echo soundings and pinger signals are recorded. (Copyright Institute of Oceanographic Sciences.)

signals are stopped for a short time so that the correspondence between transmitting and receiving pulses may be established.

The velocity of sound in sea water varies with temperature, salinity and pressure. Echo sounders are usually calibrated for nominal speeds of either 800 fathoms (1463 m) or 820 fathoms (1500 m) per second. For accurate measurements, especially in deep water, correction should be applied to the depth read on the sounder. Matthews (1939) has divided the oceans into 52 areas, for each of which tables of the necessary corrections are given.

Depth fixing of gear

Failure to take a sample when deep sea trawling or dredging is common (Menzies, 1964, 1972). Very often this seems to be due to the gear not reaching the bottom, but in other instances is due to snarling and entanglement which may often be the result of paying out too much wire (see also Chapter 6, pp. 175–6).

The construction of steel cables is such that they tend to twist and untwist with varying loads; thus, when heavy gear reaches the sea bed the consequent release of strain will cause the wire to twist. A ball-bearing swivel between the end of the wire and the gear will to some extent prevent the development of twisting strains in the wire, but if excess wire is paid out so that it lies slack on the bottom, there is a danger that it will twist around itself or the gear to form an entanglement which will not clear itself when the gear is raised off the bottom.

In shallow water in reasonably calm weather it will at once be apparent on board ship when the sampler has reached bottom, but for deep-sea work it is not only essential to measure the amount of wire paid out, but also desirable to measure the strain on the wire and the position of the gear relative to the bottom.

Tension and wire-length meters

In early deep-sea expeditions an elastic accumulator was often used to dampen the effects of sudden strains on the towing rope or cable (e.g. Menzies, 1964), and from the movement of the accumulator it was possible to obtain a measure of the strains developed. In deep-sea work the margin of safety between the strain on the warp and its breaking strain is often small, and it is therefore desirable to keep a constant check on the warp loads.

Strain may be measured simply by deflection of the wire around a pulley to which a dynamometer is attached (e.g. Tydeman, 1902; Soule, 1951; Kullenberg, 1955), but in recent years strains have been recorded electronically by load-cells, the electrical resistance of which alters as load is applied. Such a system lends itself to indication remote from the deck, to repeat indicators, graphic recording of changing loads on a chart recorder and to an overload alarm system which gives warning when a predetermined strain is exceeded.

Tension recorders not only minimize the possibilities of overstraining and parting the warp but are also an aid to determining when the sampling gear has reached bottom. Shortly before the gear is expected to reach bottom, the winch is braked and a reading of the static load recorded. It will then be apparent from the chart record when the gear later reaches bottom, or in the case of a dredge, when it starts to bite into and sample the bottom. Examples of the use of a dynamometer for deep-sea dredging are given by Menzies (1964) for a spring-accumulator system, and by Sachs (1964) for load-cell recordings.

Pingers

Pingers (or other devices employing sonar) are essential for precise location of corers, grabs and cameras in relation to the bottom. The pinger itself consists of a cylindrical pressure-tight casing containing electronic equipment

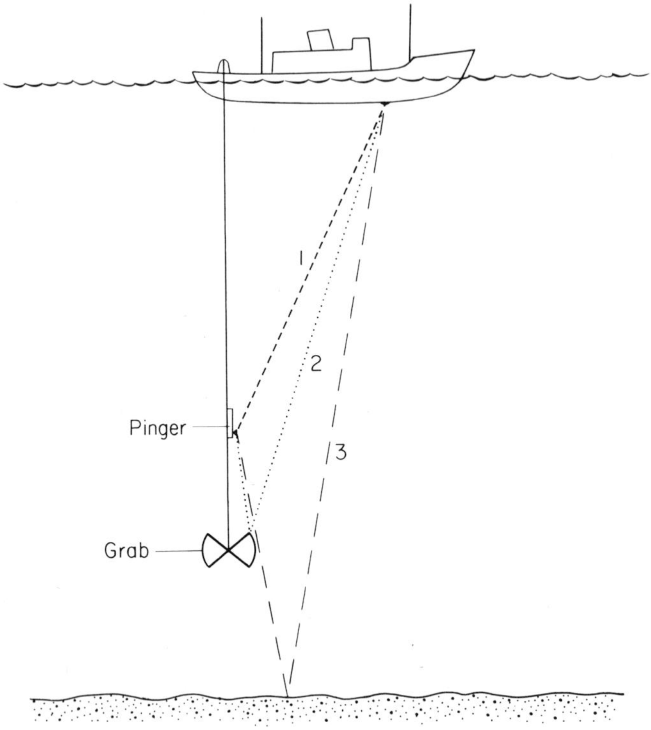

Fig. 2.2. Diagram of pinger in use for grab sampling. For explanation see text. Sound signals are normally received on the ship's echo sounder transducer.

transmitting brief sound signals, with a frequency of 10 or 12 kHz, at the rate of precisely one per second. Signals from the pinger are recorded on a Precision Depth Recorder.

When used with an instrument lowered vertically the pinger is attached to the wire a few metres above the instrument (Fig. 2.2). Three sound signals are received from the pinger:

1. A direct signal transmitted up through the water from the pinger.
2. An echo, or echoes from the sampler, which will remain a constant interval from 1 throughout the lowering.
3. An echo off the sea bed, which may not appear until the pinger approaches bottom.

As lowering proceeds, the echo off the sea bed will converge towards signals 1 and 2, so enabling precise location of the instrument in relation to the bottom. It can also be used to position an underwater camera at the correct distance above the bottom in order to take photographs of the sea bed. Sometimes a counterweight mechanism is used to alter the pinging rate as bottom is

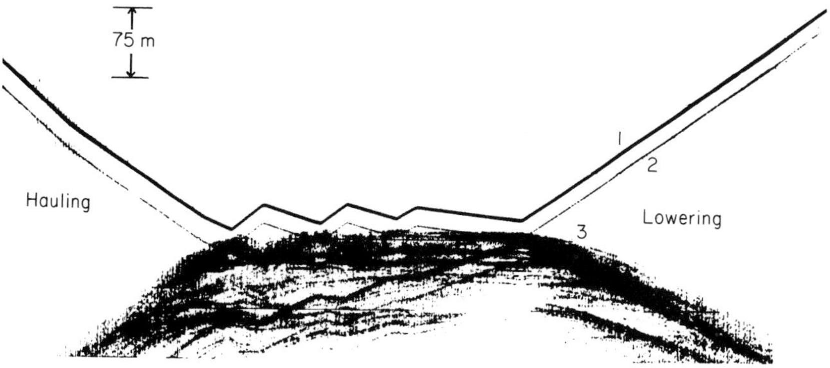

Fig. 2.3. Record of pinger signals when working the Holme grab in deep water. The grab had to be bounced on the bottom several times until an increased distance between pinger and grab echo showed that the grab had closed. 1, direct signal from pinger; 2, echo off grab; 3, sea-bed echo.

reached, giving even greater precision (Laughton, 1957; Hersey, 1967; Southward *et al.*, 1976).

The pinger not only gives precise control over lowering, but, in some instances, it is also possible to see on the record whether the sampler has operated successfully. For example, Hersey (1960) has shown that after the free-fall corer has released the component echoes off the gear change, the distance between the direct pinger signal and the echo off the body of the corer increasing. Similarly, when operating the Holme grab, which unwinds cable to close, the increased distance between pinger and grab echoes show up clearly (Fig. 2.3). Premature release of a grab or corer in mid-water while lowering would also be detectable in the same way. Bandy (1965) has shown how the echo off the large Campbell grab changes after closure and similar effects may perhaps be noticeable with other instruments (e.g. Hessler & Jumars, 1974).

Pingers are also useful for deep-sea dredging and trawling (see the section on deep-sea sampling in Chapter 6).

Position fixing by sonar

Sometimes positioning of the ship is best made in relation to sonar reflections off the sea bed. Sea floors, especially those swept by strong currents, can be remarkably varied in relief and composition, and the recognition of such features on a side-scan sonar trace can be a valuable aid to locating suitable sampling grounds. On the continental slope and in canyons the final manoeuvres of the ship are likely to rely more on echo soundings and side-scan plots than on more conventional means of navigation.

The side-scan sonar transducer is typically mounted either in the hull or in

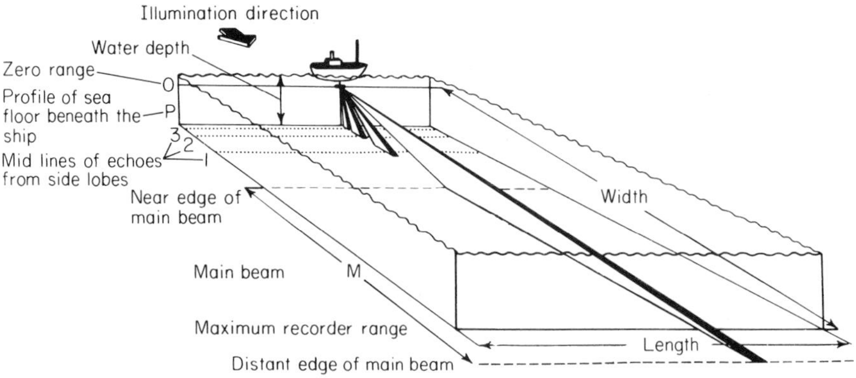

Fig. 2.4. Diagram showing operation of IOS side-scan sonar system. (From Belderson *et al.*, 1972.) Note that many side-scan systems are without side lobes.

a towed 'fish' (Fig. 2.4), the returning echoes being recorded graphically in the same way as echo sounder traces (Fig. 2.5). Owing to the oblique angle of the beam there is some distortion of scale across the near range part of the chart, however, an automatic slant-range correction system is now available for interpretation of sonographs. An outline of the techniques used for sea-bed

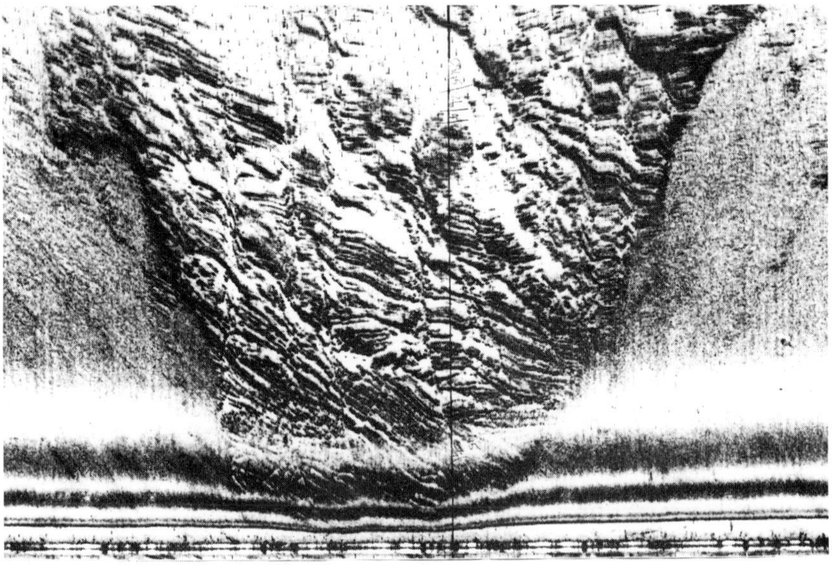

Fig. 2.5. Side-scan sonar record. An acoustic picture of a patch of sea floor about $3\frac{1}{2}$ km long and 750 m wide. The different bands towards the lower edge correspond to the lobes of the sonar beam that at the bottom correspond to a vertical echo sounding. The record shows a nearly flat sandy floor surrounding a ridge of slates. The continuity of individual beds is interrupted by fractures. (Copyright Institute of Oceanographic Sciences.)

search in relation to bottom features is given by Yules & Edgerton (1964), and the interpretation of sonographs is given in Belderson *et al.* (1972).

Special sonar equipment (GLORIA), with transducers of low frequency and high power output installed in a fish towed well below the surface, has been used for long-range examination of the continental slope and deep-sea floor (Kenyon *et al.*, 1978).

Acknowledgements

We are indebted to the Institute of Oceanographic Sciences, Wormley, England for permission to reproduce Figs 2.1 and 2.5, also to Dr A.H. Stride, formerly of that Institute, for advising on the section in this chapter relating to sonar, and for permission to reproduce Figs 2.1 and 2.4.

References

Admiralty, 1948. *Admiralty Manual of Hydrographic Surveying*, 2nd ed. Hydrographic Department, Admiralty, London, 572 pp.

Admiralty, 1977, 1973. *Manual of Navigation*, vol. I (1977) 544 pp.; vol. II (1973), 328 pp. H.M.S.O., London.

Bandy, O.L., 1965. The pinger as a deep-water grab control. *Undersea Technology*, **6**, 36.

Belderson, R.H., Kenyon, N.H., Stride, A.H. & Stubbs, A.R., 1972. *Sonographs of the Sea Floor. A Picture Atlas.* Elsevier, Amsterdam, London and New York, 185 pp.

Fifield, L.W.J., 1980. *Navigation for Watchkeepers.* Heinemann, London, 413 pp.

Hersey, J.B., 1960. Acoustically monitored bottom coring. *Deep-Sea Research*, **6**, 170–2.

Hersey, J.B., 1967. The manipulation of deep-sea cameras. Chapter 4, pp. 55–67, in J.B. Hersey (ed.) *Deep-Sea Photography.* The Johns Hopkins Oceanographic Studies, **3**, 310 pp.

Hessler, R.R. & Jumars, P.A., 1974. Abyssal community analysis from replicate box cores in the central North Pacific. *Deep-Sea Research*, **21**, 185–209.

Kenyon, N.H., Belderson, R.H. & Stride, A.H., 1978. Channels, canyons and slump folds on the continental slope between south-west Ireland and Spain. *Oceanologica Acta*, **1**, 369–80.

Kullenberg, B., 1955. Deep-sea coring. *Reports of the Swedish Deep Sea Expedition*, **4**, 35–96.

Laughton, A.S., 1957. A new deep-sea underwater camera. *Deep-Sea Research*, **4**, 120–5.

Matthews, D.J., 1939. *Tables of the Velocity of Sound in Pure Water and in Sea Water*, 2nd ed. Hydrographic Department, Admiralty, London, 52 pp.

Menzies, R.J., 1964. Improved techniques for benthic trawling at depths greater than 2000 metres. *Antarctic Research Series*, **1**, 93–109.

Menzies, R.J., 1972. Current deep benthic sampling techniques from surfaces vessels. pp. 164–9 in R.W. Brauer (ed.) *Barobiology and the experimental biology of the deep sea.* University of North Carolina, 428 pp.

Sachs, P.L., 1964. A tension recorder for deep-sea dredging and coring. *Journal of Marine Research*, **22**, 279–83.

Sonnenberg, G.J., 1978. *Radar and Electronic Navigation*. Newnes-Butterworths, London, 376 pp.

Soule, F.M., 1951. Physical oceanography of the Grand Banks region, the Labrador Sea and Davis Strait in 1949. *Bulletin of the U.S. Coast Guard*, **35**, 49–116. (*Collected Reprints, Woods Hole Oceanographic Institute*, **550**.)

Southward, A.J., Robinson, S.G., Nicholson, D. & Perry, T.J., 1976. An improved stereocamera and control system for the close-up photography of the fauna of the continental slope and outer shelf. *Journal of the Marine Biological Association of the United Kingdom*, **56**, 247–57.

Tydeman, G.F., 1902. Description of the ship and appliances used for scientific exploration. *Siboga Expedition*, **2**, 32 pp.

U.S. Navy Hydrographic Office, 1962. *American Practical Navigator. An epitome of Navigation. Originally by Nathaniel Bowditch, LLD.* U.S. Navy Hydrographic Office, Washington, 1524 pp.

Willerton, P.F., 1980. *Shiphandling for Masters, Mates and Pilots*. Stanford Marine, London, 152 pp.

Yules, J.A. & Edgerton, H.E., 1964. Bottom sonar search techniques. *Undersea Technology*, **9(3)**, 22–25, 43.

Chapter 3. Sediment Analysis

J. B. BUCHANAN

Introduction

The acquisition of large quantities of bottom sediment is an unavoidable consequence of benthic sampling. This, together with the fact that sediment grain size is easy to quantify, has often led to the gross over-simplification of considering the granulometric characteristics of the sediment to be almost synonymous with the physical environment and any correlation between the animals and the granulometry to be the key to benthic ecology. Such a correlation, when demonstrated, is far from an answer but is rather a statement of an ecological problem, which only an intimate knowledge of the biology of the individual species will resolve. Some animals, such as unselective mud swallowers living wholly within the sediment, may show a direct dependence on the nature of the sediment. To others, such as the suspension feeders, the sediment may represent nothing more than a convenient support. The distribution of both sorts of animals may nevertheless show a clear correlation with sediment type. In the first case, the correlation is direct, but in the second the correlation could be wholly indirect and the true determinants of the distribution would be found in that complex of factors, such as water movement, turbulence and suspended load, which, as a whole acting together, represent the sedimentary environment. An analysis of a sediment is a means of quantifying the result of an essentially dynamic process from a sample taken at a moment in time. The nature of a sediment is determined by the complex interaction of a large number of factors which are conveniently classified into 4 categories:

1 Factors determining source and supply of sedimentary material.
2 Factors determining transportation.
3 Factors determining deposition.
4 Post-depositional changes of mainly biogenic origin (e.g. biodeposition and bioturbation).

If the interaction of the various factors remains stable over a period of time, then it can be expected that the sedimentary environment, and, therefore, the nature of the sediment, will continue substantially unchanged. If, on the other hand, any short-term or long-term change takes place in any one of the categories of factors, there will be a corresponding alteration in the sediments. It is beyond the scope of a handbook of this sort to give full treatment to the dynamic aspects of sedimentation but a number of concise accounts may be found in text-books of marine geology (e.g. Kuenen, 1965). The results of sediment analyses will be much more meaningful biologically if an attempt is made to relate them to the causative environment of sedimentation.

Many of the techniques of sediment analysis which are used by biologists have been borrowed from the geological discipline of sedimentary petrography. Although there may be a considerable overlap of interest, it should be remembered that the questions asked and the answers sought by the two disciplines may be widely divergent. If the results of sediment analyses do not furnish a recognizable correlation of ecological significance, the uncritical use of unmodified geological techniques may prove a time-consuming and wasteful exercise. Since few measurable sediment properties vary independently, the interpretation of a correlation, in terms of the causation of the distribution of organisms, presents a considerable problem. Two sediments with disparate grain size characteristics will, in most cases, also show demonstrable and related differences in many other physical and biological properties, such as bulk density, porosity, capillarity, thixotrophy, permeability, oxygenation, plasticity, content and nature of organic matter and bacterial count. Morgans (1956) provides useful notes on the treatment and analysis of marine sediments, and discusses these problems with particular reference to the biologist. Other relevant techniques may be found in Krumbein & Pettijohn (1938), Trask (1955), Ackroyd (1964), Griffiths (1967) and Folk (1974), while Webb (1969) deals in particular with porosity and permeability and their relationship to animal distributions.

The traditional reliance on the techniques of sedimentary petrography has in recent years given rise to a tide of criticism which questions its appropriateness to purely biological investigations. The criticism is largely centred around the true nature of raw sediments with respect to the 'elements' present. A naturally occurring sediment may be regarded as being made up of two distinct types of particles—the 'mechanical elements' and the 'structural elements'. The mechanical elements are essentially the individual mineral grains, each one considered as a discrete entity, which together make up the bulk of the sediment. It is largely with this category that sedimentary petrology is concerned and steps are taken to ensure that the mineral grains are clean and completely separated one from another before the traditional analysis is commenced. The analysis is then undertaken on a sediment which is in a state of ultimate dispersion. In natural sediment, however, even a cursory microscopic examination will often reveal that there are structural elements present where the mineral grains are bound together to varying degrees into larger aggregates. Much of this structuring is due to activities of the fauna living within the sediment and would come under the category of post-depositional change. The most commonly found structural elements, particularly in more silty sediments, are undoubtedly the faecal pellets of both suspension and deposit feeders. These pellets can be persistent in the sediment over long periods of time and are often surprisingly well consolidated. In effect, then, a considerable bulk of the finer grades of mechanical elements in natural sediments may be structured by being bound together to form particles of very much greater dimensions than the original material, so that varying quantities of silt and clay may be transformed into what are essentially sand grade particles from the point of view of linear dimensions. These post-depositional changes are often vertically stratified within the sediment and, as a consequence of this structuring, the environment within the sediment can change profoundly with respect to a number of physical factors such as water content, density, permeability and porosity. This in turn produces demonstrable effects on the nature of the fauna and has given rise to a specialized branch of benthic ecology, organism–sediment relationships, a subject which has been extensively reviewed by Rhoads (1974).

There can be no doubt that the traditional methods of sedimentary petrography do not take cognizance of the possible structuring of sediments, but rather that they go to considerable lengths to destroy aggregates and organically bound elements in order to achieve ultimate dispersion. In the light of this it is not unreasonable to suggest that traditional methods may give results which are misleading in their interpretation of the true nature of the sediment. With clean sand sediments the problem of structuring is probably minimal but with increasing silt and clay content there is a tendency for structural elements to become increasingly important, to the extent that, in some sediments, $>50\%$ of the surface sedimentary material may be vested in faecal pellets. At the same time it must be admitted that traditional methods have given sound service in the past and that analyses carried out by these methods have produced powerful correlations between sediment and such aspects as the structure of the animal communities, feeding types and diversity measurements. Rather than abandoning traditional methods as being totally inappropriate, it is suggested here that they be retained, but that where structuring is suspected, a duplicate analysis be undertaken employing one of the methods which attempt to assess the degree of structuring.

Collection and storage of samples

The quantity of the sample required, and the means of obtaining it, must depend largely on the nature of the analysis contemplated and details will be considered under the heading of the various techniques. In some cases it may be sufficient to retain a small amount of sediment from the grab sample before sieving off the animals. On a patchy bottom this has the advantage of ensuring that both the animals and the sediment have come from the same locality. It has disadvantages, however, in that the sediment will often be disturbed, the natural stratification will be destroyed and some of the fine material may have been winnowed out of the surface layers by water movement during the ascent of the grab. If the distribution of a property with depth in the sediment is to be studied and the natural stratification observed, then a comparatively undisturbed core sample should be obtained. Sediment samples are best temporarily stored on shipboard or the field in convenient wide-mouthed glass jars, labelled internally as well as externally. The subsequent storage again depends on the techniques contemplated, but for some analyses the jars together with their contents can be oven dried at 100 °C, covered, and stored indefinitely. For other techniques, the storage of air dried sediment is recommended. If techniques require the use of raw sediment and there is a substantial lapse of time between collection and analysis the samples should be stored under refrigeration. Raw sediment, especially if taken from a highly organic mud, will rot if kept for any length of time at room temperature.

Particle size analysis

The essential equipment for analysis is a graded series of standard sieves suited to the intervals of the Wentworth scale. Much of the other equipment is to be found in the standard apparatus of most laboratories. A mechanical sieve shaker is useful but by no means essential. An electric beverage mixer is an ideal instrument for the mixing and dispersal of sediment samples. This should have a cup holding ≥500–600 ml with side baffles fitted and a close lid to prevent loss of material. As an alternative, the apparatus can be contrived from an ordinary laboratory stirrer and a glass beaker fitted with baffles.

Choice of grade scale

The object of size analysis is to obtain numerical, statistical and possibly graphical data which will serve to characterize a sediment in terms of the frequency distribution of grain size diameters. Size is here considered as the independent variable. The particles which make up a sediment belong to that class of frequency distribution known as a continuous distribution. That is to say that there is a continuous distribution of sizes from the largest to the smallest with the independent variable increasing by infinitesimals along its range of values. Such a distribution does not lend itself to frequency analysis unless some arbitrary series of finite increments is imposed on the independent variable to convert it into a discrete series with no gradations between. The series of finite increments is known as a grade scale. Each grade class has an upper and lower limit of size, and analysis determines the abundance of frequency within the class intervals. Many arbitrary grade scales have been suggested and the choice should suit the convenience of the analysis. Most marine sediments may be expected to contain particles ranging from a fraction of a micrometre to several thousand micrometres. Since it is generally

Table 3.1. Wentworth grade classification.

Name		Grade Limits	
		mm	μm
Boulder		>256	
Cobble		256–64	
Pebble		64–4	
Granule		4–2	
Very coarse sand	Sand	2–1	2000–1000
Coarse sand	Sand	$1-\frac{1}{2}$	1000–500
Medium sand	Sand	$\frac{1}{2}-\frac{1}{4}$	500–250
Fine sand	Sand	$\frac{1}{4}-\frac{1}{8}$	250–125
Very fine sand	Sand	$\frac{1}{8}-\frac{1}{16}$	125–62
Silt		$\frac{1}{16}-\frac{1}{256}$	62–4
Clay		$<\frac{1}{256}$	<4

Table 3.2. The Wentworth scale with the 2, $\sqrt{2}$ and ϕ notation.

	2 scale (mm)	$\sqrt{2}$ scale (mm)	ϕ
Sand	2	2	−1
		1·41	−0·5
	1	1	0
		0·71	+0·5
	0·50	0·50	+1·0
		0·351	+1·5
	0·250	0·250	+2·0
		0·177	+2·5
	0·125	0·125	+3·0
		0·088	+3·5
Silt	0·062	0·062	+4·0
		0·044	+4·5
	0·031	0·031	+5·0
		0·022	+5·5
	0·0156	0·0156	+6·0
		0·0110	+6·5
	0·0078	0·0078	+7·0
		0·0055	+7·5
	0·0039	0·0039	+8·0
Clay	<0·0039	<0·0039	

considered desirable to have closer grade intervals at the smaller end of the spectrum, an arithmetic scale would result in an impossibly large number of grade intervals. This difficulty is largely overcome by adopting a geometric grade scale. There has been a tendency in recent years for increasing numbers of benthic ecologists and marine geologists to use the Udden scale as modified by Wentworth. The Wentworth scale is geometric, based on 1 mm and a ratio of 2 (Table 3.1). The class intervals can be decreased if required by using the ratio $\sqrt{2}$ or $\sqrt[4]{2}$ instead of 2 (Table 3.2). It is a fact, which will be shown later to have considerable convenience in graphical treatment, that in a geometric series, the logarithms of the numbers to any base will form an arithmetic series. A logarithmic transformation was applied by Krumbein to the Wentworth scale in order to produce an arithmetic series of integers. This is the so-called phi notation where $\phi = -\log_2$ of the particle diameters in millimetres. The advantages of this procedure are to be found in graphical and statistical treatment. Only the Wentworth scale and the phi notation will be considered in the treatment of techniques which is to follow, but workers who wish to familiarize themselves with other scales should consult a text-book of sedimentary petrography (Krumbein & Pettijohn, 1938).

Analyses of mechanical elements

Clean beach sands—dry sieving

Sands with less than 5 % silt and clay are usually simple to deal with in the dry state. They show little tendency to inter-grain cohesion and contain only small amounts of organic matter, and pre-washing is not generally necessary. The analysis requires a stacked set of Wentworth grade sieves within the range 2000–62 μm. Finer resolution will be obtained by interpolating the $\sqrt[2]{}$ sieves to give $\frac{1}{2}\phi$ intervals. The stacks should be closed at the bottom end (after the 62 μm sieve) with a metal pan and the top closed with a cover. Undoubtedly a mechanical sieve shaker is of great benefit if a considerable number of analyses is to be undertaken. A preweighed sample (25 g is usually convenient) of oven-dried sand is introduced into the 2000 μm sieve at the top of the stack and the stack transferred to the mechanical shaker, where it should be agitated for 15 minutes. After shaking, the material on each individual sieve is weighed and noted together with any material < 62 μm which may have passed into the closing pan at the bottom of the sieve stack.

Silty sediments—rapid partial analysis by wet sieving

In conjunction with more extensive benthic surveys, sufficient information can often be derived from a measurement of the combined silt and clay content of the sediment. This involves an initial splitting of the sediment into a sand fraction (particles > 62 μm) and a silt–clay fraction (particles < 62 μm). The sand fraction may be further divided through a series of graded sieves, but the initial splitting is achieved with the 62 μm sieve employing a wet sieving method. If several 62 μm sieves are available the method is rapid and large numbers of samples can be processed in a short time. Oven dried sediment can be used for this method.

1 Accurately weigh 25 g of oven dry sediment.
2 Place sediment with about 250 ml of tap water in the dispersal beaker and add 10 ml of aqueous sodium hexametaphosphate $(NaPO_3)_6$ (6·2 g/l). Break up the sediment with a glass rod and then stir mechanically for 10–15 minutes. Allow the sediment to soak overnight and restir for 10–15 minutes.
3 Wash the sediment suspension on to a 62 μm sieve placed in a white basin, adding water until the sieve surface is submerged. Sieve by 'puddling' the sieve in the basin of water. From time to time, the material passing the sieve into the basin may be discarded and the basin filled with clean tap water until no further material is seen to pass, and the water remains clear against the white background of the basin.

4 Transfer the sieve and contents to a drying oven and dry rapidly at 100 °C.
5 Gently remove the sieve from the oven and place over a large sheet of glazed white paper. Vigorously knock and agitate the sieve over the white paper; discard any material which passes and continue dry sieving till no further material can be observed on the white paper. If the material on the sieve has tended to 'cake' during oven drying it should be gently broken up with a dry clean camel hair brush.
6 The initial splitting is now complete. Any material remaining on the 62 μm sieve may be regarded as the sand fraction. The weight of this material subtracted from original sample weight will give the silt–clay fraction by difference. If required, the sand fraction may be graded further through the sand sieve series.

Full analysis of the silt–clay fraction

After a sediment sample has been split on a 62 μm sieve, the sand fraction may be further graded by dry sieving. The material of the silt–clay fraction, which passes through the 62 μm sieve, is generally of a size which is below the practical level for further sieving. If the silt–clay fraction has been retained and its further grading is considered essential, then it will be necessary to employ some form of sedimentation analysis. It should be noted that both the preliminary preparation and the actual mechanical process of sedimentation analyses can be very time-consuming. The principle of sedimentation analysis is simply that large particles will fall faster and farther than small particles through a column of distilled water in a given time. By first assuming that all of the particles are spheres and that all have the specific gravity of quartz (2·65), it is possible to quantify the distance of fall in a given time for any particular diameter of particles by using the classic formula for settling velocities provided by Stokes' Law. The times of settling, for the Wentworth grade scale, computed from Stokes' Law, are given in Table 3.3 for a temperature of 20 °C.

Because of its simplicity and since it does not require any sophisticated apparatus, pipette analysis is generally regarded as a suitable method for

Table 3.3. Settling times in distilled water at 20 °C.

Diameter (mm)	Settling distance (cm)	Time Hours	Minutes	Seconds
0·0625	20			58
0·0312	10		1	56
0·0156	10		7	44
0·0078	10		31	0
0·0039	10	2	3	0

benthic ecology. The essential apparatus for the analysis is a 1-litre measuring cylinder, preferably of the stoppered type, and an ordinary 20 ml pipette which has a stem length below the bulb of >20 cm. The stem should be etched and well marked with two rings at distances of 10 cm and 20 cm, respectively, from the tip. Approximately 30 cm of flexible rubber tubing should be fixed to the mouth end of the pipette. For many problems of benthic ecology, bearing in mind the time factor, it should be necessary only to divide the fine fraction into three grades: coarse silt (62–15·6 μm), fine silt (15·6–3·9 μm) and clay ($<3·9$ μm). This can be achieved by taking only three pipette samples. If, however, the sediment is predominantly silt and clay, then a full analysis will be necessary at each phi interval.

The mechanics of the analysis are simple. The dispersed silt–clay fraction is transferred to the cylinder with distilled water to a volume of exactly 1 litre. The sediment is suspended by shaking and turning the stoppered cylinder, and when the sediment is judged to be uniformly dispersed throughout the suspension the cylinder is placed upright and a stop-watch started. From the moment of placing the cylinder upright, all of the particles will start to fall through the water column under the action of gravity and at the settling velocities appropriate to their individual diameters. If, as an example, particles of 15·6 μm are considered, Table 3.3 indicates that such particles will fall a distance of 10 cm in 7 min 44 sec. After this length of time, it can be confidently expected that the part of the water column between the surface and a depth of 10 cm will be free of all particles $>15·6$ μm but that at the level of exactly 10 cm there will still be a representative amount of all particles $<15·6$ μm, occurring in the same proportions as they did when the sediment was originally homogeneously suspended throughout the water column. If a 20 ml pipette sample is taken at the level of 10 cm after a lapse of time of 7 minutes 44 seconds and if the material in the pipette is subsequently dried and weighed it should represent the weight of material $<15·6$ μm contained in a volume of 20 ml of the original suspension. This figure can easily be extrapolated by multiplying by fifty to obtain an estimate for the whole suspension. Provided that both the weight of material <62 μm and the weight of that $<15·6$ μm is known, it is simply a matter of subtraction to find the weight of material lying between the grade limits 62–15·6 μm. The analysis proceeds in this manner from the coarsest to the finest grades.

When taking a sample, the pipette is lowered very gently into the cylinder (preferably by mechanical means) to the appropriate mark just before the expiry of the time interval. The pipette can be held steadily at this depth by one hand resting on the top of the cylinder. To take a sample an even suction is applied by the mouth through the rubber tube attached to the pipette. When the pipette is filled to the mark the rubber tube is pinched with the free hand, the pipette is removed from the cylinder and the contents of the pipette transferred to a tared 50 ml crystallizing dish. The sample is dried at 100 °C, but care must be taken to avoid boiling and possible spattering. The dried sample, cooled in a desiccator, should be weighed accurately to the third decimal place, preferably on an aperiodic balance. Some workers advocate the use of more elaborate suction devices, usually involving a suction pump and an evacuated aspirator, for taking samples, but after some preliminary practice, mouth suction produces adequately accurate results. In order to be confident of his results, each worker should first familiarize himself with the technique by comparing a series of replicate pipette samples taken from the same sedimentation grade.

Variation in temperature can lead to serious errors in pipette analysis. If the table of settling velocities computed for 20 °C is employed, it is essential that the entire analysis is carried out at this temperature. A quartz particle of 50 μm diameter will show an increase of settling velocity of approximately 2·3 % for each Centigrade degree rise in temperature. For this reason the sedimentation cylinder should ideally be immersed in a thermostatically-controlled water bath to within a few centimetres of its mouth. Even if such a bath is not available, it is still good practice to immerse the cylinder in a water bath at the ambient temperature in order to buffer the effects of room temperature fluctuations. If the analysis is carried out in a room temperature which differs from 20 °C, it will be necessary to recalculate a new table of settling velocities from the Stokes' Law formula. The formula may be expressed as:

$$V = Cr^2 \tag{3.1}$$

where V = the settling velocity in cm sec^{-1}, C is a constant, and r = radius of the particle in cm.

The constant $C = 2(d_1 - d_2)g/9z$, where d_1 = density of the particle = 2·65 (quartz), d_2 = density of water, g = acceleration of gravity = 980 cm sec^{-2}, and z = viscosity of water in dyne seconds cm^{-2} (1 dyne sec cm^{-2} = 1 poise).

In the range of temperature 15–30 °C, the density of water can be regarded as 1 without great loss of accuracy, although the figure for viscosity varies considerably. The numerator of the equation for this temperature range may be conveniently regarded as 3234, so that $C = 3234/9z$.

The values for z over a wide temperature range can be extracted from physical tables of water viscosity. Most tables given the value in centipoises (0·01 poise), so that it will be necessary to divide the value by 100 before applying to the equation. It should also be noted that although most tables for settling velocities are calculated for the diameter of the particle in millimetres, Stokes' Law is calculated from the radius of the particle in centimetres. The values of C for 15 °C and 20 °C are $3{\cdot}14 \times 10^4$ and $3{\cdot}57 \times 10^4$, respectively.

When choosing the weight of sediment to start a pipette analysis, the aim should be to arrive at a suspension of silt and clay weighing approximately 15 g. That is to say, if the sediment contains 50 % sand and 50 % silt and clay, an initial sample of 30 g should be chosen to start the analysis. Because of this, no universal starting weight can be recommended and it is necessary to carry out a rapid partial analysis for silt and clay before embarking on the pipette analysis. The complete analysis is outlined below in seven stages.

1 Pretreatment

(a) Weigh an appropriate amount of air-dried sediment to produce approximately 15 g of silt and clay fraction.

(b) Transfer the sediment to a litre beaker with 100 ml of 6 % hydrogen peroxide. Stand overnight and then heat gently on a water bath. Add further small quantities of peroxide until there is no further reaction. This process should effectively remove the organic matter.

(c) Wash the contents of the beaker on to a filter paper (Whatman No. 50) in a Buchner funnel. Wash thoroughly under gentle suction with distilled water to remove any electrolytes.

(d) Wash the sediment from the filter paper into the cup of the mechanical stirrer using a jet of distilled water and a camel hair brush. About 200–300 ml of water should be used. Add 10 ml of sodium hexametaphosphate solution (6·2 g l^{-1} aqueous), and stir for 10–15 minutes. Leave the sediment to soak overnight.

2 Initial splitting of silt–clay fraction

Stir the sediment again for 10–15 minutes and then transfer to the surface of a clean 62 μm sieve placed in a flat-bottomed white basin. Add about 300–400 ml distilled water, sufficient to flood the sieve surface. The volume of water within the basin should on no account exceed 1 litre. Wet sieve the sediment by agitating and puddling in the basin of water until most of the fine fraction has passed. Lift the sieve and allow to drain over the basin. Transfer the sieve and its contents to a drying oven at 100 °C.

3 Dry sieving of the sand fraction

Although the process of wet sieving will have removed most of the material < 62 μm, a certain amount will have been retained by the water film on the sieve. It is, therefore, necessary to subject the sieve to a period of dry sieving. The first stage of this can be carried out by agitating the dried 62 μm sieve over a large sheet of white glazed paper and transferring any fine material which passes through to the suspension in the basin. When no further fine material is seen to pass, the contents of the sieve can be transferred carefully to the uppermost (coarsest) of a stacked series of graded sand sieves, taking care to gently brush all of the material from both surfaces and the pores of the 62 μm sieve with a fine sieve brush. Although the bulk of the silt–clay fraction will now have been removed it is still advisable to have a fresh 62 μm sieve at the bottom of the stack of sieves and care should be taken to use a pan below this finest sieve to catch the last of any fine material which may still pass. The stacked column of sieves may now be transferred to an automatic sieve shaker for a period of 10–15 minutes. If an automatic sieve shaker is not available, the column of sieves should be rocked and tapped with the flat of the hand until sieving on the topmost sieve is judged to be complete. This sieve may then be detached and any doubts about the finality of sieving may be checked by agitating and tapping over a sheet of white glazed paper. Any material passing through should be immediately transferred to the next finer sieve. When the finality of sieving has been checked, the material on the sieve should be emptied on to the sheet of glazed paper and any grains lodged in the sieve should be removed with a sieve brush. The fraction can then be transferred to the balance pan for weighing. The analysis continues sieve by sieve through the series until, finally, any material passing the last (62 μm) sieve into the pan is transferred to the suspension of silt and clay in the basin.

4 First pipette sample

The grading of the silt–clay fraction may now proceed. It will be assumed that the analysis takes place at 20 °C. Wash the fine material in the basin into the litre cylinder using a large filter funnel and a wash bottle of distilled water, then make up the volume to exactly 1 litre with distilled water. Place the cylinder in the thermostatic water bath until the temperature has equilibrated at 20 °C. Remove the cylinder, shake and turn to suspend the sediment evenly throughout the water column. Place upright and immediately withdraw a 20 ml pipette sample from a depth of 20 cm. Transfer the pipette sample to a tared crystallizing dish and dry in the oven at 100 °C. The weight of this material will represent the total amount of sediment <62 μm in the suspension.

5 Second pipette sample

A few seconds before the expiry of 7 min 44 sec, lower the pipette tip to a depth of exactly 10 cm below the surface of the suspension. At exactly 7 min 44 sec withdraw the 20 ml sample and transfer to a tared crystallizing dish. Dry at 100 °C.

6 Third pipette sample

Take the third sample after a time interval of 2 h 3 min. Dry the sample. It is not necessary to resuspend pipette samples.

7 Calculation of results

Wt. of first pipette sample + dish	31·799
Wt. of dish	31·317
∴ Wt. of material <62 μm in 20 ml suspension	0·482
∴ Wt. of material <62 μm in 1 litre suspension = 0·482 × 50 = 24·100 g	
Wt. of second pipette sample + dish	30·247
Wt. of dish	30·126
∴ Wt. of material <15·6 μm in 20 ml suspension	0·121
∴ Wt. of material <15·6 μm in 1 litre suspension = 0·121 × 50 = 6·050 g	
Wt. of material finer than 62 μm	24·100
Wt. of material finer than 15·6 μm	6·050
Difference: amount in 62–15·6 μm grade	18·050

Similar calculations are carried out for the third pipette sample to give the weight of material in the 15·6–3·9 μm grade together with the amount $<3{\cdot}9\ \mu$m (i.e., the clay content). Finally the results of the pipette analysis are combined with the sieve analysis and the weights in each grade expressed as a percentage of the dry weight of the total sample.

Analysis of structural elements—natural aggregates

Although there is now a degree of consensus and standardization with respect to the analysis of the mechanical elements of a sediment, no such agreement exists in the treatment of the structural elements and the subject is open to innovation. In the full analysis of mechanical elements, the assumption that the sediment consists of fully dispersed quartz spheres is made. This ideal might be approximated in some clean sands, but it is very far from the truth in sediments containing substantial quantities of silt and clay. Silty sediments, as a rule, contain varying quantities of natural aggregates in the form of faecal pellets and also more amorphous organic-mineral particles with variable degrees of aggregation (Johnson, 1974). In addition, such sediments generally contain a highly diverse selection of particle 'species' of biogenic origin (Whitlach, 1981)—diatoms, echinoderm spines, polychaete tubes, crustacean exuvia, etc. Two possible treatments commend themselves:

1 Gentle wet sieving of raw untreated sediment. This technique has proved useful in assessing the degree of 'pelletization' of silty sediments.
2 Direct microscopical observation of small representative samples of sediment in order to measure and categorize the particle species.

Wet sieving of untreated sediment

Since many of the natural aggregates are vertically stratified, it is usual to treat slices of extruded cores (Moore, 1931). For the analysis, the portion of wet untreated sediment is placed on the largest sand sieve (2 mm) and the sieve plus contents is 'puddled' gently in a white basin of sea water which conveniently fits the sieve diameter. The technique requires two basins, since the material which passes the coarse sieve into the basin is transferred with the water to the next finer sieve in another basin, and so on. The material on each sieve can be washed into petri-dishes and the aggregates counted, measured and, if desired, picked out for dry weight measurement. If the weight of aggregates is to be compared with the total weight of the sample, it will be necessary to dry and weigh all residues after the removal of the aggregates. This will naturally include all of the material in the water which has passed the finest sieve. This can be recovered by filtration under gentle suction in a Buchner funnel using a washed, tared, hardened filter paper (e.g., Whatman No. 50), which is subsequently dried and weighed.

Direct microscopic observation

Varying degrees of elaboration have been employed in the microscopic observation of small representative samples of raw sediment:

1 Measurement of particle size by eyepiece graticule, with or without categorizing particle species. A magnification of 200× is usually recommended and particles <5 μm (on occasion 10 μm) are not measured.
2 Measurement of particles after ethanol fixation and histochemical staining in order to distinguish organic material from inorganic (Johnson, 1974, 1977; Whitlach & Johnson, 1974; Hughes, 1979). Periodic and Schiff (PAS) appears to be the most useful stain in this respect (Whitlach, 1981).
3 The use of fluorescence microscopy to assess the relative importance of adsorbed bacteria and other micro-organisms (Hughes, 1979, Meadows & Anderson, 1968).

There is no doubt that, especially for silty sediments, microscopic observations should be regarded as an essential accompaniment to conventional analysis of the mechanical elements.

Graphical presentation and statistics of analysis data

Presentation of two variables

Two simple devices are available for depicting size frequency distribution data:

1 The frequency distribution histogram.
2 The cumulative frequency curve.

Of these, the second is preferable since it allows the direct derivation of statistical data. The size of the particle grades should be laid off on the horizontal axis using the phi notation and arithmetic graph paper. By convention, the phi values should increase positively to the right (i.e., particle diameters should decrease towards the right). At the upper limit of the first (largest diameter) class interval, an ordinate is erected equal in height to the frequency percentage in that class. At the end of the second class interval another ordinate is erected equal in height to the cumulative sum of the frequencies of the first two classes. At the end of the third, the ordinate should equal a height of the sum of the first three classes, and so on. The cumulative curve is in effect similar to setting each block of a histogram above and to the right of its predecessor. When complete, the ordinates are joined with a smooth curve. If the analysis of fine sediment has been incomplete at the small diameter end of the spectrum, the curve can be left open-ended, but for

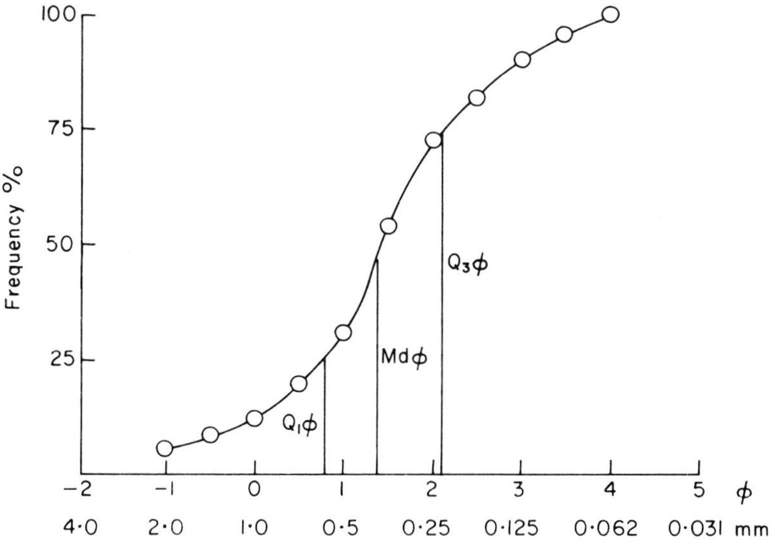

Fig. 3.1. Cumulative curve of beach data sand from Table 3.4: $Md\phi = 1{\cdot}36$ $Qd\phi = (2{\cdot}10 - 0{\cdot}80)/2 = 0{\cdot}65$. $Q_1\phi = 0{\cdot}80$. $Q_3\phi = 2{\cdot}10$ $Skq\phi = [(0{\cdot}80 + 2{\cdot}10)/2] - 1{\cdot}36 = +0{\cdot}09$.

statistical purposes it is essential that $\geq 75\%$ by weight of the sediment should have been graded.

An example of a cumulative curve derived from sieving analysis of a beach sand is shown in Fig. 3.1 and the relevant analysis data are presented in Table 3.4. It should be noted that the cumulative frequency curve allows the

Table 3.4. Sieve analysis data—beach sand.

mm	ϕ	% wt. on sieve	Cumulative %	Grade
4·00	−2·0	0	0	4·00
2·83	−1·5	0	0	4·00–2·83
2·00	−1·0	5	5	2·83–2·00
1·41	−0·5	3	8	2·00–1·41
1·00	0	4	12	1·41–1·00
0·71	+0·5	8	20	1·00–0·71
0·50	+1·0	11	31	0·71–0·50
0·351	+1·5	23	54	0·50–0·351
0·250	+2·0	19	73	0·351–0·250
0·177	+2·5	9	82	0·250–0·177
0·125	+3·0	8	90	0·177–0·125
0·088	+3·5	6	96	0·125–0·088
0·062	+4·0	4	100	0·088–0·062

comparison of sediments analysed by different sieve series, provided that the aperture interval between successive sieves is not too great.

Statistical treatment of two variables

It is often convenient to express the characteristics of the size frequency curve in terms of numbers. Three attributes of the curve are generally considered:

1 A measure of central tendency.
2 A measure of degree of scatter.
3 A measure of the degree of asymmetry.

The median diameter is a good measure of central tendency and this can be easily determined from the cumulative curve by reading the phi value which corresponds to the point where the 50 % line crosses the cumulative curve. It is defined as the phi value which is > 50 % of the phi values in the distributions and smaller than the other 50 %. It should be abbreviated as Mdϕ.

The median diameter gives an average value and represents a central point, but it does not indicate the degree of spread of the data about this central tendency. A second measure, the quartile deviation, measures the number of phi units lying between the first and third quartile diameters, that is, between the 25 % and 75 % points on the cumulative curve, where:

$$QD\phi = (Q_3\phi - Q_1\phi)/2$$

A sediment with a small spread between the quartiles is regarded as being 'well sorted'.

The quartile deviation, which measures the spread, does not give any indication of the symmetry of the spread on either side of the average. If there is a tendency for the data to spread on one side more than the other, this asymmetry is called skewness. The phi quartile skewness may be calculated from the equation:

$$Sk_q\phi = (Q_1\phi + Q_3\phi)/2 - Md\phi$$

A positive value will indicate that the mean of the quartiles lies to the right of the Mdϕ and should be prefixed '+', while a negative value would lie to the left and should be prefixed '−' to indicate negative skewness. The values for the various statistical measures have been worked out for the curve in Fig. 3.1.

The three measures—Mdϕ, QDϕ and $Sk_q\phi$ have been criticized by Folk (1974), on the grounds that they deal only with 50 % of the distribution and ignore the 'tails' of the curve: in the case of a skewed sediment or a bimodal distribution they would cease to be sensitive measures. Folk developed three measures which utilize up to 90 % of the curve:

1 Graphic Mean,

2 Inclusive Graphic Standard Deviation

$$\sigma_I = \frac{\phi 84 - \phi 16}{4} + \frac{\phi 95 - \phi 5}{6 \cdot 6}$$

Verbal classification

<0·35	very well sorted
0·35–0·50	well sorted
0·50–0·71	moderately well sorted
0·71–1·00	moderately sorted
1·00–2·00	poorly sorted
2·00–4·00	very poorly sorted
>4·00	extremely poorly sorted

3 Inclusive graphic skewness

$$Sk_I = \frac{\phi 16 + \phi 84 - 2\phi 50}{2(\phi 84 - \phi 16)} + \frac{\phi 5 + \phi 95 - 2\phi 50}{2(\phi 95 - \phi 5)}$$

Verbal classification

+1·00 to +0·30	strongly fine skewed
+0·30 to +0·10	fine skewed
+0·10 to −0·10	symmetrical
−0·10 to −0·30	coarse skewed
−0·30 to −1·00	strongly coarse skewed

Since these measures make use of the 'tails' of the curves, the cumulative curve should be drawn on probability paper, which has the effect of straightening and extending the 'tails' to give greater accuracy. The use of probability paper also allows an immediate assessment of departure from the normal distribution, i.e., a departure from a straight line in the cumulative curve. This can be quantified using a measure of kurtosis. If the central portion of the frequency distribution is excessively peaked, the curve is leptokurtic. If the curve is flat peaked, it is platykurtic.

Graphic kurtosis:

$$K_G = \frac{\phi 95 - \phi 5}{2 \cdot 44(\phi 75 - \phi 25)}$$

Verbal classification

<0·67	very platykurtic
0·67–0·90	platykurtic
0·90–1·11	mesokurtic (nearly normal)
1·11–1·50	leptokurtic
1·50–3·00	very leptokurtic

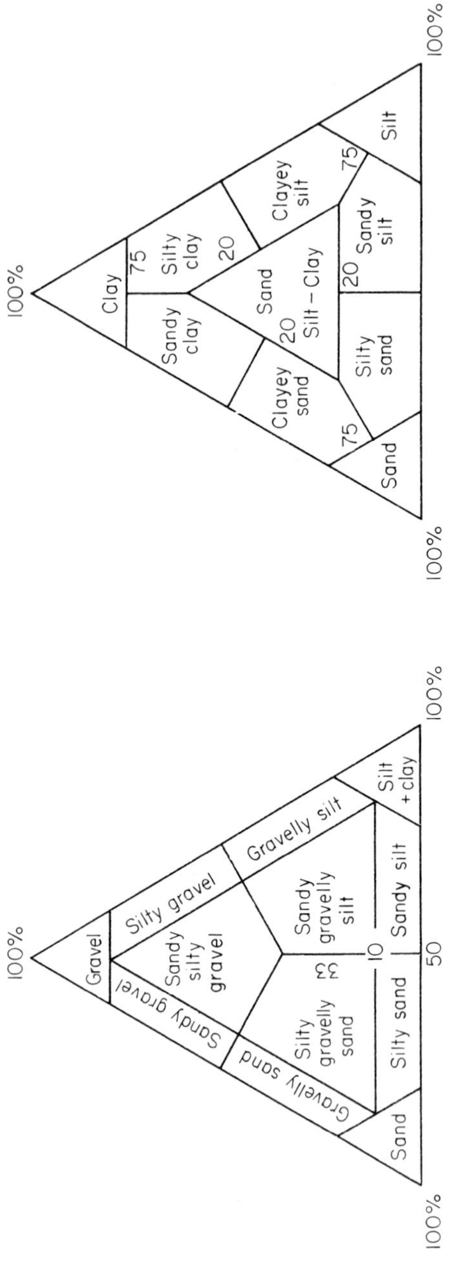

Fig. 3.2. Possible methods of presenting sediment data as triangular graphs.

Presentation of three variables

In a rather different approach, a sediment can be characterized with respect to its percentage content of three variables, for instance sand, silt and clay. This can be done by utilizing the grade nomenclature of the Wentworth scale. If triangular graph paper is used, it is possible to represent the percentage content of three variables by a single graph point. When a number of different sediments are represented by points on the same graph, it is often possible to classify the sediments into groups which have regional ecological significance.

In practice, each of the three vertices of the triangular graph are labelled for one of the three variables. The distance from each vertex to its corresponding base is considered to be 100%. If, for instance, a sediment contains 30% sand, 60% silt and 10% clay, the plotting would proceed as shown in Fig. 3.2. Move a ruler 30 units up the sand axis and draw a faint line parallel to the base. Repeat the process moving 60 units up the silt axis and draw a second faint line parallel to the silt base. The transection of the two lines fixes the point which will automatically also be correct for the clay percentage.

Another common function for the triangular graph is to subdivide the entire field into classes for descriptive purposes. Two of many possible subdivisions are shown in Fig. 3.2 the one for sand, silt and clay, and the other for coarser sediments using gravel (particles >2 mm), sand and silt + clay.

Measurement of some other physical properties

Porosity

Porosity is the percentage volume of pore space in the total volume of sediment. In a submerged sediment it measures, in effect, the volume occupied by water. Although the concept is simple it is, in fact, difficult to measure with accuracy. Since porosity varies markedly with depth in the sediment, it is essential to obtain undisturbed cores for depth sectioning, avoiding core compaction during sampling. The determination of porosity involves the measurement of the volume of water in the wet sediment and the volume of the sediment grains. Both volumes may be measured indirectly by gravimetric techniques. The weight of water in the sediment can be obtained by washing the salt out of the sediment, drying, and subtracting the dry weight of washed sediment from the original weight of wet sediment. Assuming that the salinity of the interstitial water is similar to the overlying sea water, the weight can be converted to volume using sea water density tables. The dry weight of washed sediment can be converted to approximate volume by assuming a mean grain specific gravity of 2·65.

$$\text{Porosity } (P) = \frac{V_{\text{water}} \times 100}{V_{\text{water}} + V_{\text{solids}}}$$

$$= \frac{\dfrac{M_{\text{water}}}{D_{\text{water}}} \times 100}{\dfrac{M_{\text{water}}}{D_{\text{water}}} + \dfrac{M_{\text{solids}}}{D_{\text{solids}}}}$$

M = mass and D = density. $D_{\text{water}} = 1{\cdot}025\text{ g cm}^{-3}$ (at salinity 35‰), $D_{\text{solids}} = 2{\cdot}65$ (quartz).

In some sediments which are highly pelletized, the amount of water in the upper layers may well be $> 80\%$, and it is advisable to correct the M values for salt (Swan *et al.*, 1982).

Permeability

For sediments, permeability is usually understood as the rate at which water passes through a cylindrical section of core. The coefficient of permeability P can be conveniently measured using a constant head permeameter (Fig. 3.3), P being calculated as follows:

$$P = \frac{QL}{hAt}$$

where Q = volume of water collected, L = length of core, h = head of water, A = cross sectional area of core, t = time.

If all measurements are in c.g.s. units, P will be in cm sec^{-1}. A full treatment of permeability may be found in Frazer (1935) and some aspects of biological significance are covered in Webb (1958, 1969).

Capillary rise

Of particular importance in beach deposits, capillary rise was studied by Bruce (1928) for both graded and ungraded sands. The procedure is to arrange a number of drained sediment cores in a frame with their lower ends resting in a shallow dish of sea water and to measure the capillary rise against time. The temperature of the sea water must be kept constant in order to avoid viscosity changes.

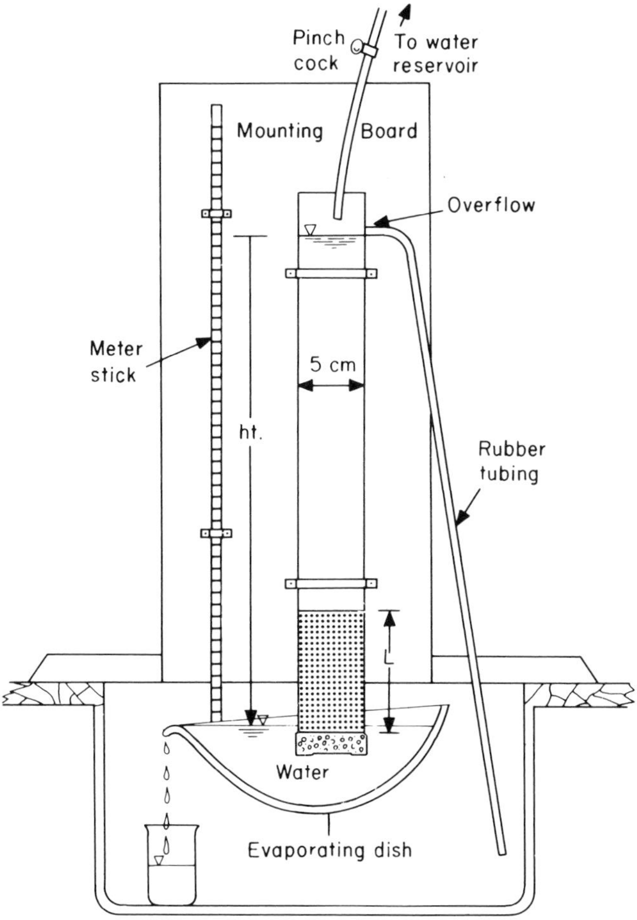

Fig. 3.3. Constant head permeameter. (After Means & Parcher, 1964.)

pH and Eh

A concise account of the measurement and biological significance of pH and redox profile can be found in Fenchel (1969) and Pearson & Stanley (1979). In both cases, undisturbed core samples are required.

Temperature

Thermoprobes are available to geologists for measuring the temperature of sediments *in situ* but most biologists will find that adequate results can be obtained by pushing a thermometer into the middle of a grab sample as soon as it is brought up. Sediments act quite effectively as insulating water bottles.

Other factors

A number of other physical factors may be measured, such as shear strength, plasticity and thixotropy. A good general account will be found in Emery (1960), but the biological significance of such measurements has not as yet been fully demonstrated.

Determination of organic matter in sediments

For the routine determination of organic matter in sediments, a rough estimate may be obtained by measuring the loss of weight on ignition at 600 °C of a dried sediment sample from which the carbonates have been previously removed by acid treatment.

For more accurate determinations, the organic nitrogen or organic carbon should be estimated. The most commonly adopted method involves the estimation of organic carbon using modifications of the Schollenberger chromic acid oxidation technique. In this, the sediment sample is digested with a chromic acid–sulphuric acid mixture and the excess of chromic acid not reduced by the organic matter is titrated with a standard ferrous salt. Two routine methods are available:

1 The method of Walkley & Black (1934).
2 The method of El Wakeel & Riley (1956).

There is little to choose between the methods, both give similar results and both are easy to operate.

The analytical procedure for the Walkley & Black method is as follows.

Reagents

N Potassium dichromate. Dissolve 49·04 g of reagent grade $K_2Cr_2O_7$ in water and dilute to 1 litre.

Sulphuric acid. Not less than 96 % with 1·25 g of silver sulphate added for every 100 ml of acid. (The silver sulphate removes the interference of chlorides.)

Phosphoric acid. At least 85 %.

Diphenylamine. Dissolve 0·5 g diphenylamine in 20 ml water and add 100 ml of conc. sulphuric acid.

N Ferrous sulphate. Dissolve 278 g of reagent grade $FeSO_4 . 7H_2O$ in water, add 15 ml of conc. sulphuric acid and dilute to 1 litre. Standardize by titrating against 10·5 ml potassium dichromate as described below.

Method

The sediment sample should be ground to pass a 0·5 mm screen. Transfer a weighed quantity of sediment (≤ 10 g) containing 10–25 mg organic carbon to a 500 ml conical flask. The most suitable quantity of sediment for the estimation can only be judged by experience of the benthic area being sampled. In practice a weight of about 1·5 g is often suitable. If, however, the sediment is very highly organic, it is possible that this amount will reduce all of the dichromate and leave none to be titrated. This situation will be immediately obvious at the commencement of the titration and can be corrected by a reduction in the weight of sediment used. If the titration value is <1 ml, the sediment weight should be reduced to one-third of the original weight. If, on the other hand, the titration value is >9 ml, the weight of sediment should be trebled. Add 10 ml of normal potassium dichromate followed by 20 ml of conc. sulphuric acid. Shake by hand for one minute, then place in a boiling water bath for 30 min. Cool, add 200 ml water, 10 ml phosphoric acid and 1 ml diphenylamine indicator solution. Titrate by adding ferrous sulphate from an automatic burette until the solution is purple or blue. Continue to add the ferrous sulphate in small lots of about 0·5 ml until the colour flashes to green. This occurs with little or no warning. Then add 0·5 ml dichromate to restore an excess of dichromate and complete the titration by adding ferrous sulphate drop by drop until the last trace of blue colour disappears.

The end point can easily be recognized to within one drop of ferrous sulphate. The colour is not always purple on adding the indicator at the beginning of the titration but the purple or blue always appears just before the end point. The original blue frequently fails to reappear on the addition of 0·5 ml excess potassium dichromate but soon redevelops after the addition of the first drop or two of ferrous sulphate.

Calculation

One ml of normal dichromate is equivalent to 3 mg of carbon. The amount of carbon present in the sediment is, therefore, given by the expression:

$$\frac{V_1 - V_2}{W} \times 0{\cdot}003 \times 100$$

where V_1 equals the volume of normal potassium dichromate (10·5 ml), V_2 equals the volume of ferrous sulphate in ml, W equals the weight of soil taken.

The percentage recovery by this method varies according to the sediment type and is 75–90 %. For this reason it is undesirable to use a conversion factor to correct the results; rather, the results are expressed as single value determinations and are designated 'Organic Carbon, chromic acid oxidation values'.

Nitrates interfere only if present in amounts in excess of one twentieth of the carbon content. Carbonates, even when they constitute 50 % of the soil, do not affect the results. Manganese dioxide may also be present (three or four times in excess of the carbon content does not introduce serious error). Interference due to chlorides is overcome by the addition of excess silver sulphate to the sulphuric acid. Elementary carbon is almost unattacked in this method so this source of error is eliminated. The presence of coal will invalidate the estimations, since it reacts with both organic carbon and nitrogen estimations to give highly inflated values. To overcome this difficulty in Northumberland sediments, Buchanan & Longbottom (1970) resorted to protein analysis.

References

Ackroyd, T.N.W., 1964. *Laboratory Testing in Soil Engineering*. Soil Mechanics Ltd., London, 233 pp.

Bruce, J.R., 1928. Physical factors on the sandy beach. Part I. Tidal, climatic and edaphic. *Journal of the Marine Biological Association of the United Kingdom*, **15**, 535–565.

Buchanan, J.B. & Longbottom, M.R., 1970. The determination of organic matter in marine muds: the effect of the presence of coal and the routine determination of protein. *Journal of Experimental Marine Biology and Ecology*, **5**, 158–169.

Emery, K.O., 1960. *The Sea off Southern California. A modern Habitat for Petroleum*. John Wiley & Sons, New York, 366 pp.

Fenchel, T., 1969. The ecology of marine microbenthos. IV. Structure and function of the benthic ecosystem, its chemical and physical factors and the micro-fauna communities with special reference to ciliated Protozoa. *Ophelia*, **6**, 1–182.

Folk, R.L., 1974. *Petrology of Sedimentary Rocks*. Hemphill Publishing Co., Austin, Texas, 182 pp.

Frazer, J.H., 1935. Experimental study of porosity and permeability of clastic sediments. *Journal of Geology*, **43**, 910–1010.

Griffiths, J.C., 1967. *Scientific Method in Analysis of Sediments*. McGraw Hill, New York, pp. 109–173.

Hughes, T.G., 1979. Studies on the sediment of St Margaret's Bay, Nova Scotia. *Journal of the Fisheries Research Board of Canada*, **36(5)**, 529–536.

Johnson, R.G., 1974. Particulate matter at the sediment–water interface in coastal environments. *Journal of Marine Research*, **32(2)**, 313–330.

Johnson, R.G., 1977. Vertical variations in particulate matter in the upper twenty centimetres of marine sediments. *Journal of Marine Research*, **35(2)**, 273–282.

Krumbein, W.C. & Pettijohn, F.J., 1938. *Manual of Sedimentary Petrography*. Appleton-Century Crofts Inc., New York, 549 pp.

Kuenen, P.H., 1965. Geological conditions of sedimentation. pp. 1–21 in J.P. Riley & G. Skirrow (eds) *Chemical Oceanography*, vol. 2. Academic Press, London, 508 pp.

Meadows, P.S. & Anderson, J.G., 1968. Micro-organisms attached to marine sand grains. *Journal of the Marine Biological Association of the United Kingdom*, **48**, 161–175.

Means, R.E. & Parcher, J.V., 1964. *Physical Proportions of Soils*. Constable, London, 464 pp.

Moore, H.B., 1931. The muds of the Clyde Sea area. III. Chemical and physical conditions; rate of sedimentation; and fauna. *Journal of the Marine Biological Association of the United Kingdom*, **17**, 325–358.

Morgans, J.F.C., 1956. Notes on the analysis of shallow-water soft substrata. *Journal of Animal Ecology*, **25**, 367–387.

Pearson, T.H. & Stanley, S.O., 1979. Comparative measurement of the redox potential of marine sediments as a rapid means of assessing the effect of organic pollution. *Marine Biology*, **53**, 371–379.

Rhoads, D.C., 1974. Organism–sediment relations on the muddy sea floor. *Annual Reviews, Oceanography and Marine Biology*, **12**, 263–300.

Swan, D.S., Baxter, M.S., McKinley, I.G. & Jack, W., 1982. Radiocaesium and ^{210}Pb in Clyde sea loch sediments. *Estuarine and Coastal Shelf Science*, **15**, 515–536.

Trask, P.D. (ed.), *Recent Marine Sediments; a Symposium.* American Association of Petroleum Geologists, Tulsa, Oklahoma, 736 pp. (1955 edition: Dover Publications Inc., New York.)

El Wakeel, S.K. & Riley, J.P., 1956. The determination of organic carbon in marine muds. *Journal du Conseil International pour l'Exploration de la Mer*, **22**, 180–183.

Walkley, A. & Black, I.A., 1934. An examination of the Degtjareff method for determining soil organic matter, and a proposed modification of the chromic acid titration method. *Soil Science*, **37**, 29–38.

Webb, J.E., 1958. The ecology of Lagos Lagoon. V. Some physical properties of lagoon deposits. *Philosophical Transactions of the Royal Society, London, B*, **241**, 393–419.

Webb, J.E., 1969. Biologically significant properties of submerged marine sands. *Proceedings of the Royal Society, London, B*, **174**, 355–402.

Whitlach, R.B., 1981. Animal–sediment relationships in intertidal marine benthic habitats: some determinants of deposit-feeding species diversity. *Journal of Experimental Marine Biology and Ecology*, **53**, 31–45.

Whitlach, R.B. & Johnson, R.G., 1974. Methods for staining organic matter in marine sediments. *Journal of Sedimentary Petrology*, **44**, 1310–1312.

Chapter 4. Photography and Television

N. A. HOLME

Introduction

Both photography and television have been increasingly employed in recent years as a means of recording data on the structure, ecology and behaviour of marine organisms. These media have the important advantage of being non-destructive (so allowing repeat descriptions of the same site over a period of time), non-selective with regard to the information recorded, and a ready means of description and demonstration of phenomena which cannot easily be portrayed in any other way.

In-air photography

Seashore photography

Photography, particularly in colour, is an invaluable means of recording the life forms and ecology of marine organisms, especially on rocky shores. General aspects of photography on the seashore are described by Angel (1975) and George (1980). For many purposes a reflex 35 mm camera is ideal, the principal problem being protection of the equipment from salt spray. The fitting of a UV or 1A filter in front of the lens is a standard means of protection against splashes and minor bumps. Under more extreme conditions the whole camera can be placed inside a plastic bag which incorporates a flat glass disc to fit over the lens—a modification of this equipment, with glove inserts for operation of the controls, can be used in shallow depths below the sea surface. Alternatively a Nikonos camera (Fig. 4.1) can be used, either in air or under water.

Bright light reflected from the sea, sand or coral may require adjustment to exposure settings, to avoid under-exposure, while reflections from wet surfaces can be minimized by the use of a Polaroid filter in front of the lens.

Fig. 4.1. Nikonos underwater camera with close-up attachment and with flashgun fitted with dome to spread beam over an angle of 94°. (From George, 1980.)

For quantitative work, photographs can be taken of the area defined by a quadrat frame, or by a circular ring (H. Littler, 1971; M.M. Littler, 1980), and further information on procedures in the intertidal environment are given by Gonor & Kemp (1978).

Where repeat photographs, usually of a quantitative nature, are required, there is a need to standardize positioning of the camera, and here the problems are similar to those described for underwater photography of hard substrata (p. 75).

Aerial photography

Aerial photography is useful for defining the limits of the shoreline, mapping of intertidal and submerged reefs and banks, and in the assessment of the distribution of vegetation. Although the taking of high quality isometric photographs suitable for map-making requires specialized equipment and techniques, much information can be gained by simpler means. An oblique photograph is often more descriptive than a vertical shot, and these can be taken fairly satisfactorily through the windows of aircraft on commercial flights. Where an aircraft or helicopter is available for a survey much can be achieved in a short time, and Kelly & Conrod (1969) give useful information on taking photographs from a small plane. Other methods include photography from balloons and kites, and some success has been achieved with radio-controlled model aircraft carrying a camera (e.g. Anon, 1980; Hoer, 1981). In some circumstances satellite photographs may provide additional information, particularly on hydrography.

Applications of aerial photography for coastal surveys are described by Kumpf & Randall (1961); Ellis (1966); Kelly & Conrod (1969); Kelly (1969, 1970); Steffensen & McGregor (1976); Kirkman (1978); relevant chapters in Stoddart & Johannes (1978); Coulson *et al.* (1980); and George (1980). The advantages of having a continuous video record when surveying coastlines in relation to oilspill countermeasures are described by Owens & Robilliard (1980).

Apart from ordinary monochrome and colour photography, there are applications for false colour, particularly infra-red, to show differences between actively growing and stressed or polluted vegetation in coastal regions.

Underwater photography

In recent years there have been rapid advances in the development of underwater photographic and television gear, the impetus for which has come not only from marine biologists, fisheries workers, and SCUBA divers, but also from the growing commercial requirements for equipment for examination of underwater engineering structures, particularly in relation to offshore oil and gas exploitation.

The first underwater photographs were taken towards the end of the last century by Louis Boutan (1893, 1900) in the Bay of Banyuls, but the development of the underwater camera as an oceanographic tool dates from the pioneer work of Maurice Ewing and associates in the United States in the 1940s, the history of which is described by Ewing *et al.* (1967); see also Stephens (1967).

Following the development of reliable underwater cameras, many thousands of photographs have been taken at all depths, particularly as an aid to geological and sedimentological studies of the sea bed. Many of these incidentally provide information on the occurrence, abundance, and habits of organisms living on the sea floor.

For biological studies, photography has usually been employed as an adjunct to trawl, dredge or grab surveys rather than as the prime means of investigation. This is because it has often proved difficult to make positive species identifications from underwater photographs, particularly where actual specimens have not been taken as part of the sampling programme. In addition, only those species living on the surface of sediment grounds are portrayed, the many burrowing animals being out of sight or at most revealed only by the presence of holes, or by siphons or other parts protruding from the sediment. However, in certain circumstances photography or TV may be the only practical means of investigation—an example being the estimation of stocks of scallops (*Pecten maximus*) below diving depths, as described by Franklin *et al.* (1980).

Improvements in the quality of underwater photographs have been aided by the development of photographic films of greater speed and resolution and with better colour rendering, through advances in lens design of both in-air and underwater cameras, and in the development of powerful and compact electronic flash units capable of giving large numbers of flashes from a single set of batteries.

The use of photography and television underwater raises a number of problems, which are described in specialized works such as Hopkins & Edgerton (1962), Thorndike (1959, 1967), Mertens (1970), Glover *et al.* (1977), and more elementary descriptions are given in popular articles written for SCUBA divers. The problems relate to:

1 Refraction of light at the air/glass/water interfaces situated at or in front of the camera lens.
2 Transmission of light through sea water, which is subject to absorption, and to the effects of suspended particles, and which is selective for different wave lengths.

3 Provision of a suitable watertight underwater case, with optical port, and associated controls.
4 Provision of a light source.
5 Orientation and control of the camera to provide the required information.

The optical problems are now well understood, and equipment available today is capable of producing satisfactory results under most circumstances. However, the problems of orientation and control of the camera are peculiar to each investigation, and even after the acquisition of a purpose-designed commercial camera considerable development and modification of ancillary equipment may be necessary to fulfil the purpose in hand.

Optical problems

Air–glass–water interface

The simplest situation is that in which an ordinary in-air camera is mounted in an underwater case having a flat glass port (Thorndike, 1967; Richter, 1968; Mertens, 1970; Strickland, 1980). Rays of light are bent due to the relative refractive indices of water and air, with the following results:

1 The apparent distance of the object is reduced, so that the camera must be focussed to approximately $\frac{3}{4}$ of the actual distance in water.
2 The field of view in water is narrowed, giving the effect of a lens of longer focal length, in the ratio of 4:3. This means that a wide angle lens must be used to obtain the same angle of view as a standard lens in air (Table 4.1). The use of a wide angle lens is desirable because it minimizes the effect of light absorption and turbidity by reducing the distance required between camera and subject.

Table 4.1. Angle of view in air (θ_a) and water (θ_w) for lenses of different focal lengths, behind a plane optical part. The values are for a standard 35 mm film frame (diagonal 43·3 mm). The formula for finding the angle under water is:

$$\sin \theta_w/2 = 0{\cdot}75 \times \sin \theta_a/2$$

Focal length (mm)	θ_a (focus = ∞)	θ_a (focus = 1 m)	θ_w (focus = 1 m)
24	84°	83°	60°
28	75°	74°	54°
35	63°	62°	45°
40	57°	55°	41°
50	47°	45°	33°
55	43°	41°	30°

3 Spherical aberration arises, particularly towards the periphery, so that the image of a square grid shows characteristic 'pillow' or 'pincushion' distortion.
4 Chromatic aberration is marked towards the edge of the field, even with a lens corrected for this fault in air. This causes blurring of the image towards the edge of the field.

In spite of these inherent defects, an arrangement with an ordinary in-air lens behind a flat glass port is frequently employed in both photographic and TV systems, where it provides a reasonably satisfactory image for many biological purposes, particularly where a relatively narrow angle of field is acceptable. However for photogrammetric work or where a wide angle is required, it is nececessary to improve the optical system by:

1 Use of a curved port in front of an in-air lens. The port can be either hemispherical with the lens positioned near its centre, or the lens may be behind a plano-concave port having the concave side towards the lens, or other similar optical arrangement. The construction of a simple hemispherical port, the dimensions and curvature of which are not critical, is described by Glover *et al.* (1977, p. 96), and such a system allows a wider angle of coverage to be obtained with a diver-held camera. An optically corrected domed port is used with a standard in-air lens in some deep-sea camera systems (e.g. UMEL, see Southward *et al.*, 1976).
2 Use of a specially designed camera lens behind an optically flat glass port. For deep-sea cameras the Hopkins f11 and f4·5 lenses are often employed for this purpose where wide-angle coverage is not important (Hopkins & Edgerton, 1962).
3 Using a camera lens corrected for use in direct contact with the water. The best example is the Nikonos camera (Fig. 4.1) used by divers, which is designed for use under water without an additional water-tight case.

Light absorption and turbidity

Except under the most favourable conditions (bright sunlight, clear water, shallow depth), artificial light is needed to illuminate the subject. It is then only necessary to consider the passage of light from the artificial light source to the subject and back to the camera.

Light passing through water is selectively absorbed for different wavelengths. Red light is rapidly absorbed, while blue and green penetrate farther, giving the characteristic blue-green cast to underwater scenes. There are differences in the selective absorption of different spectra in coastal as opposed to oceanic waters, resulting in the differing colours of sea water. Where an artificial light source is used—and the electronic flash used for photography has a spectral composition comparable to that of sunlight—similar effects are obtained, although the distances involved are much shorter. Formulae have been developed to show the amount of additional exposure required under different conditions; in addition some colour correction may be needed for underwater scenes (Schenk & Kendall, 1954; Richter, 1968). At short subject-to-camera distances ($\leq 0{\cdot}75$ m), colour rendering of objects appears to be satisfactory, i.e. it is similar to that of the object when brought to the surface, and pictures of objects of this range, which are largely used for identification, requires no colour correction. At greater distances colour-compensating filters (e.g. a CC30 red filter) may be required. To compensate fully for the reduction in red light a heavy degree of filtering, resulting in a need for greatly increased exposure, may be required, so it may prove more satisfactory to compensate only partially for the colour bias. In any case it would seem more natural for a distant underwater scene to have the characteristic blue-green cast seen by a diver.

Suspended particles are always present to a greater or lesser extent, and show up as bright specks where they are illuminated by the light source. It is therefore necessary to achieve lateral separation of the light from the camera to minimize illumination of the water between camera and subject. In particular, it is important to avoid illuminating the water close to the camera lens, and, in some instances, shading of the light beam may be necessary. Some diver-held cameras (both photographic and TV) have the light incorporated into the same casings as, and close to, the camera; while this makes the equipment easy to handle it accentuates the undesirable effects of suspended matter.

In estuaries and harbours suspended matter often reduces visibility to such an extent that underwater photography and television become virtually impossible. In coastal waters, and on the continental shelf generally, there may be considerable quantities of suspended matter at certain times, so that it is necessary to select suitable periods for carrying out surveys. Calm weather and neap tides result in reduction in the suspended load, but underwater visibility is not necessarily better in summer because there may be large quantities of living or moribund plankton in the bottom waters during this season.

In the deep oceans the bottom water is often exceptionally clear, as evinced by the many high definition photographs obtained in this environment.

Photographic cameras

For general purposes, 35 mm film cameras are often employed, although cameras taking 70 mm film are sometimes used, and 16 mm ciné cameras may be employed for motion pictures, time-lapse work, and other purposes where many frames are to be exposed. Cameras used by SCUBA divers and for remotely-controlled systems on the continental shelf are often standard in-air cameras placed in specially constructed cases, while those for deep-sea use are purpose-built.

For SCUBA divers, plastic cases have been manufactured to house many types of still and ciné cameras (e.g. Glover *et al.*, 1977; George, 1980), the controls being actuated by the diver, although exposure, stop, and focus are often preset, distance for still subjects being standardized by a rod or frame placed up to the subject (Fig. 4.1). A wide-angle lens is used, and cases often have domed ports to increase the angle of coverage. The Nikonos camera has already been mentioned as a specially designed underwater camera not needing a separate case.

For work on the continental shelf it has been customary to use standard 35 mm cameras, with motor wind, and a magazine back with film capacity of 250–500 frames. These are mounted in purpose-built underwater cases which, because of small user demand, are not generally available commercially. Some workers have used deep-sea cameras, modified by the provision of a shutter (e.g. Carrothers *et al.*, 1974). Sometimes the flash is mounted within the same case, which, if sufficiently long, provides the necessary lateral separation. There is now much accumulated experience in the design of underwater cases (Edgerton & Hoadley, 1955; Edgerton, 1963, 1967) for operation on the shelf and at greater depths.

For deep-sea work a number of specially designed cameras, mostly employing long lengths of 35 mm film, have been produced. They are shaped to fit into a narrow cylindrical casing, of sufficient thickness to withstand great pressure, and have an optically corrected port and lens system at one end of the case (Edgerton, 1963, 1967). Such cameras normally have no shutter (so must be tested after dark when in shallow depths), and film wind-on is relatively slow. Each picture has time and other parameters displayed in a corner of the frame, although the design is such that there is a separation of several frames between the data display and the frame to which it refers.

Lighting

Electronic flash is used almost exclusively for still photography underwater, and high speed strobe flash or continuous lighting for ciné photography. As with the camera system, the use of a flat port restricts the beam angle under water, so it is necessary to check that the flash gives adequate and even illumination of the field. It may be necessary to use a domed port (Fig. 4.1). Complete flash units are available commercially for both diver-operated and deep-sea systems; alternatively, an ordinary flash unit may be mounted in an underwater case. Flash units with a capacity sufficient for up to 500 exposures from disposable or rechargeable batteries are readily obtainable.

Positioning of the light source is all-important. For a near-vertical view of the sea bed, the light may be placed to one side to illuminate the scene at about 45°. Sometimes two lights are used, but these should not be positioned so as to eliminate all shadows, as these are often essential for interpretation of the photograph. On sledge-mounted equipment both camera and lights are normally pointed forward at an angle of between 30° and 45° to the horizontal. The light is typically placed to the side or below the camera, providing oblique or low angle illumination (Fig. 4.5). Where the camera is pointing at a low angle it may be difficult to illuminated the scene uniformly because of the great range of subject distances from the light source.

Many modern flash units are designed so that the duration of flash is determined by the light reflected back from the subject. Because reflection from suspended matter would be likely to affect such a unit, it is unlikely that these would be suited to underwater photography.

Cable connections

With a vertically-lowered photographic camera, there are typically two connections:

1 An electric cable from the switch unit controlled by a counterweight mechanism (p. 77), for firing the shutter.
2 A synchronizing cable from camera to flash.

If the units are in separate underwater cases, cable connections are made via special plugs inserted into sockets fitted to the cases. A number of commercial connectors are available, including the Electro Watermate® Connectors made by Electro-Oceanics Inc., 15146 Downey Avenue, Paramount, California 90723, USA. All connectors should be carefully washed, cleaned and dried after use as they are subject to corrosion, and many of the faults associated with underwater cameras are caused by bad cable connections. Before use, the connectors should be lightly smeared with silicone grease.

Where the camera is attached to a towed system the shutter may be fired by an electronic timing device, with an initial delay or other means of switching on when bottom is reached. It is not necessary to have an electrical connection to the surface for firing the camera but this may be provided in conjunction with other equipment, such as underwater TV.

Deployment and control of photographic cameras

Manual operation

This includes use by divers and from manned submersibles where the

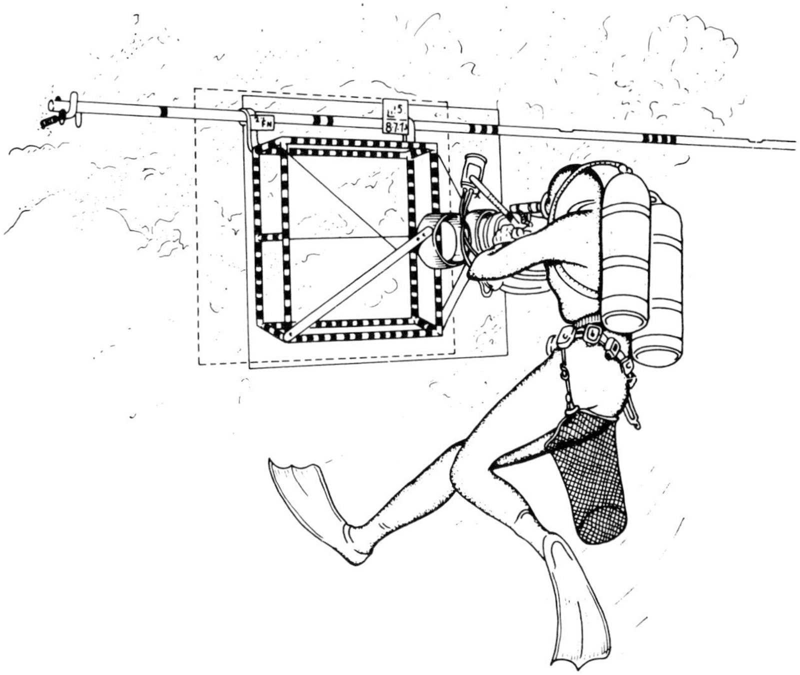

Fig. 4.2. Diver making stereophotographic recording on permanently marked, vertical rock face. (From Lundälv, 1976).

positioning and firing of the camera is under visual control. The greatest single problem is in determining the distance from camera to subject which, for monitoring purposes, should be standardized. Diver-operated cameras can be held at a known distance from the subject by the use of a rod or frame attached to the camera (Fig. 4.1), and the use of a frame allows quantitative counts of organisms on rocky substrata to be made. Lundälv (1971, 1974, 1976) describes a diver-operated system for taking repeat stereo photographs on near-vertical rock faces (Fig. 4.2). Two holes are drilled in the rock 3 m apart and pegs are inserted on to which a rigid rod carrying the camera frame is attached . The frame is arranged so that pairs of photographs can be taken by a single camera at a constant distance (20 cm) apart, which are mounted to form a stereo pair. From these, photogrammetric measurements can be made of size, area, or volume of organisms with a high degree of accuracy (Torlegård & Lundälv, 1974). Since the two frames forming the stereo pair are not taken simultaneously the technique is only suitable for measurement of sedentary organisms.

A similar system, but using twin 35 mm cameras rather than the single Hasselblad with 70 mm film used by Lundälv, is described by Rørslett *et al.*

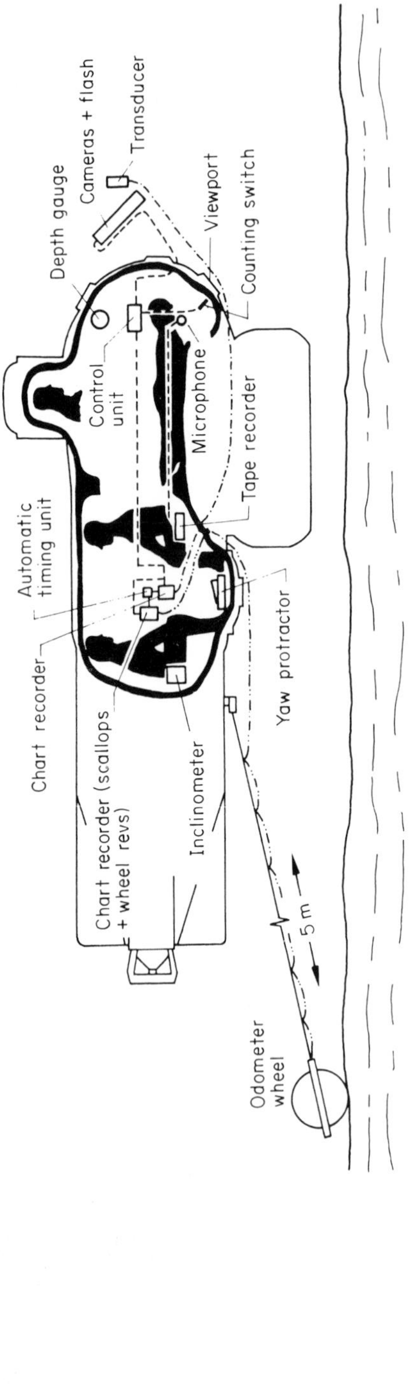

Fig. 4.3. Submersible fitted for quantitative surveys of scallop populations. (From Caddy, 1976.)

(1978). Gulliksen (1980) describes another method for taking repeat photographs on rocky substrata (Fig. 5.3), and Johnston *et al.* (1969) describe a quick technique for three-dimensional studies.

When working from a submersible, constant distance off bottom is attained by visual means aided by echo sounding (cf. methods for positioning remotely controlled cameras, p. 79), but workers have reported difficulties in estimating precise distances at which photographs (as well as direct observations) have been made. Caddy (1970, 1976) found that the odometer wheel attached to the rear of a submersible (Fig. 4.3) aided the maintenance of a constant distance off bottom, which was necessary for the quantification of visual observations. Uzmann *et al.* (1977) used a chain trailed from the submersible to define the lateral width of field of view and distance off bottom, while Grassle *et al.* (1975) had a ski attached to the equipment basket by means of which distance off bottom was standardized. Brundage *et al.* (1967) utilized the degree of overlap of two stereoscopically-mounted cameras to estimate distance. In addition, these authors (also Barham *et al.*, 1967), used animals of known size (urchins, brittle-stars) in the field of view as a scale for the photographs.

Remote operation

Triggering of the camera may be controlled in a number of ways. If lowered vertically, the camera is normally fired by a switch actuated on contact with the sea bed; if towed, the camera may be controlled by an electronic timer set to fire the shutter at regular intervals, following an initial delay; or the camera control may be linked with a TV viewing system allowing photographs to be taken manually or at constant intervals as required. Another technique is to use the camera on a tripod or other fixed support to take photographs at regular intervals by means of an electronic timing mechanism.

Vertical lowering. The camera case is typically mounted within a protective framework, from which a trigger weight which activates a switch on touching bottom is suspended (Craig & Priestley, 1963; Hersey, 1967; Ewing *et al.*, 1967; Owen, 1967; Southward *et al.*, 1976). Alternatively a rigid foot switch may be employed (Vevers, 1951).

Control of the camera in shallow water can be exercised by hauling in on the lowering cable as soon as bottom is detected, but a more precise system involving two bottom-detecting sensors is described by Eagle *et al.* (1978). In deep water or where the design of the frame makes it undesirable for it to touch bottom, a pinger can be used for precise control of lowering (Hersey, 1967—see also Fig. 2.2). Using such a system it is possible to 'bounce' the frame just off bottom to take a series of photographs as the ship drifts slowly over the

Fig. 4.4. Distribution of brittle-stars in patches or scattered on rocks in a way which makes estimation by grab unreliable. The camera trigger weight is seen at the top. (From McIntyre, 1956.)

ground. Southward *et al.* (1976) have described a sophisticated control system where the rate of pinging changes after each firing of the camera, the signal being displayed both visually and audibly.

The trigger weight often forms a conspicuous object in the photograph (Fig. 4.4) where it not only provides a scale, but evidence of the grain size of the deposit (Athearn, 1967). In the set-up used by Kanneworff (1979) for surveying shrimp stocks the bottom sensor was outside the field of view.

An underwater camera mounted within the jaws of a Campbell Grab is described by Menzies *et al.* (1963) and Emery *et al.* (1965). This provided a photograph of the sea bed just before it was sampled by the grab. The degree of correspondence between the two methods is considered below (p. 90).

A method of photographing a profile of a silt–clay sediment *in situ*, using a slowly-descending wedge-shaped plexiglass chamber filled with water and having a mirror placed at 45° is described by Rhoads & Cande (1971). A further development of this system, incorporating a video camera, is described by Rhoads & Germano (1982).

Towed bodies. Vertical or oblique photographs have been taken from equipment towed 1–2 m above the sea bed, the camera being fired either automatically at predetermined intervals or by acoustic or cable controls from the ship. Distance off bottom can be determined by an echo sounding transducer attached to the towed body, but this distance will tend to fluctuate and greater precision for interpretation of individual shots (both from vertical lowering or from towed bodies) can be obtained by use of the Benthos Height Recording Deep Sea Camera System which provides an altitude readout for each film frame.

Blacker & Woodhead (1965) describe a towed underwater camera mounted in a high speed townet body, the camera pointing down to cover an area 60 × 48 cm. Depth off bottom was measured by a cable-linked headline transducer, sounding to the bottom, and the device could be towed at speeds of up to 5 knots.

A more elaborate towed instrument with a camera, using transponders linked to a shipboard computer for position fixing, is described by Phillips *et al.* (1979), and other unmanned free-floating or towed bodies carrying TV cameras are mentioned on p. 87.

Sledges. Attachment of a camera to a sledge towed along the sea bed removes much of the uncertainty as to distance between camera and subject. On the other hand there is a risk of the sledge and its equipment being damaged or lost through collision with rocks or other obstructions on the sea bed. The sledge must, therefore, be sturdily constructed and capable of negotiating rocks, even if it is normally employed on smooth level grounds. Such an instrument is the 'troika' described by Laban *et al.* (1963), and subsequently used by Vaissière & Fredj (1964) and by Chardy *et al.* (1980) (Fig. 4.5) for benthic surveys. Other sledges have been described by Thiel (1970), Wigley & Theroux (1970), Wigley *et al.* (1975), and Uzmann *et al.* (1977). Sledges carrying TV as well as photographic cameras are described by Machan & Fedra (1975), Bascom (1976a,b) and Holme & Barrett (1977) (Fig. 4.7).

The camera is usually positioned to point ahead of the sledge at an angle of $<45°$ to the horizontal. For general purposes 35° may be a suitable angle, but the 'troika' has the camera mounted at a more acute angle. Holme & Barrett (1977) positioned a camera to point vertically downwards from a distance of 60 cm, where it photographed a small area of 0·2 m^2; on the other hand Wigley *et al.* (1975) used a sideways-pointing camera for red crab surveys in deep water, each picture covering 31·8 m^2 of sea bed. Interpretation of results from this survey are given by Patil *et al.* (1980).

Fig. 4.5. Troika equipped for bottom photography. (From Chardy *et al.*, 1980. Photograph CNEXO.)

An epibenthic sledge carrying a forward-facing camera is described by Aldred *et al.* (1976), and by Rice *et al.* (1982) (Fig. 6.5). Apart from such purpose-made sledges, photographic and TV cameras have been attached to trawls and dredges used for fishery and benthic surveys (e.g. Craig & Priestley, 1963; Hemmings, 1972; Dyer *et al.*, 1982). These provide an opportunity for quantitative comparisons between estimates of population density by photography and actual catches in the net (p. 89). Menzies (1972) describes the use of a forward-facing camera in the mouth of a deep-sea trawl to monitor its behaviour on the sea bed (p. 177 and Fig. 6.25), and photographic evidence of the fishing behaviour of the epibenthic sledge is given by Aldred *et al.* (1976).

Fixed cameras. Time-lapse cameras, usually employing 16 mm ciné film, have been used for recording long-term changes on the sea bed. Fedra & Machan (1979) describe such a system capable of taking up to 4000 single frames over a period of up to a week. Observations were made on the behaviour and distribution of species living on or close to the bottom, including fish, crustaceans, anemones and echinoderms. A camera of similar film capacity is described by Jogansen & Propp (1980). Rumohr (1979) describes a fixed camera arrangement, capable of taking up to 1000 frames on 35 mm film, used for recording fish around an artificial 'benthosgarten'. An inexpensive system for shallow water, using a 16 mm film camera, is described by Mitchell (1967).

In the deep sea, time-lapse photographs and observations of benthos from the bathyscaph *Trieste* are described by LaFond (1967). Photographic records covering much longer periods are described by Paul *et al.* (1978), who set up a time-lapse camera at 4873 m depth to take pictures every four hours over a period of several months.

Time-lapse cameras have also been used to record the sequence of predators at a food source (Isaacs & Schwartzlose, 1975), while the use of TV cameras for long-term observations is considered on p. 86.

Stereophotography

Interpretation of photographs of both rock and sediment substrata is greatly aided by stereophotography. This not only improves recognition of bottom features (for example by enabling a small stone to be distinguished from a burrow), but measurement of bottom features and organisms is possible where the camera system is precisely aligned.

Stereophotography of rock surfaces using either one or a pair of cameras on a framework is described above (Lundälv, 1971, 1974, 1976; Rørslett *et al.*, 1978), and for sediment grounds Moore (1976, 1978) describes a framework which allows a diver to take stereo pairs for photogrammetric evaluation of ripples and sandwaves.

Stereophotography has often been employed on remotely-controlled cameras for the shelf and deep sea. Owen (1967) describes the use of two 35 mm cameras mounted in line within a single underwater case, but where cameras are in separate cases they can be mounted with axes converging to provide a better overlap of the subject. Thus Southward *et al.* (1976) inclined two cameras at a mutual angle of 15° for close-up work. A number of workers have employed black and white film in one camera and colour in the other (Grassle *et al.*, 1975; Owen, 1967; Lemche *et al.*, 1976; Southward *et al.*, 1976). Southward *et al.* describe a method for viewing these by transparency as a stereo pair, and also consider the technique of triggering deep-sea cameras for stereophotography.

Photogrammetric techniques, following those used for interpretation of aerial photographs, are described by the above authors and also by Atkinson & Newton (1968) and Schuldt *et al.* (1967).

Interpretation of photographs

In some instances there will be a frame number, time, or other record in the corner of each film frame; in other cases it is necessary to carefully label each negative or colour transparency so that there can be no possibility of confusion. There may be advantages in keeping lengths of film uncut, keeping the pictures in their correct sequence, so eliminating the risk of losing individual frames.

Use of colour film is often an advantage for identification and interpretation (Trumbull & Emery, 1967), as is comparison of stereo pairs, and wherever possible photographs should be examined in conjunction with actual material collected from the sea bed.

When interpreting sets of photographs for quantitative studies it is essential to take account of every frame, care being taken to avoid over-emphasis of the perhaps few frames containing material of particular interest.

Photographs are of value not only for identification of species and for estimating density or percentage cover, but also in showing the effects of organisms on their surroundings, for example through grazing of rock surfaces or bioturbation of sediments. They also can be used for showing the nature of the sea bed, the direction and strength of bottom currents, and for assessing effects of pollution and waste dumping. Because the information displayed is unselective, photographs are invaluable as evidence of 'before and after' conditions related to pollution or other environmental change. Discussion of the information revealed by underwater photographs is given *inter alia* by Laughton (1963), Heezen & Hollister (1964), Fell (1967), Wigley & Emery (1967), Horikoshi & Ohta (1975), Dangeard *et al.* (1978) and George (1980).

For quantitative studies the area covered by each frame is usually known with some degree of accuracy, although the importance of standardizing subject distance to give uniformity of scale and area must again be emphasized. For many purposes, total numbers of individuals per frame are all that may be required, but for more detailed analysis of sessile organisms the use of randomly positioned points, as described by Bohnsack (1979) can be employed, or other methods used for vegetation quadrats on land may be applicable. For more precise measurement photogrammetric mapping (p. 75) from stereo pairs, or the simpler method described by Johnston *et al.* (1969) may be employed.

The scale may vary within the picture due to distortion of the image near the periphery (particularly when a plane optical port has been used), or because the picture is not normal to the surface of the subject. Pictures are often taken at an inclined angle, and a theoretical consideration of perspective in such situations is given by Mertens (1970)—see also Barham *et al.* (1967). However, it is usually best to take an underwater photograph of a board, marked into squares and placed in the same position as the subject. This photograph can then be made into a transparent overlay for analysis of photographic prints or marked on to a screen on to which a transparency is projected. Examples of the employment of such grids are given by Uzmann *et al.* (1977) and Rice *et al.* (1979) for towed sledges, and also by Grassle *et al.* (1975).

Consideration must also be given to the proportion of the total benthic population visible in the photographs. While surface-living sessile or slowly moving organisms should be well portrayed, and some success has been achieved in estimating populations of scallops and queens which swim when disturbed (Caddy, 1970; Franklin *et al.*, 1980), fast-moving organisms such as fish and crustacea are seldom seen in photographs. Vaissière and Fredj (1964) noted that fauna was rarely seen above the bottom in front of the 'troika', but on the other hand Wigley & Theroux (1970) recorded a fair number of fish from a sledge-mounted camera. Kanneworff (1979) was able to use a camera for surveys of shrimp populations. Some organisms, particularly larger burrowing crustacea such as *Nephrops*, tend to come out of their burrows at night and so are more likely to be recorded at this time (Chapman & Rice, 1971).

The validity of results obtained by photography as compared with other survey methods are considered on p. 89.

Underwater television

The use of television under water dates from the late 1940s, and an account of its early development for this purpose is given by Barnes (1963). The rapid advances in technology which are still taking place render it particularly necessary to consult up-to-date sources of information from manufacturers and the technical and scientific literature when planning a research programme.

Television has the advantage of providing a continuous and instantaneous view, and can serve as a 'viewfinder' for the operation of still and ciné cameras. Besides giving information on the occurrence, distribution and behaviour of benthic organisms, it can be used for defining the limits of communities and grounds, and for studying the operation of gear under water (Barnes, 1955). Its limitations are due to poorer definition than still photography, the greater expense and complexity of the equipment, and the need for a special cable linked to the monitor and recording gear in the ship or on shore. Studies of behaviour may be affected if an artificial light source is required.

Problems of underwater housings, optics, and the positioning of lighting so as to minimize the effects of suspended matter are similar to those described above for photography. However, the special features of TV pick-up tubes need to be considered in relation to particular requirements.

The most widely used pick-up tube is the Vidicon which gives good results under normal daylight levels of illumination. However it suffers from 'burn-in' (long-term retention of a brightly-lit scene) and 'lag' (short-term retention,

Table 4.2. Comparison of performance of TV camera pick up tubes at different light intensities. The 'low-light' SIT and ISIT tubes incorporate image intensifiers. (From Harris, 1980.)

1. PLAIN VIDICON (sulphide type)					Fading into amplifier noise & vidicon defects	Good, very high definition pictures by target control, marred by excessive 'memory' at lower light levels				
2. NEWVICON CHALNICON SILICON VIDICON				Fading into amplifier noise	Good, high definition pictures with low lag		N.D.* filter or iris needed to prevent 'blooming'			
3. SIT (16 mm) TUBE		Fading into photo-electron and amplifier noise	Good, very clear pictures		Very good pictures can be maintained with N.D.* filter on lens					
4. ISIT (16 mm) TUBE	Fading into photo-electron noise	Good clear pictures			Good clear pictures with N.D.* filter on lens					
SCENE ILLUM. LUX	10^{-4}	10^{-3}	10^{-2}	10^{-1}	1	10	10^2	10^3	10^4	10^5
VIEWING CONDITIONS	STARLIGHT OVERCAST	CLEAR	MOONLIGHT $\frac{1}{4}$ MOON	FULL		TWILIGHT	ROOM LIGHTING		DAYLIGHT OVERCAST	FULL SUN

The above conditions are assumed in all cases (1) Cameras are fitted with T2 lens; (2) Scene reflectance of 50%.
* Neutral density.

causing blurring and comet-tailing of a moving image), which may render it unsuitable in certain situations. Other tubes such as the Newvicon, Chalnicon and Silicon-Vidicon are better in the above respects, and are more sensitive at lower light levels (Harris, 1980). Low light tubes such as SIT and ISIT, using image intensifiers, are sensitive to much lower light intensities (Table 4.2). Although these are more costly they have advantages in that they can sometimes be used without artificial lighting for behaviour studies, and for viewing at ambient light levels in turbid water, where the use of an artificial light source would result in deterioration of the picture through illumination of suspended particles.

Pick-up tubes can, to some extent, compensate for changing light intensities, but a motorized iris may be provided to widen the range of light levels to which the camera can accommodate. A useful review of the characteristics of underwater TV cameras is given by Harris (1980), and a general account of closed-circuit TV is given by Shackel & Watson (1968). Underwater colour TV systems are discussed by Robinson (1978).

Specialized uses of underwater TV include a slow-scan system for transmission of still pictures from great depths (Flato, 1965; Brundage *et al.*, 1967), and the use of two interlacing signals to create a three-dimensional display, as described by Berry (1979).

Lighting

The most suitable sources for underwater lighting are mercury vapour or thallium iodide rather than incandescent lamps, since the first two more nearly match both the transmission characteristics of sea water and the spectral sensitivity of Vidicon tubes. A review of the uses of different forms of lighting for both TV and photographic applications is given by Strickland (undated). Bascom (1976a) considers that the emerald green light from a thallium iodide source disturbs fish less than white light.

Cables

The TV camera is linked to its control unit on ship or shore by a coaxial cable which both powers the camera and transmits signals to the surface. Special TV cables, incorporating additional conductors for focussing, lights, and control functions, are available. These are normally of low tensile strength and need to be combined with heavy duty ropes or cables needed for lifting or towing the gear. TV signals can be transmitted over several kilometres of cable, given suitable boosters, but for operation at sea the deployment of long lengths of cable may present problems. High tensile armoured cable is available, but is likely to prove too costly, except in short lengths, for other than industrial applications. For deep-sea use deployment from a submersible may prove a more practicable alternative.

Videorecording

Successful recording is dependent upon an AC supply free from electrical interference and with a frequency stability better than $\pm 1\%$. For small boat operation a portable generator is unlikely to provide sufficient stability—this problem was overcome by Machan & Fedra (1975) by feeding the AC supply from a portable generator into a rectifier used to charge a 24 volt battery. The 24 volts DC was subsequently converted to 220 volts AC at a constant 50 Hz by a thyristor-inverter, thus providing a constant frequency source for videorecording, and a reserve of power should the generator fail.

Mounting of TV cameras

Hand-held cameras

The commonest use is by divers investigating underwater structures, where there are advantages in the view being transmitted to the surface for others to see, and for making a videorecording. Here the camera is under the direct control of the diver, but it has been noted that when used with an image intensification tube under turbid conditions more could be seen on videotape replay than had been visible to the diver holding the camera (Rees, 1976; see also Machan & Fedra, 1975). TV cameras are frequently deployed from manned submersibles, where, again, direct control of the camera and lighting is possible.

Remote operation

Vertical lowering and fixed cameras. Because of the requirement for continuous viewing, TV cameras lowered vertically to within a metre or two of the bottom have met with only limited success—the motion of the ship makes it difficult to obtain a steady picture. TV cameras suspended just above the bottom or resting on a framework have been used by Schwenke (1965), Farrow *et al.* (1979), and Goeden (1980), and fixed cameras have been employed for longer-term observation by Nishimura (1966), Chapman & Rice (1971) and Eleftheriou & Basford (1983). A TV viewing system mounted permanently on the sea bed at 20 m depth is described by Kronengold & Loewenstein (1965), Stevenson (1967), and Myrberg (1973). As well as giving valuable information on fish behaviour and sound production, the effect on the bottom of varying hydrographic conditions, including storms, was observed.

Fig. 4.6. Television and photographic camera instrument package TRIP 2, well protected within pipe frame. (From Goeden, 1981.)

When a camera is used for long-term observations, particularly in shallow water, the likelihood of the viewing system becoming obscured through settlement of fouling organisms must be considered.

Towed bodies. TV cameras, with associated lighting, are frequently mounted as part of the control system in towed bodies (Winterhalter *et al.*, 1970; Holmes & Dunbar, 1978; Phillips *et al.*, 1979). Goeden (1981), describes a remote instrument package, with TV and photographic cameras, which is 'flown' above the bottom (Fig. 4.6). Views obtained with cameras suspended

Fig. 4.7. Sledge with television and photographic cameras, as described by Holme & Barrett (1977). The forward-facing TV camera is at the top, below which is the transversely mounted case containing the still camera. Below this is the case for the electronic flash, with the television light on the right. The odometer wheel, mounted centrally on trailing arms near the back of the sledge, transmits an electrical impulse on each revolution, which is recorded on the sound track of the videotape.

above the bottom are likely to pose the same problems of scale and distance as were noted for underwater photography.

Towed sledges. Some towed bodies may also be used on the sea bed; others, such as the sledges described by Machan & Fedra (1975) and Holme & Barrett (1977), have been specifically designed for this purpose. Both of these sledges also carry photographic cameras, and that of Holme & Barrett has a wheel which measures distance over the bottom (Fig. 4.7). Each revolution of the wheel is recorded as a signal on the video sound track, allowing quantification of the observations made by TV. The rotation of the wheel can also be used to trigger the camera, enabling photographs to be taken at constant distance intervals along the sea bed. Bascom (1976a) describes a sledge which is normally towed along the bottom, but which can be lightened to float a few metres above. He details the kinds of faunistic observations which can be made, and concludes that 'we can think of no other single marine instrument that produces so much information'.

Comparison with other survey techniques

Those who have made observations under water, whether by diving or from a submersible, would agree that there can be no substitute for direct visual inspection for obtaining a real understanding of conditions of life on the sea bed. Commentary on this aspect of the use of submersibles is given by Saunders (1972), Heirtzler & Grassle (1976), Treadwell (1977), Uzmann *et al.* (1977), and Chapman (1979). Exceptionally, a low light TV camera can record more than an underwater observer (p. 86), but under most conditions direct observation will enable a better interpretation to be made.

The exacting nature of continuous observation under water, the need for simultaneous recording of different attributes of the scene (nature of the bottom, occurrence of different species, estimates of density), and the desirability of obtaining a permanent record for further study and for demonstration to others, make the employment of photographic and videorecording techniques essential in conjunction with any exploration by submersible.

For reasons of cost, most investigations below diving depths are carried out by remotely-operated instruments rather than from manned submersibles, and here there are differences of opinion as to the value of photographic and TV records. These views reflect, on the one hand, the varying results obtained on different types of ground, and on the other, the objectives of particular investigations. Thus, photographic or TV surveys have been considered as quite unsuitable for obtaining a quantitative estimate of the total fauna of a soft mud (cp. Owen *et al.*, 1967), but eminently suitable for survey of particular species of large size living on the surface of the sediment. Wigley *et al.* (1975) successfully employed a sledge-mounted photographic camera for surveying stocks of red crabs, and Franklin *et al.* (1980) used a TV camera for estimating stocks of scallops and queens. McIntyre (1956) considered that a camera could be the only means of obtaining quantitative data on epifauna species on hard grounds.

A number of comparisons have been made between estimates of species density obtained by trawl, dredge and camera. McIntyre (1956) compared the sampling ability of an Agassiz trawl, grab, and camera for brittle-stars, starfish and queen scallops in a Scottish loch (Table 4.3). Wigley & Theroux (1970) compared scallop-dredge samples with bottom photographs, Uzmann *et al.* (1977) compared visual observations from a submersible with otter trawl and camera sled results, and Rice *et al.* (1979, 1982) compared results from a camera mounted in the mouth of an epibenthic sledge with species taken in the net, in samples from the deep sea. Owing to the low efficiency of most trawls and dredges (Dickie, 1955, found efficiencies as low as 5% for a scallop dredge—see also p. 181), such comparisons are more a commentary on the

Table 4.3. Numbers of the brittle-star, *Ophiothrix fragilis*, on 5 stretches of sea bed, as estimated by trawl, camera and grab. (From McIntyre, 1956.)

Stretch	Trawl	Camera	Grab
A	1093	39 542	19 418
B	1992	32 702	41 919
C	215	15 577	(No samples)
D	524	61 330	79 686
E	290	4916	10 455

functioning of the towed nets than on the absolute validity of camera survey techniques.

A better comparison is possible between grab and photographic surveys: McIntyre (1956) found that results were of the same order (Table 4.3) for members of the epifauna, the reliability of either method depending on the density and patchiness of the species to be compared. More precise comparisons have been made from photographs taken by a camera mounted between the jaws of a Campbell grab: Menzies *et al.* (1963) found that there was good correspondence between the two methods for larger members of the epifauna, although the situation was less satisfactory for smaller animals, most of which lived below the surface of the sediment. For this reason Emery *et al.* (1965) and Owen *et al.* (1967) concluded that photographs provided a poor quantitative estimate of the organisms present. However, Wigley & Emery (1967) considered that photographs are a desirable supplement to grab samples for biomass studies. Furthermore the photographic method was judged to be superior for determining the biomass of larger surface-living organisms. The usefulness of a photographic survey will naturally depend on the area covered by each frame and the degree of resolution of fine detail in the photographs: Barham *et al.* (1967), also Grassle *et al.* (1975), have commented on the better resolution obtained by still as opposed to ciné camera surveys.

In conclusion, photography can be a valuable adjunct to any benthic survey, and should be employed whenever practicable. Because of the greater expense and complexity of TV (including the need for a cable link to the ship, and a supply of stabilized frequency AC), combined with a requirement for greater care in operation and maintenance, the deployment of such equipment for a particular survey requires careful consideration and a firm decision must be made at the outset as to whether to devote a significant proportion of time and resources to the development and use of television gear. For the future, the increasing availability of compact and reliable underwater TV systems is likely to lead to greater use of this medium to supplement other means of investigation.

Acknowledgements

I am indebted to the following for permission to reproduce material, and for supplying the appropriate photographs or drawings: Dr J.D. George, Trustees of the British Museum (Natural History) (Fig. 4.1); Dr T. Lundälv, Kristineberg, Academic Press (Fig. 4.2); Dr J.F. Caddy, Rome, Academic Press (Fig. 4.3); M.P. Chardy, Centre Océanologique de Bretagne (CNEXO) (Fig. 4.5); Dr G.B. Goeden, Bungalow, Queensland (Fig. 4.6); Mr R.J. Harris, Wells, Somerset (Table 4.2).

References

Aldred, R.G., Thurston, M.H., Rice, A.L. & Morley, D.R., 1976. An acoustically monitored opening and closing epibenthic sledge. *Deep-Sea Research*, **23**, 167–74.

Angel, H., 1975. *Photographing Nature: Seashore*. Fountain Press, Kings Langley, England, 96 pp.

Anon, 1980. (Aerial pictures from a model aircraft.) *Water Research News*, **4**, 1–2.

Athearn, W.D., 1967. Estimation of relative grain size from sediment clouds. Chapter 16, pp. 173–6, in J.B. Hersey (ed.) *Deep-Sea Photography*. The Johns Hopkins Oceanographic Studies, **3**, 310 pp.

Atkinson, K.B. & Newton, I., 1968. Photogrammetry. Chapter 6, pp. 273–97, in C.E. Engel (ed.) *Photography for the Scientist*. Academic Press, London, 632 pp.

Barham, E.G., Ayer, N.J. Jr. & Boyce, R.E., 1967. Macrobenthos of the San Diego trough: photographic census and observations from bathyscaphe, *Trieste*. *Deep-Sea Research*, **14**, 773–84.

Barnes, H., 1955. Underwater television and research in marine biology, bottom topography and geology. Part 2. Experience with the equipment. *Deutsche Hydrographische Zeitschrift*, **8**, 213–36.

Barnes, H., 1963. Underwater television. *Oceanography and Marine Biology. An Annual Review*, **1**, 115–28.

Bascom, W., 1976a. An underwater television system. *Southern California Coastal Water Research Project. Annual Report for the year ended 30 June 1976*, 171–4.

Bascom, W., 1976b. Underwater color photography. *Southern California Coastal Water Research Project. Annual Report for the year ended 30 June 1976*, 175–8.

Berry, H.W., 1979. An underwater three-dimensional (3D) television system. *Proceedings of the 11th Annual Offshore Technology Conference*, **2**, 893–900.

Blacker, R.W. & Woodhead, P.M.J., 1965. A towed underwater camera. *Journal of the Marine Biological Association of the United Kingdom*, **45**, 593–7.

Bohnsack, J.A., 1979. Photographic quantitative sampling of hard-bottom benthic communities. *Bulletin of Marine Science*, **29**, 242–52.

Boutan, L., 1893. Mémoire sur la photographie sous-marine. *Archives de Zoologie Expérimentale et Generale*, Ser. **3, (1)**, 281–324.

Boutan, L., 1900. *La Phótographie Sous-Marine et les Progrès de la Photographie*. Schleicher Frères, Paris, 332 pp.

Brundage, W.L. Jr., Buchanan, C.L. & Patterson, R.B., 1967. Search and serendipity. Chapter 6, pp. 75–87, in J.B. Hersey (ed.) *Deep Sea Photography*. The Johns Hopkins Oceanographic Studies, **3**, 310 pp.

Caddy, J.F., 1970. A method of surveying scallop populations from a submersible. *Journal of the Fisheries Research Board of Canada*, **27**, 535–49.

Caddy, J.F., 1976. Practical considerations for quantitative estimation of benthos from a submersible. pp. 285–98 in E.A. Drew, J.N. Lythgoe and J.D. Woods (eds) *Underwater Research*. Academic Press, London, 430 pp.

Carrothers, P. J. G., Polar, S. M. & Foulkes, T. J., 1974. The EG & G, 200/210, Automatic, Underwater, Photographic System adapted to fisheries research. *International Council for the Exploration of the Sea. Gear & behaviour committee*, C.M. 1974/B:4, 14 pp.

Chapman, C.J., 1979. Some observations on populations of Norway lobster, *Nephrops norvegicus* (L), using diving, television and photography. *Rapports et Procès-Verbaux de Réunions, Conseil International pour l'Exploration de la Mer*, **175**, 127–33.

Chapman, C.J. & Rice, A.L., 1971. Some direct observations on the ecology and behaviour of the Norway lobster *Nephrops norvegicus*. *Marine Biology*, **10**, 321–9.

Chardy, P., Guennegan, Y. & Brannellec, J., 1980. Photographie sous-marine et analyse des peuplements benthiques. *Centre National pour l'Exploitation des Océans, Rapports Scientifiques et Techniques*, **41**, 32 pp.

Coulson, M.G., Budd, J.T.C., Withers, R.G. & Nicholls, D.N., 1980. Remote sensing and field sampling of mudflat organisms in Langstone and Chichester Harbours, southern England. Chapter 9, pp. 241–63, in J.H. Price, D.E.G. Irvine and W.F. Farnham (eds) *The Shore Environment Vol. 1 Methods*. Systematics Association Special Volume, 17(a). Academic Press, London, 321 + 41 pp.

Craig, R.E. & Priestley, R., 1963. Undersea photography in marine research. *Department of Agriculture and Fisheries for Scotland, Marine Research*, 1963. No. **1**, 24 pp.

Dangeard, L., Deniaux, B. & Le Notre, N., 1978. Apports de la photographie sous-marine à l'étude de l'interface eau-sediment. *Océanis*, **4**, 613–21.

Dickie, L. M., 1955. Fluctuations in abundance of the giant scallop *Placopecten magellanicus* (Gmelin), in the Digby area of the Bay of Fundy. *Journal of the Fisheries Research Board of Canada*, **12**, 797–857.

Dyer, M.F., Fry, W.G., Fry, P.D. & Cranmer, G.J., 1982. A series of North Sea benthos surveys with trawl and headline camera. *Journal of the Marine Biological Association of the United Kingdom*, **62**, 297–313.

Eagle, R.A., Norton, M.G., Nunny, R.S. & Rolfe, M.S., 1978. The field assessment of effects of dumping wastes at sea: 2. Methods. *Fisheries Research Technical Report, MAFF, Directorate of Fisheries Research*, **47**, 24 pp.

Edgerton, H.E., 1963. Underwater photography, pp. 473–9 in M.N. Hill (ed.) *The Sea. Ideas and Observations on Progress in the Study of the Seas*, *3*. Interscience Publishers, New York, 963 pp.

Edgerton, H.E., 1967. The instruments of deep-sea photography. Chapter 3, pp. 47–54 in J.B. Hersey (ed.) *Deep-Sea Photography*. The Johns Hopkins Oceanographic Studies, **3**, 310 pp.

Edgerton, H.E. & Hoadley, L.D., 1955. Cameras and lights for underwater use. *Journal of the Society of Motion Picture and Television Engineers*, **64**, 345–50. (*Collected Reprints, Woods Hole Oceanographic Institution*, 1955, No. 817.)

Eleftheriou, A. & Basford, D.J., 1983. The general behaviour and feeding of *Cerianthus lloydi* Gosse (Anthozoa, Coelenterata). *Cahiers de Biologie Marine*, **24**, 147–158.

Ellis, D.V., 1966. Aerial photography from helicopter as a technique for intertidal surveys. *Limnology and Oceanography*, **11**, 299–301.

Emery, K.O., Merrill, A.S. & Trumbull, V.A., 1965. Geology and biology of the sea floor as deduced from simultaneous photographs and samples. *Limnology and Oceanography*, **10**, 1–21. (*Collected Reprints, Woods Hole Oceanographic Institution*, 1965(1), No. 1508.)

Ewing, M., Worzel, J.L. & Vine, A.C., 1967. Early development of ocean-bottom photography at Woods Hole Oceanographic Institution and Lamont Geological Observatory. Chapter 1, pp. 13–41 in J.B. Hersey (ed.) *Deep-Sea Photography*. The Johns Hopkins Oceanographic Studies, **3**, 310 pp.

Farrow, G., Scoffin, T., Brown, B. & Cucci, M., 1979. An underwater television survey of facies variations on the inner Scottish shelf between Colonsay, Islay and Jura. *Scottish Geological Journal*, **15**, 13–29.

Fedra, K. & Machan, R., 1979. A self-contained underwater time-lapse camera for *in situ* long-term observations. *Marine Biology*, **55**, 239–46.

Fell, H.B., 1967. Biological applications of sea-floor photography. Chapter 19, pp. 207–21, in J.B. Hersey (ed.) *Deep-Sea Photography*. The Johns Hopkins Oceanographic Studies, **3**, 310 pp.

Flato, M., 1965. Long cable telemetry and command system. *Proceedings of the National Telemetering Conference*, Session 10, 176–9.

Franklin, A., Pickett, G.D., Holme, N.A. & Barrett, R.L., 1980. Surveying stocks of scallop (*Pecten maximus*) and queens (*Chlamys opercularis*) with underwater television. *Journal of the Marine Biological Association of the United Kingdom*, **60**, 181–91.

George, J.D., 1980. Photography as a marine biological research tool. pp. 45–115 in J.H. Price, D.E.G. Irvine and W.F. Farnham (eds) *The Shore Environment, Vol. 1: Methods*, Systematics Association Special Volume, 17(a). Academic Press, London, 321 + 41 pp.

Glover, T., Harwood, G.E. & Lythgoe, J.N., 1977. *A Manual of Underwater Photography*. Academic Press, London, 219 pp.

Goeden, G.B., 1980. Reef survey finds commercial fish habitats. *Australian Fisheries*, **39(6)**, 8–9.

Goeden, G.B., 1981. A towed instrument package for fisheries research in Great Barrier Reef waters. *Fisheries Research*, **1**, 35–44.

Gonor, J.J. & Kemp, P.F., 1978. Procedures for quantitative ecological assessments in intertidal environments. *U.S. Environmental Protection Agency, Office of Research and Development, Ecological Research Series*, **EPA-600/3-78-087**, Corvallis, Oregon, 103 pp.

Grassle, J.F., Sanders, H.L., Hessler, R.R., Rowe, G.T. & McLellan, T., 1975. Pattern and zonation: a study of the bathyal megafauna using the research submersible *Alvin*. *Deep-Sea Research*, **22**, 457–81.

Gulliksen, B., 1980. The macrobenthic rocky-bottom fauna of Borgenfjorden, North-Tröndelag, Norway. *Sarsia*, **65**, 115–38.

Harris, R.J., 1980. Improving the design of underwater TV cameras. *International Underwater Systems Design*, **2**, 7–11.

Heezen, B.C. & Hollister, C., 1964. Deep-sea current evidence from abyssal sediments. *Marine Geology*, **1**, 141–74.

Heirtzler, J.R. & Grassle, J.F., 1976. Deep-sea research by manned submersibles. *Science, New York*, **194**, 294–9.

Hemmings, C.C., 1972. Underwater photography in fisheries research. *Journal du Conseil International pour l'Exploration de la Mer*, **34**, 466–84.

Hersey, J.B., 1967. The manipulation of deep-sea cameras. Chapter 4, pp. 55–67 in J.B. Hersey (ed.) *Deep-Sea Photography*. The Johns Hopkins Oceanographic Studies, **3**, 310 pp.

Hoer, J., 1981. Remote control survey. *Water Research News*, **5**, 4.

Holme, N.A. & Barrett, R.L., 1977. A sledge with television and photographic cameras for quantitative investigation of the epifauna on the continental shelf. *Journal of the Marine Biological Association of the United Kingdom*, **57**, 391–403.

Holmes, R.T. & Dunbar, R.M., 1978. Operational experience with the remotely manned A.N.G.U.S. submersible. *Offshore Technology Conference*, **OTC 3234**, 1577–83.

Hopkins, R.E. & Edgerton, H.E., 1962. Lenses for underwater photography. *Deep-Sea Research*, **8**, 312–19.

Horikoshi, M. & Ohta, S., 1975. Underwater photography in ecological investigations of benthic animals including demersal fishes. *Pacific Science Association Special Symposium on Marine Sciences. Hong Kong: Committee for Scientific Coordination*, 29–31.

Isaacs, J.D. & Schwartzlose, R.A., 1975. Active animals of the deep-sea floor. *Scientific American*, **233(4)**, 85–91.

Jogansen, V.S. & Propp, M.V., 1980. Automatic underwater camera for registration of slow processes in marine environment. (In Russian; English Summary). *Biologia Moria*, **1**, 86–8.

Johnston, C.S., Morrison, I.A. & MacLachan, K., 1969. A photographic method for recording the underwater distribution of marine benthic organisms. *Journal of Ecology*, **57**, 453–9.

Kanneworff, P., 1979. Density of shrimp (*Pandalus borealis*) in Greenland waters observed by means of photography. *Rapports et Procès-verbaux des Reunions, Conseil International pour l'Exploration de la Mer*, **175**, 134–8.

Kelly, M.G., 1969. Aerial photography for the study of near-shore ocean biology. pp. 347–55 in *New Horizons in Color Aerial Photography*. Proceedings of Seminar jointly presented by the American Society of Photogrammetry and the Society of Photographic Scientists and Engineers, June 9–11, 1969.

Kelly, M.G., 1970. Aerial photography and coastal ecology. *Marine Pollution Bulletin*, **1** (NS), 11–13.

Kelly, M.G. & Conrod, A., 1969. Aerial photography. *Bioscience*, **19**, 352–3.

Kirkman, H., 1978. Decline of seagrass in northern areas of Moreton Bay, Queensland. *Aquatic Botany*, **5**, 63–76.

Kronengold, M. & Loewenstein, J.M., 1965. Cinematography from an underwater television camera. *Research Film*, **5**, 242–8.

Kumpf, H. E. & Randall, H.A., 1961. Charting the marine environments of St John, U.S. Virgin Islands. *Bulletin of Marine Science of the Gulf and Caribbean*, **11**, 543–51.

Laban, A., Pérès, J-M. & Picard, J., 1963. La photographie sous-marine profonde et son exploitation scientifique. *Bulletin de l'Institut Océanographique, Monaco*, **60**, **No. 1258**, 32 pp.

LaFond, E.C., 1967. Movements of benthonic organisms and bottom currents as measured from the bathyscaph *Trieste*. Chapter 25, pp. 295–301, in J.B. Hersey (ed.) *Deep-Sea Photography*. The Johns Hopkins Oceanographic Studies, **3**, 310 pp.

Laughton, A.S., 1963. Microtopography. pp. 437–72 in M.N. Hill (ed.) *The Sea. Ideas and Observations on Progress in the Study of the Seas. 3*, Interscience Publishers, New York, 963 pp.

Lemche, H., Hansen, B., Madsen, F.J., Tendal, O.S. & Wolff, T., 1976. Hadal life as analysed from photographs. *Videnskabelige Meddelelser fra Dansk Naturhistorisk Forening*, **139**, 263–336.

Littler, H., 1971. Standing stock measurements of crustose coralline algae (Rhodophyta) and other saxicolous organisms. *Journal of Experimental Biology and Ecology*, **6**, 91–9.

Littler, M.M., 1980. Southern California rocky intertidal ecosystems: methods, community structure and variability. Chapter 6, pp. 565–608, in J.H. Price, D.E.G. Irvine & W.F. Farnham (eds) *The Shore Environment, Vol. 2. Ecosystems*, Systematics Association Special Volume 17(b). Academic Press, London, 945 + 100 pp.

Lundälv, T.L., 1971. Quantitative studies on rocky bottom biocoenoses by underwater photogrammetry. A methodological study. *Thalassia Jugoslavica*, **7**, 201–08.

Lundälv, T.L., 1974. Undervattensfotogrammetri-nytt hjälpmedel för biodynamiska studier i marin miljö. *Svensk Naturvetenskap*, **1974**, 222–229.

Lundälv, T.L., 1976. A stereophotographic method for quantitative studies on rocky-bottom biocoenoses, pp. 299–302 in E.A. Drew, J.N. Lythgoe & J.D. Woods (eds) *Underwater Research*. Academic Press, London, 430 pp.

Machan, R. & Fedra, K., 1975. A new towed underwater camera system for wide-ranging benthic surveys. *Marine Biology*, **33**, 75–84.

McIntyre, A.D., 1956. The use of trawl, grab and camera in estimating marine benthos. *Journal of the Marine Biological Association of the United Kingdom*, **35**, 419–29.

Menzies, R.J., 1972. Current deep benthic sampling techniques from surface vessels. pp. 164–9 in R.W. Brauer (ed.) *Barobiology and the Experimental Biology of the Deep Sea*. University of North Carolina, 428 pp.

Menzies, R.J., Smith, L. & Emery, K.O., 1963. A combined underwater camera and bottom grab: a new tool for investigation of deep-sea benthos. *Internationale Revue der Gesamten Hydrobiologie*, **48**, 529–45.

Mertens, L.E., 1970. *In-water Photography. Theory and Practice*. Wiley-Interscience, New York, 391 pp.

Mitchell, C.T., 1967. An inexpensive, self-contained underwater data recording camera. *California Fish and Game*, **53**, 203–08.

Moore, E.J., 1976. Underwater photogrammetry. *Photogrammetric Record*, **8**, 748–63.

Moore, E.J., 1978. Underwater photogrammetry. *Progress in Underwater Science*, **3**, 101–110.

Myrberg, A., 1973. Underwater television—a tool for the marine biologist. *Bulletin of Marine Science*, **23**, 824–36.

Nishimura, M., 1966. A study of the application of underwater television (Report No. 2). *Technical Report of Fishing Boat*, **20(4)**, 18 pp.

Owen, D.M., 1967. A multi-shot stereoscopic camera for close-up ocean-bottom photography. Chapter 8, pp. 95–105 in J.B. Hersey (ed.) *Deep-Sea Photography*. The Johns Hopkins Oceanographic Studies, **3**, 310 pp.

Owen, D.M., Sanders, H.L. & Hessler, R.R., 1967. Bottom photography as a tool for estimating benthic populations. Chapter 21, pp. 229–34, in J.B. Hersey (ed.) *Deep-Sea Photography*. The Johns Hopkins Oceanographic Studies, **3**, 310 pp.

Owens, E.H. & Robilliard, G.A., 1980. Shoreline aerial video-tape surveys for spill countermeasures. *Spill Technology Newsletter*, **5**, 126–30.

Patil, G.P., Taillie, C. & Wigley, R.L., 1980. Transect sampling methods and their application to the deep-sea red crab. pp. 493–504 in F.P. Diemer, F.J. Vernberg & D.Z. Mirkes (eds) *Advanced Concepts in Ocean Measurements for Marine Biology*. University of South Carolina Press, 572 pp. (The Belle W. Baruch Library in Marine Science, No. 10).

Paul, A.Z., Thorndike, E.M., Sullivan, L.G., Heezen, B.C. & Gerard, R.D., 1978. Observations of the deep-sea floor from 202 days of time-lapse photography. *Nature, London*, **272**, 812–14.

Phillips, J.D., Driscoll, A.H., Peal, K.R., Marquet, W.M. & Owen, D.M., 1979. A new undersea geological survey tool: ANGUS. *Deep-Sea Research*, **26A**, 211–25.

Rees, E.I.S., 1976. Notes on benthos visible using underwater TV and in cores. Appendix M, pp. 183–88 in *Out of Sight Out of Mind*. Report of a Working Party on sludge disposal in Liverpool Bay. Department of the Environment, 188 pp.

Rhoads, D.C. & Cande, S., 1971. Sediment profile camera for *in situ* study of organism–sediment relations. *Limnology and Oceanography*, **16**, 110–14.

Rhoads, D.C. & Germano, J.D., 1982. Characterization of organism–sediment relations using sediment profile imaging: an efficient method of remote ecological monitoring of the sea floor (REMOTS™ system). *Marine Ecology—Progress Series*, **8**, 115–28.

Rice, A.L., Aldred, R.G., Billett, D.S.M. & Thurston, M.H., 1979. The combined use of an epibenthic sledge and a deep-sea camera to give quantitative relevance to macrobenthos samples. *Ambio Special Report*, **6**, 59–72.

Rice, A.L., Aldred, R.G., Darlington, E. & Wild, R.A., 1982. A quantitative estimation of the deep-sea megabenthos; a new approach to an old problem. *Oceanologica Acta*, **5**, 63–72.

Richter, H.U., 1968. Underwater Photography. Chapter 9, pp. 385–419 in C.E. Engel (ed.) *Photography for the Scientist*. Academic Press, London, 632 pp.

Robinson, J., 1978. An underwater color viewing system—what it can and will do. *Marine Technology Society Journal*, **12**, 28–33.

Rørslett, B., Green, N.W. & Kvalvågnaes, K., 1978. Stereophotography as a tool in aquatic bilogy. *Aquatic Botany*, **4**, 73–81.

Rumohr, H., 1979. Automatic camera observations on common demersal fish in the Western Baltic. *Meeresforschung*, **27**, 198–202.

Saunders, W., 1972. Submersibles as sample-collecting devices and as possible platforms for *in situ* experiments. pp. 188–192 in R.W. Brauer (ed.) *Barobiology and the Experimental Biology of the Deep Sea*, University of North Carolina, 428 pp.

Schenk, H. & Kendall, H., 1954. *Underwater Photography*. Cornell Maritime Press, Cambridge, Maryland.

Schuldt, M.D., Cook, C.E. & Hale, B.W., 1967. Photogrammetry applied to photography at the bottom. Chapter 5, pp. 69–73 in J.B. Hersey (ed.) *Deep-Sea Photography*. The Johns Hopkins Oceanographic Studies, **3**, 310 pp.

Schwenke, H., 1965. Über die Anwendung des Unterwasserfernsehens in der Meeresbotanik. *Kieler Meeresforschungen*, **21**, 101–06.

Shackel, B. & Watson, G.R., 1968. Closed-circuit television. Chapter 14, pp. 553–606 in C.E. Engel (ed.) *Photography for the Scientist*. Academic Press, London and New York, 632 pp.

Southward, A.J., Robinson, S.G., Nicholson, D. & Perry, T.J., 1976. An improved stereocamera and control system for close-up photography of the fauna of the continental slope and outer shelf. *Journal of the Marine Biological Association of the United Kingdom*, **56**, 247–57.

Steffensen, D.A. & McGregor, F.E., 1976. The application of aerial photography to estuarine ecology. *Aquatic Botany*, **2**, 3–11.

Stephens, W.M., 1967. Sophisticated underwater cameras to bring depths into sharp focus. *Oceanology International*, **2**(**3**), 37–40.

Stevenson, R.A., 1967. Underwater television. *Oceanology International*, **2**(**7**), 30–35.

Stoddart, D.R. & Johannes, R.E. (eds), 1978. Coral reefs: research methods. *Monographs on Oceanographic Methodology*, *UNESCO*, *Paris*, **5**, 581 pp.

Strickland, C. (undated). *Facts on Underwater Illumination*. Hydro Products guide, 37 pp.

Strickland, C.L., 1980. The design and use of still cameras underwater. *Underwater Systems Design*, **2**, 13–17.

Thiel, H., 1970. Ein Fotoschlitten für biologische und geologische Kartierungen des Meeresbodens. *Marine Biology*, **7**, 223–9.

Thorndike, E.M., 1959. Deep-sea cameras of the Lamont observatory. *Deep-Sea Research*, **5**, 234–7.

Thorndike, E.M., 1967. Physics of underwater photography. Chapter 3, pp. 43–5 in J.B. Hersey (ed.) *Deep-Sea Photography*. The Johns Hopkins Oceanographic Studies, **3**, 310 pp.

Torlegård, A.K.I. & Lundälv, T.L., 1974. Underwater analytical system. *Photogrammetric Engineering*, **40**, 287–93.

Treadwell, T.K., 1977. The rationale for submersibles. Chapter 1, pp. 13–21 in R.A. Geyer (ed.) *Submersibles and their use in oceanography and ocean engineering*. Elsevier, Amsterdam, 383 pp.

Trumbull, J.V.A. & Emery, K.O., 1967. Advantages of colour photography on the continental shelf. Chapter 12, pp. 141–3 in J.B. Hersey (ed.) *Deep-Sea Photography*. The Johns Hopkins Oceanographic Studies, **3**, 310 pp.

Uzmann, J.R., Cooper, R.A., Theroux, R.B. & Wigley, R.L., 1977. Synoptic comparison of three sampling techniques for estimating abundance and distribution of selected megafauna: submersible vs camera sled vs otter trawl. *Marine Fisheries Review*, **39**(**12**), 11–19.

Vaissière, R. & Fredj, J., 1964. Étude photographique préliminaire de l'étage bathyal dans la région de Saint-Tropez (ensemble A). *Bulletin de l'Institut Océanographique*, *Monaco*, **64**, **No. 1323**, 70 pp.

Vevers, H.G., 1951. Photography of the sea floor, *Journal of the Marine Biological Association of the United Kingdom*, **30**, 101–11.

Wigley, R.L. & Emery, K.O., 1967. Benthic animals, particularly *Hyalinoecia* (Annelida) and *Ophiomusium* (Echinodermata), in sea-bottom photographs from the continental slope. Chapter 22, pp. 235–49, in J.B. Hersey (ed.) *Deep-Sea Photography*. The Johns Hopkins Oceanographic Studies, **3**, 310 pp.

Wigley, R.L. & Theroux, R.B., 1970. Sea-bottom photographs and macrobenthos collections from the continental shelf off Massachusetts. *United States Fish and Wildlife Service, Special Scientific Report—Fisheries*, **613**, 12 pp.

Wigley, R.L., Theroux, R.B. & Murray, H.E., 1975. Deep-sea red crab, *Geryon quinquedens*, survey off northeastern United States. *Marine Fisheries Review*, **37(8)**, 1–21.

Winterhalter, B., Niemistö, L. & Voipio, A., 1970. Kauko-ohjattu vedenalainen tutkimuslaitteisto. *Eripainos Teknikkalehden*, **2**, 3 pp.

Chapter 5. Diving

J. C. GAMBLE

Diving systems

Modern diving techniques can be classified according to type of breathing mixture and whether the diver is free-swimming or tethered to the surface by a life-support umbilicus. At present, almost all diving associated with benthic research is carried out with air as the breathing mixture and in most cases with the free-swimming 'aqualung' system (frequently known as SCUBA: self-contained underwater breathing apparatus). However, surface demand and traditional 'basic air diving' (helmet diving) have applications in situations where, for example, diver endurance takes precedence over mobility. The 'basic air diving' system, as was used for instance by Kitching *et al.* (1934), was the only system available to the diving scientist prior to the invention of the aqualung by Cousteau and Gagnan in 1943. Riedl (1966, 1980) reviews the development of diving as a tool in littoral research.

It will be assumed that, for the most part, air breathing SCUBA is being used, but with rapid and recent commercial developments the modern diving scientist should be prepared to consider other systems. Closed circuit systems based traditionally on oxygen and, more recently, on an electronically monitored and maintained helium:oxygen mixture are available (NOAA, 1979).

Diving on oxygen alone is strictly limited to depths <10 m and is banned in some countries, while the depth limitation of $He:O_2$ mixture sets depends mostly on gas carrying capacity. Except for the early work of Hass (1949) little benthic marine biological research has been done with these sets, although the absence of expiratory gas bubbles should be favourable for *in situ* behaviour studies, and it is possible to get closer to fish when using the bubble-free $He:O_2$ set compared with a conventional aqualung (J. Main, personal communication). Chapman *et al.* (1974), also Okamoto *et al.* (1981) have observed the reactions of fish to divers.

Limitations of the diving technique

SCUBA diving using air is the simplest and cheapest means of conducting *in situ* benthic research underwater but it is essential that the scientist is initially proficient in the diving technique. The usual routes to proficiency are through the training schedule of a recognized national sport diving organization, through training courses set up within marine institutes or through specialized courses run at diving schools. Further post-training diving experience is desirable before the diving scientist can expect to become competent underwater. International standardization of diver competence has yet to be achieved, although the scientific diver brevet of the Confederation Mondaile des Activités Subaquatiques (CMAS) assists the recognition of scientific diving qualifications between countries. Legislation has recently been passed in Britain that specifies the standards of proficiency of different categories of working diver (Health and Safety, 1981) and requires potential divers to attend an approved training course leading to certification. Details of diving equipment and specification of training schedules, including theoretical as well as practical aspects of diving, are available in many excellent manuals, e.g. British Sub-Aqua Club (1978), Ministry of Defence (Navy) (1972), US Naval Ship Systems Command Supervisor of Diving (1973) as well as in popular texts (Poulet, 1962; Terrell, 1965). Aspects of scientific diving are specifically dealt with in the NOAA (1979) *Diving Manual* and in the Underwater Association *Code of Practice for Scientific Diving* (Underwater Association for Scientific Research Ltd, 1979).

In addition to the time limit imposed by the volume of breathing mixture carried by a SCUBA diver, three main physical factors limit efficiency: depth, temperature and visibility. Other factors such as sea state, currents and dangerous animals will also affect diving activities but these tend to be localized or short-term effects.

Although the three major factors are often interrelated, depth is the universal limitation for all diving activity. Human hyperbaric physiology has been extensively investigated and is discussed in many texts such as Miles & Mackay (1976), Shilling *et al.* (1976), and Strauss (1976). The time of a dive to >10 m on air is limited according to depth by the risk due to inert gas absorption on decompression. Figure 5.1 illustrates the time–depth decompression relationship based on US Navy Diving Tables. Normal decompression (Zone II, Fig. 5.1) can be a very lengthy procedure involving ascent stops at specified depths for prolonged periods. This is usually exceptionally unpleasant in cold waters and is not desirable in the tropics. In consequence, unless the diver has immediate access to an onshore or shipboard decompression facility, most scientific dives should be organized within the depth–time limits set for a 'non-stop' decompression schedule (Zone I, Fig. 5.1).

Nitrogen narcosis progressively affects diver performance as diving depth increases (Baddeley *et al.*, 1968) and effectively limits air diving to about 50 m. Individual diver susceptibility to nitrogen narcosis varies and some acclimatization to moderate narcotic effects can be acquired by careful dive planning involving incremental increases in the maximum depth of successive dives (Underwater Association for Scientific Research Ltd, 1979). Helium is not narcotic (Shilling *et al.*, 1976) and is substituted for nitrogen in all commercial deep diving systems. To date, because of the complexity of the helium decompression schedules, and probably the expense of helium itself, little benthic scientific research has been carried out taking advantage of $He:O_2$ breathing mixtures. However, Flemming & Stride (1967) used self-contained $He:O_2$ sets at 50 m depth to study sand and gravel patches.

Dive duration is also affected by temperature and this is not only a problem in higher latitudes, since water even a few degrees lower than body temperature can rapidly cool a stationary working diver. Hence, an insulating suit is essential except for a mobile diver in tropical water. The foam neoprene rubber wet suit provides the greatest mobility for the diver but is not as efficient an insulator as the constant volume dry suits which are now available. Being more cumbersome, these are more suitable for working with sleds,

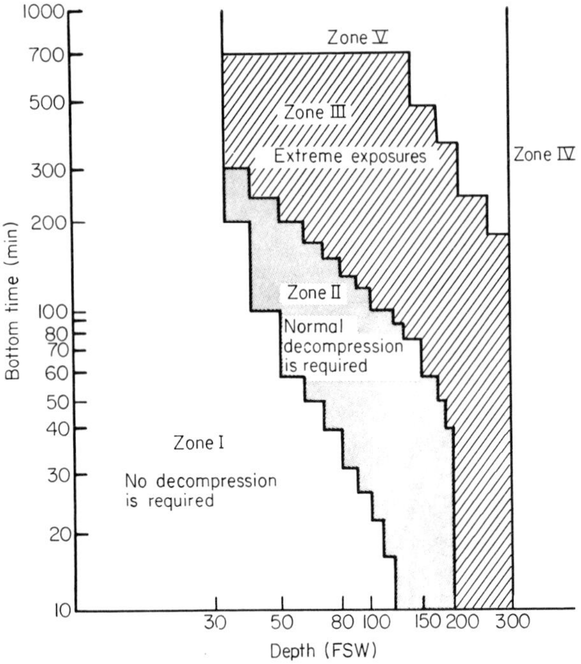

Fig. 5.1. Decompression risk associated with a dive on air at selected depths (FSW = feet of seawater, 1 m = 3·28 ft) and bottom time (min). Decompression schedules for Zones II and III are taken from US Navy standard schedules. Zone IV diving is *dangerous* (too deep) on air. Zone V dives are covered by US Navy saturation decompression schedules. (From Shilling *et al.*, 1976.)

towed wet submersibles or other transportation devices. The best temperature control is achieved by hot water circulation within a suit. This is usually provided through an umbilicus from the surface, although a self-contained system has been developed (NOAA, 1979). Where temperature and decompression considerations are not paramount the surface umbilicus system is ideal for long duration dives in a limited area. It is particularly favoured by archaeologists working on a specific site (Muckelroy, 1978) and has great potential for detailed *in situ* shallow water benthic research.

The depth limitation of air diving techniques hence confines scientific benthic diving activity to the sublittoral fringe where ambient levels are often drastically reduced by suspended particle load, although in clear water there is sufficient light to work by even at 50 m depth. The particles can be biogenic, such as a spring phytoplankton bloom, or inorganic in origin. Many kinds of underwater light are available to the diver, but reflectance and light scatter from suspended particles limit their effectiveness (see also Chapter 4). Light loss due to particle suspension is a serious problem when working on very soft sediments which are inevitably disturbed by the diver himself. Various devices which may be of assistance in these conditions are discussed later, but anyone diving in these circumstances should be prepared to carry out the allotted underwater task in complete darkness. Relocation of experimental sites and apparatus in conditions of very low visibility usually necessitates marking the site with a buoyed line. Where this is impossible 'pinger' beacons for diver-operated locators are commercially available (NOAA, 1979). An underwater compass is useful as an aid in crude direction finding but is too unreliable for accurate work.

Logistic limitations on diving are imposed by the necessary safe minimum size of the diving team. It is foolhardy, and in some countries illegal, to dive alone. The size of the team depends on many factors and guidance should be taken from reputable manuals (Underwater Association for Scientific Research Ltd, 1979; NOAA, 1979). Some countries have legislation procedures which affect working divers, including those engaged in scientific research. It is essential to check whether such regulations exist before embarking on a scientific programme.

Data recording

Written

Psychological investigations by Godden (1975) and by Godden & Baddeley (1975) have demonstrated that recall to memory of observations made during a dive is worse than the recall of similar observations made in a terrestrial situation. Consequently, apart from the briefest collection dive, the diving scientist should carry some means of directly recording or communicating his *in situ* observations.

A writing board of white plastic laminate (or other rigid waterproof surface) with a graphite pencil is the simplest method of recording information underwater. The pencil line is not easily erased underwater without the aid of a domestic abrasive cleanser. At its simplest, the board need only have a wrist strap and pencil attached by a short lanyard, although it can be readily elaborated to meet the needs of a specific exercise, e.g. with an attached plumb line and angular gradations for measuring inclination (Woods & Lythgoe, 1971), or the writing area can be increased by incorporating a number of sheets of plastic 'paper' (Fager *et al.*, 1966). A measuring scale along one edge is also useful. However, it is best to limit the board size since large boards drag during a dive; 20 × 30 cm is a useful size. Prepared layouts can greatly increase dive efficiency during specific tasks. Waterproof marking pen lines will resist several pencil line erasures but etched laminate boards provide a permanent proforma. Nichols (1978) shows a layout for simple survey details of *Echinus esculentus*.

Spoken

Writing boards, while essential for sketch diagrams, are often a nuisance to carry and both hands are usually required during writing. Intelligible speech is possible underwater and can either be recorded directly or transmitted to the surface. Portable, battery-powered, cassette tape recorders in waterproof housings are extremely useful for rapid data recording. Fager *et al.* (1966) describe a custom-designed microphone fitted to the mouth-piece of a single hose regulator. More recent systems are based on bone conduction microphones held tightly against the diver's skull by the suit hood (Woods & Lythgoe, 1971; Byers, 1977; Main & Sangster, 1978a). Intelligible communication requires the diver to use a mouthbox instead of the usual 'biting' mouthpiece of the regulator. Nevertheless valve movements and bubble production during breathing as well as direct pressure effects on speech production often make post-dive interpretation difficult. Before using a tape recorder for important scientific recording, it is essential that some practice in transliteration has taken place. The waterproof housing for the recorders is usually strapped to the aqualung cylinder. Main & Sangster (1978a) show two housing designs (Fig. 5.2) based on an inexpensive Sanyo M48M recorder. The cylindrical housing is stronger and has a more reliable seal than the rectangular perspex (acrylic) case, although it is more buoyant. An on/off

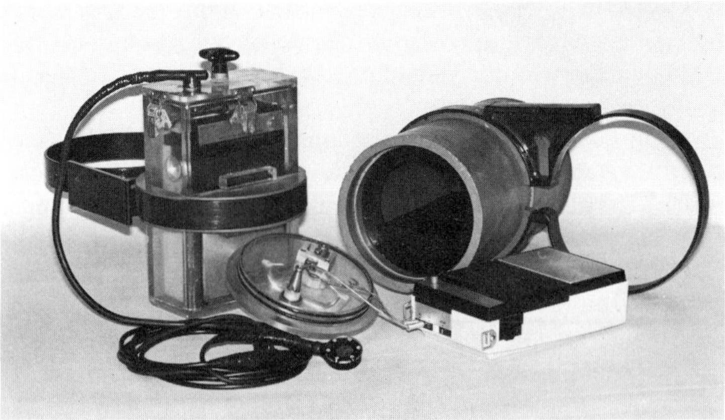

Fig. 5.2. Two types of underwater housing for a portable, battery-powered tape recorder. Housing on the left is made out of 9 mm clear acrylic sheet, that on right from 9·5 mm wall PVC tubing with a clear, O-ring sealed, acrylic lid. Both are clamped to the diver's air cylinder. (From Main & Sangster, 1978a.)

switch can be operated by the diver (Fager *et al.*, 1966) but it is usually simpler and more reliable to leave the recorder on for the duration of the dive.

Communication to the surface via a telephone system is ideal for a diver on an umbilical system but can also be used by SCUBA divers and this method gives the best speech reproduction. 'Through-water' systems (NOAA, 1979), which provide diver-to-diver communication, are available but tend to be bulky and expensive, and in some systems the efficacy can be affected by obstructions in the direct line between transmitter and receiver. Surface communications using this system are two-way and their use is most valuable when a diver is carrying out tasks such as operating TV cameras, which are monitored at the surface.

Visual

The camera is a most important item for the diving scientist and, although aspects of underwater photography are discussed in Chapter 4, further points, specific to diver-operated systems are emphasized here. A wide range of custom-built underwater housings are available for many types of camera, including ciné (see catalogues produced by Ikelite, Aquasnap and others). Many cases for SLR cameras are included in the range but, unless the housing can take an action/sports finder (such as the Nikon DA-1 Action Finder), or reflex viewfinder, as with the 70 mm Hasselblad housing, delineation of the field of view will not be available to a mask-wearing diver.

The waterproof Nikonos camera (Fig. 4.1) is the most straightforward and useful camera for the diving scientist. Nikon provide four interchangeable lenses, 15 mm, 28 mm, 35 mm and 80 mm, a close-up lens attachment kit, flashguns and viewfinders, while a range of supplementary equipment is made by other firms (Taylor, 1977). Earlier models of the Nikonos were all manual but these have been superseded by the aperture-priority automatic Mk IV-A which also has an automatic electronic flash facility.

Despite the apparent simplicity of the Nikonos camera, underwater photography requires much care to achieve success. While *in situ* practice is the best route to proficiency there are many useful publications (e.g. Church & Church, 1972; Glover *et al.*, 1977; Schulke, 1978; George, 1980). George (1980) makes many recommendations for efficient underwater photography, but two points are worth emphasizing: first it is difficult to write notes while handling a camera, so a portable tape recorder is desirable for noting information; secondly, it is often better to be overweight for benthic photography and to use an adjustable buoyancy device for control throughout a range of depths. Extreme care is needed when working over muddy bottoms if sediment resuspension is to be avoided.

Riedl (1966) beautifully illustrates the value of the diver-operated camera

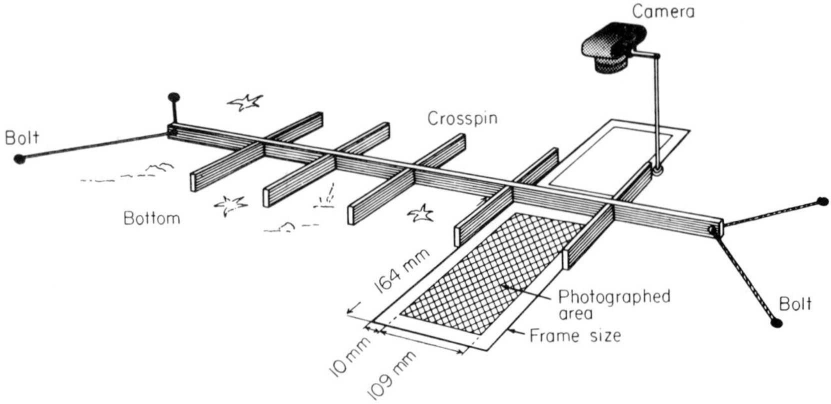

Fig. 5.3. Aluminium 'test-rod' fixed permanently to a specific rocky test site and used to determine position of area for repeated photographing. Note that the photographed area is considerably larger than the aluminium guides. (From Gulliksen, 1980.)

in his studies on the biota of clefts, crannies, caves and other hard surface features. More recently, precise determination of epibenthic cover has been made using the fixed focus close-up Nikonos system (Hiscock & Hoare, 1975; Gulliksen, 1980). Gulliksen, who visited the same site over a $3\frac{1}{2}$ year period, used a permanently fixed aluminium guide to ensure exact replication of the area of his photographs (Fig. 5.3). In consistently clear waters, as for instance in coral reef regions, descriptive photography of large structures and extensive features is commonplace (Stoddart & Johannes, 1978). Pearson (1981) used vertical and oblique photographs of permanent study plots to illustrate coral recovery over several years after catastrophic natural events.

Photography of permanent study plots over a period of time not only provides information on species succession and seasonal changes but can give estimates of linear growth rates of sessile organisms. However, where organisms are competing for light, masking effects, such as those caused by rapid growth of tabular *Acropora* sp. corals (Pearson, 1981), can overshadow successional developments beneath the canopy. Careful searching and sampling is the only way of determining such eventualities. Nevertheless some stereophotographic methods have been developed for studying permanent plots (see also Chapter 4). Such methods should provide more accurate non-destructive estimates of biomass and, according to Lundälv (1971), there is less concealment and superimposition in a stereo-pair of photographs than in a single shot. The single camera stereo technique described by Lundälv (1971), and by Moore (1978) necessitates accurate camera positioning in a rigidly held guide. This limits the use of the technique to suitable surfaces and, as mentioned in Chapter 4, is for sedentary organisms only. The use of two synchronous cameras as a stereo pair overcomes these limitations (Rørslett *et al.*, 1978). Johnston *et al.* (1969), however, describe a much simpler single camera method based on photographing the subject within a movable, graduated rectangular box-quadrat.

The diver-operated camera is also of great value for *in situ* behavioural studies. George (1980) lists several examples, such as interspecific aggression in corals (Lang, 1973), pumping in sponges (Reiswig, 1971), orientation and feeding in sea fans (Velmirov, 1973), ophiuroids (Warner & Woodley, 1975) and crinoids (La Touche, 1978). *In situ* photography has featured in many studies of fish behaviour and, although outside the scope of this Handbook, benthic interactions such as predation on echinoids (Fricke, 1971, 1973a) and symbiotic relationships between fish and anemones (Mariscal, 1970) have been described photographically.

In some respects ciné and TV are better media for behavioural studies than still photography, but, unless the diver is carried in a towed vehicle, the cable from a TV camera can be an enormous drag. Diver-operated ciné and TV have been used extensively for fish behaviour studies, particularly in relation to fishing gear (Main & Sangster, 1978b), and stationary TV and time-lapse ciné have been used to observe the behaviour of relatively static organisms (A. Eleftheriou, personal communication) and the diurnal emergence patterns of burrow dwellers such as *Nephrops norvegicus* (Chapman *et al.*, 1975; Chapman & Howard, 1979). Howard (1980) used a motor-driven SLR camera in a similar time-lapse set-up to study the effects of currents on foraging of *Cancer pagurus* and *Homarus gammarus*. Both of the time-lapse systems, which have the advantage of being self-contained, require bait to attract the decapods within range and need white electronic flash for each exposure. Chapman & Howard (1979) compared time-lapse and TV methods as well as low light and conventional Vidicon cameras. They used red perspex filters which restricted light for the cameras to wavelengths >600 nm, i.e. outside the lobster's visual range. There was, however, no difference between activity as determined by TV and white flash time-lapse, nor was there any difference between the red light levels (2×500 W and 1×15 W) required for the two types of TV camera. Chapman & Rice (1971) previously noticed that the nocturnal emergence of *Nephrops* was adversely affected by continuous 'white' light.

Sample collection

Large macroflora and fauna are usually collected by hand and transferred to pre-labelled polythene bags, mesh sacs or other suitable containers. Self-sealing polythene bags are particularly useful for smaller organisms. Carriage of samples is a problem underwater although it is possible to use lifting bags or electric tugs (Flemming & Stride, 1967) to raise heavy samples. Specialized sample containers, ranging from weighted crates for jars (Fager *et al.*, 1966) to buoyed lines of sacs (Bailey *et al.*, 1967) optimize the short time available for underwater work, but, when working on fine sediments, it is important to be able to operate any systematic collection system entirely by feel. Gullicksen & Derås (1975) have described an underwater sample and equipment carrier supported on a buoy and tethered to the diver.

Many benthic organisms are either too small, or are too fragile to be collected by hand. Some hidden habitats, such as macrophyte holdfasts, can be removed in their entirety with a knife or paint scraper and sealed in a polythene bag. McKenzie & Moore (1981) recommend introducing a few drops of concentrated formalin into the holdfasts as soon as possible to induce the organisms to leave their habitat. Dilute formalin or bleach can also be squirted into burrows to drive out motile organisms (Rice & Johnstone, 1972) while Russell *et al.* (1978) detail methods of using poisons or anaesthetics for collecting localized populations of reef fish (see also pp. 173–4; Chapter 6).

The simplest method of collecting small attached epibenthic organisms is to use a paint scraper within a frame with an attached mesh bag (Kain, 1960; True, 1970; Drew, 1971). These methods can be used quantitatively but low-growing encrusting organisms such as bryozoans, serpulids and barnacles can be missed or crushed (Hiscock, in preparation). Hard-shelled organisms readily fall into mesh sample bags but many delicate or soft-bodied species are easily wafted away. Several methods however, have been devised using a collecting current to carry organisms into a mesh container. Such suction samplers, which are effective for collecting infauna from sediments as well as for sampling detached epibiota, have been developed to various levels of complexity and efficiency.

Suction samplers

Manual suckers

These are essentially qualitative samplers designed to pick up slow-moving delicate organisms, e.g. nudibranchs, from the substratum. The typical design (Fig. 5.4), often called a 'slurp gun', as described by Tanner *et al.* (1977), is valved to permit multiple sampling into the same container. Simpler and cheaper devices are described by Bleakney (1969) and Clark (1971). Tanner *et al.* (1977) used their sampler for collecting delicate algal macrophytes and suggest that it could be used to sample organisms >1 mm diameter in small delimited areas of approximately $0{\cdot}1\ m^2$. A two-man manual sampler was used by Gullicksen and Derås (1975) for sessile epifauna and the fine sediment

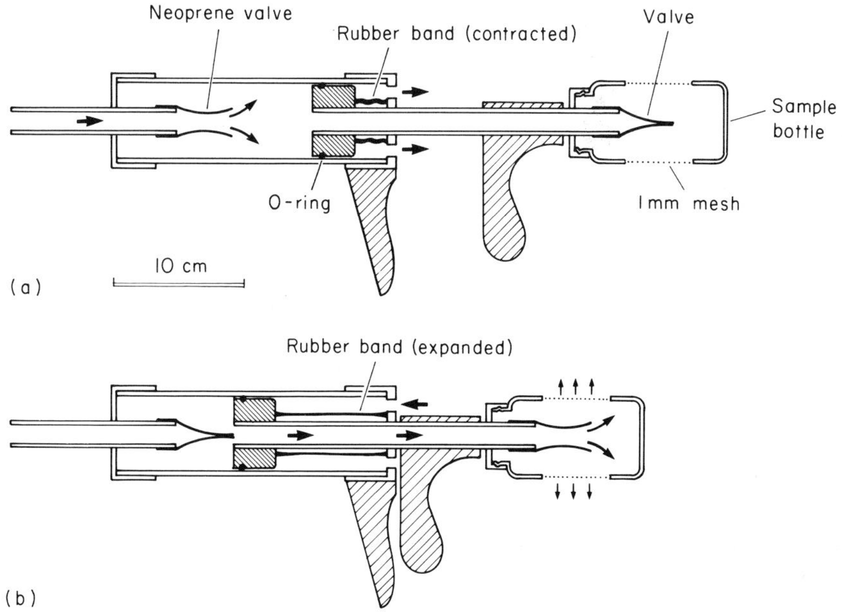

Fig. 5.4. Longitudinal section of manual suction sampler indicating mode of operation (arrows denote direction of water flow). (a) Rubber bands pull piston back to fill pump chamber. (b) Chamber evacuated through sample bottle mesh as handles are squeezed together. This sampler has been used for collecting filamentous algae, and slow moving invertebrates and fish. (From Tanner *et al.*, 1977.)

layer on rock surfaces. This could sample an area of up to 500 cm^2, depending on the rate at which the mesh filters become clogged.

Mechanical suckers—hydraulic

The limitations of mechanical suckers—pulsed action and low capacity—are overcome in mechanical, continuous high flow rate devices. Hydraulic samplers depend on the aspirator principle, using pumped water injected axially through a venturi into a larger tube. Samples are drawn through the nozzle of the larger tube and collected in rigid cone filters or a mesh bag. The original diver-operated hydraulic sampler, designed by Brett (1964) and improved upon by Massé (1967, 1970), depended upon a surface water pump and hosed supply. A self-contained version based on a 12 V pump (Fig. 5.5) is described by Emig & Lienhart (1967). When used in conjunction with a coring cylinder driven into the sediment to delimit the volume of sampled sediment and prevent caving in, these devices are most efficient quantitative macrofauna samplers (Massé, 1967, 1970).

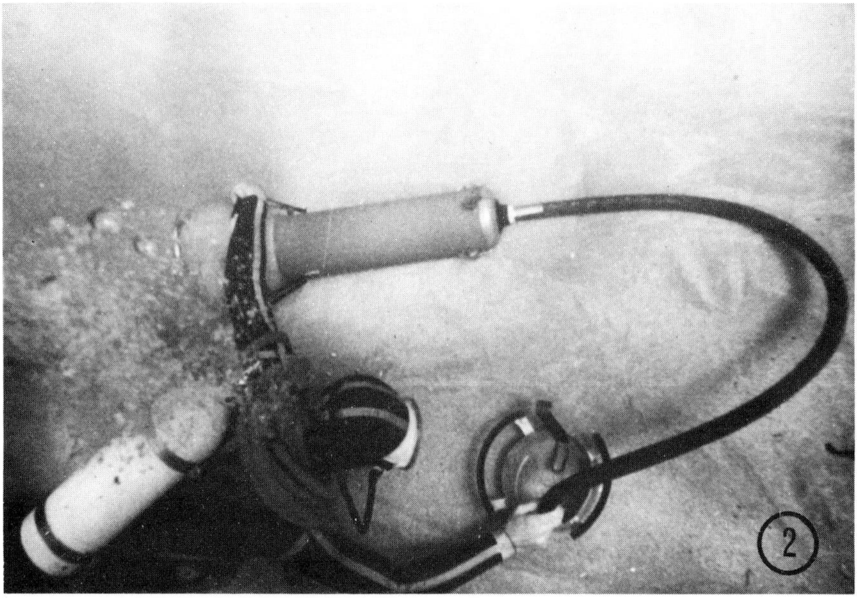

Fig. 5.5. Hydraulic suction sampler being used to excavate sediment contained within the sampling cylinder. (From Emig & Lienhart, 1967.)

Mechanical suckers—air lift

Barnett & Hardy (1967) describe a diver-operated soft sediment sampler (Fig. 5.6) which operates using suction created at the bottom end of a long, vertical tube into which a stream of fine air bubbles is continually injected. Various modifications of the original design have been made, usually incorporating a self-contained air supply, other than one at the surface (Christie & Allen, 1972; Keegan & Könnecker, 1973). Air lifts have been mostly used to excavate sediments, but the Finnish IBP-PM Group (1969) and Hiscock & Hoare (1973) describe the use of air lifts to collect epibiota scraped off hard surfaces. Barnett & Hardy's (1967) sampler incorporated a device for mechanically sucking the coring cylinder (0·1 m^2 surface area) into the sediment but Keegan & Könnecker (1973) noted that this was both time-consuming and inefficient in coarser sediments. Like Massé (1970), they used a stainless steel coring cylinder with a bevelled lower edge and circular hand grip at the top to enable the cylinder to be driven manually into the sediment.

Although very efficient at removing organisms to depths down to 60 cm, particularly in sandier sediments (Massé, 1970; Christie, 1976; Massé *et al.*, 1977), mechanical suckers have disadvantages (Hiscock, in preparation). For instance, air lifts are cumbersome instruments requiring a vertical tube up to 6 m in length, while hydraulic suckers have a tendency to clog rapidly and

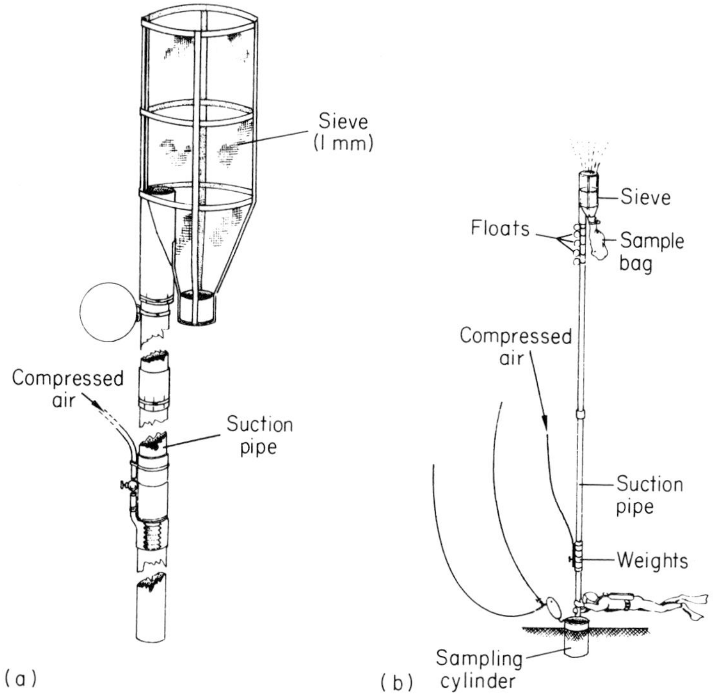

Fig. 5.6. Barnett & Hardy (1967) air-lift suction sampler. (a) Details of suction pipe and sieve; (b) method of excavation of contents of sunken sampling cylinder; sieved material falls into attached plastic bag. The plastic floats beneath the sieve are now replaced by an adjustable buoyancy cylinder fed from the system compressed air supply. (P.R.O. Barnett, personal communication.)

create a dense cloud of suspended sediment. In addition, the sample size is usually limited to about 0·1 m^2 area due to the difficulty of driving in the coring cylinder. See also remotely-operated samplers on pp. 169–71 in Chapter 6.

Corers

The corer is one of the simplest and most effective sampling tools available to the diver. Simple corers are small diameter (approximately 5 cm) metal or plastic tubes driven into the sediment by hand or, if necessary, with the aid of a hammer (Angel & Angel, 1967; Gage, 1975). A bung or tight lid prevents muddy sediments escaping from the tube as the corer is removed, but since, in these sediments, many meiobenthic organisms inhabit the flocculent upper few centimetres (McIntyre, 1971), care must be taken when sealing the top of the tube. A small 'bung within the bung' or a simple flap-valved lid (Gage, 1975) facilitates this operation. McIntyre (1971) has emphasized that a minimum core diameter of 10 cm is required for effective sampling of surface flocculent material.

When used in coarser sediments, larger diameter cores tend to lose their excavated material on removal. This can be avoided by sealing the base with a gloved hand, a bung or a detachable base-plate (Finnish IBP-PM Group, 1969). Multilevel corers, where the core is sectioned by a series of horizontal plates introduced through slots in the core tube, are described by Fager *et al.* (1966) and by Kirchner (1974). Mechanically driven corers have been designed by Walker (1967) and by Keegan & Könnecker (1973).

A later modification of the Barnett & Hardy (1967) suction sampler eliminates the *in situ* air lift sampling stage by converting the sampling cylinder into a large corer (P.R.O. Barnett, B.L.S. Hardy & J. Watson, personal communication). The coring action is controlled by a diver but the cylinder and core are withdrawn by a shipboard winch with a wire bridle attached to lugs close to the lower edge of the cylinder. The cylinder, which has been strengthened to withstand the great tensions (up to 1·25 t) involved during core extraction, is made of 3·2 mm steel and the quick-release lid clamps are replaced by screw clamps. A concentric tube at the top of the corer keeps the bridle from fouling the clamps during inversion and also serves as a handle. Two-thirds of the original suction sampling diving time is saved by sieving the contents onboard ship, although the original suction tube is still used to sample large areas qualitatively or to assist *in situ* collection of macrofauna.

Resin casting

Polyester resin casting techniques, originally devised for intertidal use by palaeontologists for verifying trace fossils (Shinn, 1968; Farrow, 1975), have been successfully applied to the investigation of the burrow architecture of certain sublittoral benthic organisms. Rice & Chapman (1971), Atkinson (1974), Pemberton *et al.* (1976) and Myers (1979) have made resin casts of the burrows of crustacea, and similar techniques have been employed by Nash (1980) for several species of burrowing fish. Rice & Chapman mixed the resin and catalyst ashore but delayed addition of the hardener until immediately before diving. They found that a small frame standing 0·75 m above the sediment was desirable as a platform to aid resin pouring and to minimize bottom resuspension.

Survey and sampling methods

It is possible, with a little ingenuity, to apply most conventional terrestrial ecological survey and sampling techniques to *in situ* studies of the benthos.

However, because of the severe restrictions of time on site and the reduced operational capacity of the diving ecologist, it is important to decide upon an optimal working strategy. Furthermore the pre-dive strategy must take account of the variety and nature of the near-shore sublittoral habitat. Substrata range from hard rock to flocculent muds, surface inclination from vertical to horizontal, faunal mobility from sessile to free-swimming, situation from surface-dwelling to cryptobiotic. Within the 40 m depth normally investigated by SCUBA there is usually a distinct zonation of the macrophytes, primarily imposed by differential light absorption (Drew, 1971). This zonation is itself affected by vertically differentiated stresses such as wave action and exposure (Hiscock, 1979).

One distinct advantage which the sub-aqua ecologist has over his terrestrial counterpart is his ability to move independently of the sea bed because of neutral buoyancy. This not only facilitates investigation of 'inaccessible' regions such as vertical or overhung faces (e.g. Hiscock & Hoare, 1975; Lundälv 1971), cave interiors (Riedl, 1966, 1976; Larkum *et al.*, 1967) and the infrastructure of giant kelp forests (Aleem, 1956), but also allows for over-view swimming of an area, either as an objective in itself or as a way to determine limits of a particular area of interest. Overall strategy generally involves an initial qualitative over-view of the region leading to precise quantification of the particular habitat of interest.

Qualitative over-view

The effectiveness of the over-view will depend upon ambient conditions, particularly visibility. The simplest method is to swim along a path determined by depth contour and/or compass bearing or by some obvious linear feature such as the edge of a reef or base of a cliff. The value of the over-view can be much enhanced if the area observed during the dive is increased by towing the diver from a surface craft or by use of an electric tug. Hiscock (in preparation) points out that a diver controlled tug is preferable to towing when in the vicinity of cliff faces or rock outcrops!

It is possible to hold onto a tow rope from a boat but various underwater hydroplanes (Fig. 5.7) have been designed for divers, ranging from simple one-man vanes (Riley & Holford, 1965; Erwin, 1977a) through two-man sleds (Sigl *et al.*, 1969) to a manoeuvrable mid-water two-man vehicle (Main & Sangster, 1978b). Since the diver is manually occupied with controlling the simple hydroplane a tape recorder is essential for data logging during the dive. Two-man vehicles, when fitted with a screen, can be towed at speeds in excess of two knots, carrying a pilot and also an observer who is able to handle recording gear such as a TV camera. They do, however, require more motive power for towing. Midwater vehicles, unlike the sled, do not depend upon a

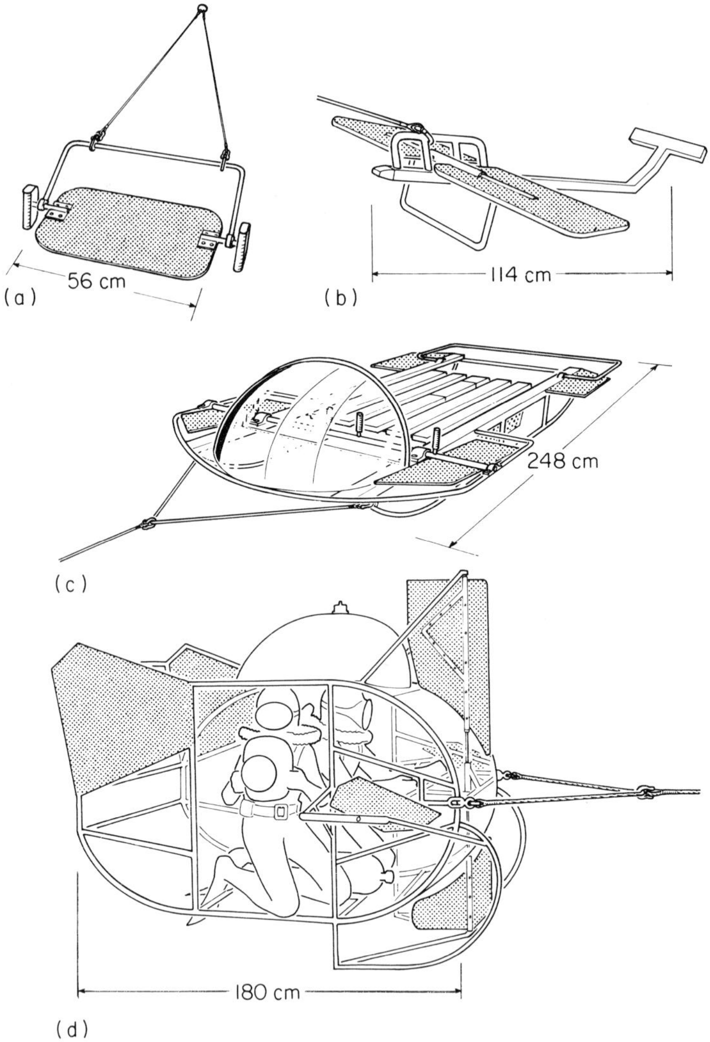

Fig. 5.7. Types of vehicle used in benthic surveys. (a) One-man hydroplane (after Riley & Holford, 1965); (b) one-man hydroplane (after Erwin, 1977a); (c) two-man benthic sled (after Sigl *et al.*, 1969); (d) two-man midwater vehicle (after Main & Sangster, 1978b).

uniform bottom and do not disturb the sediment as they pass. The value of the broad scale hydroplane survey is illustrated by Erwin (1977b), while Earll (1977) discusses primary survey strategy.

Use of amateur divers

Broad scale qualitative ecological information can be obtained in detail over a wide area and throughout a prolonged period by the systematic use of amateur divers. In Britain, projects have been organized on kelp (Bellamy *et al.*, 1968), *Echinus esculentus* (Nichols, 1978, 1981) and on general coastal habitat or sublittoral species survey (Earll & Erwin, 1979). The projects on single organisms can often be simply defined, but habitat and species recording schemes need careful design and assessment by the co-ordinators. Feedback to the participants through meetings, newsletters or the more popular journals (Bellamy & Whittick, 1968; Earll, 1979) is essential for maintenance of interest.

A concise set of simple instructions is required for all schemes which include help from amateur and dispersed groups of divers. Ideally, the instructions should either be formulated so that they can be easily transferred to a diver's writing board or be printed on waterproof paper. If interest can be sustained, such schemes should provide useful information on background species and habitat distribution patterns which could have long-term significance.

Detailed sampling and survey methods

The usual objectives of underwater ecological surveys, such as estimates of population density and biomass, assessments of spatial relationships, structural composition of communities, and characterization of habitats, are common to ecological studies in all ecosystems. Since many of the organisms, particularly those found on hard surfaces, are sessile, procedures used in plant ecology and sociobiology are frequently adopted and adapted (e.g. Goodall, 1952; Greig-Smith, 1965; Guonot, 1969). Drew (1971) has pointed out that most ecological investigations of benthic macrophytes have been based on transect and quadrat methods, while similar techniques have been applied to sessile animals (e.g. Weinberg, 1981; Hiscock, in preparation).

The more dispersed, visible, macrobenthic organisms such as echinoderms (Forster, 1959; Larsson, 1968) have been surveyed by the 'swim-line' method, where individuals are counted within an area defined by the length of a horizontal rod (usually 1 m) held normal to a measured line stretched along the bottom. Levin & Shenderov (1975) have pointed out that for small samples (<20 widely dispersed organisms, density less than $0{\cdot}2\ m^{-2}$) the total error in estimating population density is associated principally with the sample size and not with estimation of area. Hence they conclude that accurate methods of measuring area are unnecessary and have devised census methods based on approximate area estimation. Their methods, which depend on the assumption that the average travel speed of an experienced diver is constant (20 m per minute), are based on a system of multiple tacks at a set distance above the bottom, or on the total time taken to collect a number of specimens of known spatial separation. More precise plotless methods suitable for studying coral dispersion patterns, such as closest individual and nearest neighbour methods (Cottam & Curtis, 1956) are discussed by Loya (1978).

Transects

A transect normal to the shoreline running from shallow into deeper water is commonly used for *in situ* studies. The technique usually consists of laying a non-buoyant line on the substratum in a direction chosen from the surface or by means of a depth profile or underwater compass heading. When the inclination of the sea bed is excessively steep the line is best attached to natural features, algal holdfasts, or artificial devices such as rock-climbing 'pegs' or bolts in holes made by pneumatic drills (NOAA, 1979). As well as serving as a reference, the judicious use of buoys, sacs and carabiner clips allows the line to serve as a means of storing and recovering sample collections (Bailey *et al.*, 1967).

The profile of a transect is usually determined trigonometrically from differential depth and incremental transect length between adjacent points. Hiscock (in preparation) describes a simple method whereby profile determination is aided by a rough sketch of prominent features on the profile. Measurements can be made by means of a fibreglass surveying tape, and precise profiles of particular features can be determined with the aid of an inclinometer (Woods & Lythgoe, 1971).

Transects are often initially used in an investigation as a means of recording qualitative changes in distribution with increasing depth, and the flora and sessile fauna can be represented symbolically on a plot of the measured profile (e.g. Neushul, 1965; Riedl, 1966, 1980; Drew, 1971). However, line transects can also be used to quantify distribution, by means of techniques developed in particular by Schmid (1965), for aquatic vegetation, and by Loya & Slobodkin (1971) and Loya (1972) for abundance and diversity estimates of hermatypic corals. The latter authors worked a series of short transects running along selected depth contours (Fig. 5.8) each parallel to the shoreline; coral abundance estimates were based on the length of the individual colony intersections on each line. Abundance estimates can also be made by noting intercepts at set points on the transect (Weinberg, 1981), but the most frequently used method of quantifying from a transect is to sample

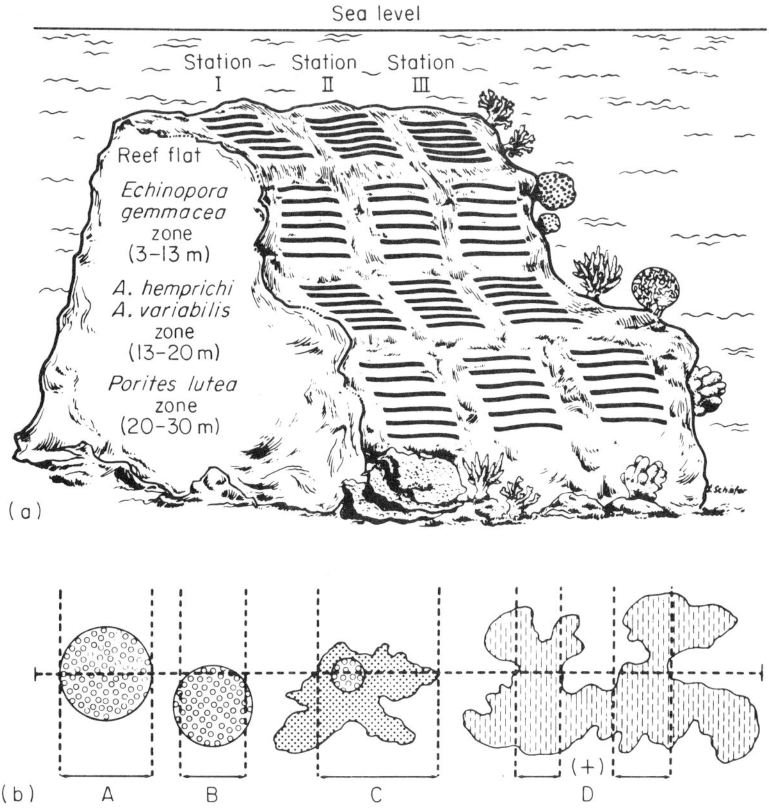

Fig. 5.8. (a) Line transect method devised by Loya & Slobodkin (1971) and Loya (1972) for quantitative studies on corals. (b) Intercept method used to quantify different colonies (from Loya, 1978).

plots at specific depths or zone intervals down a transect laid normally to the shore (e.g. Mann, 1972a; Lambert *et al.*, 1972). Sampling can be undertaken photographically (e.g. Lundälv, 1971; Bohnsack, 1979), but more frequently involves counting, visual assessment, or removal and collection of all material within the area of a quadrat.

Quadrats

Frames larger than 1 m^2 are too unwieldy to use underwater unless assembled *in situ*. However, as with terrestrial ecology, the important initial aim is to determine the most efficient, i.e. minimal, area which should be sampled. Several authors have discussed this minimal area concept in relation to diving (e.g. Boudouresque, 1971, 1974; Drew, 1971; Weinberg, 1978, 1981) and

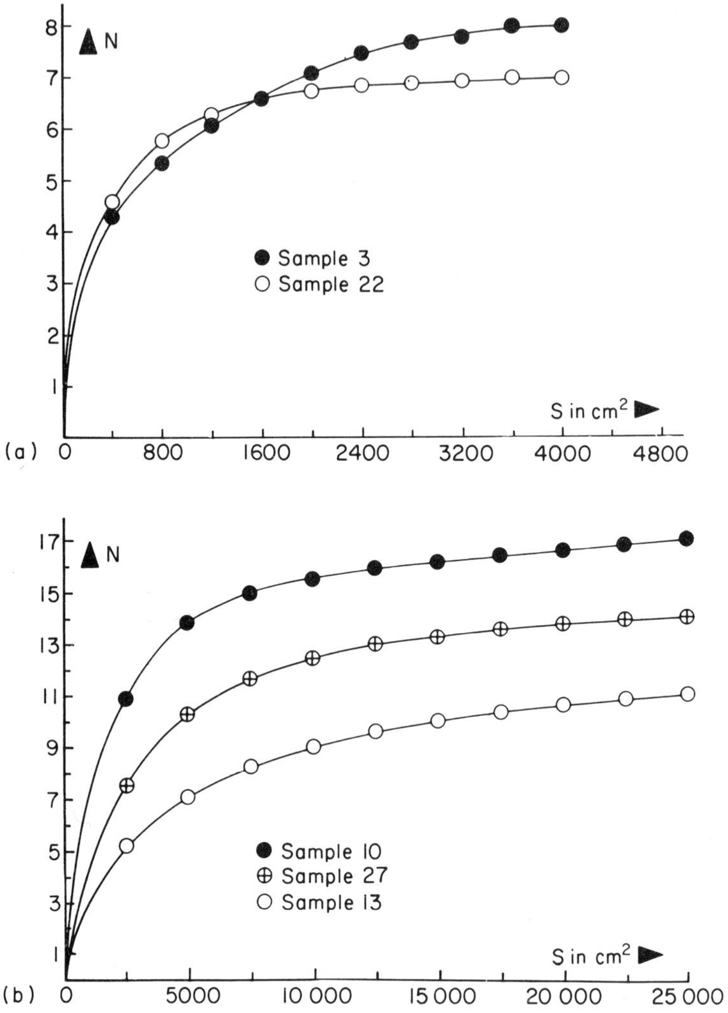

Fig. 5.9. Species-area curves for Mediterranean rock epifauna in (a) dark caves and (b) open communities. N, number of species; S, sample surface. (From Weinberg, 1978.)

generally agree that the minimal sampling area should be that which, if doubled, would yield only a 10% increase in information. Figure 5.9 shows characteristic asymptotic species–area curves used to estimate minimal area but Weinberg (1978) points out that the minimal sampling area will vary, depending on the nature of the information required. For instance, the minimal area to be sampled according to a qualitative similarity index (Sørensen) was half that defined for the same habitat by conventional species–area criteria. Weinberg (1978) also discusses a method for obtaining species–area data which is based on a series of ten equal-sized quadrats instead of the customary technique of using a succession of quadrats of increasing size; this is more suited to use by divers.

It is unwise to generalize, but Hiscock (1979) lists sizes of quadrats used on sublittoral rock surfaces which range from 0·5 m^2 for his samples at Loch Ine, Ireland to 100 cm^2 for algae in shallow shaded habitats, 250 cm^2 for circallittoral coralligenous areas of the Mediterranean (Boudouresque, 1971), and 400 cm^2 for algal vegetation off Malta (Larkum *et al.*, 1967). As emphasized by Weinberg (1978) these values can vary with habitat: Larkum *et al.* (1967) required 1 m^2 to differentiate significantly between plant assemblages at 30 and 45 m off Malta.

Choice of method and evaluation of data

Since shortage of time on site is a critical consideration for any diving survey it is important to consider whether a qualitative abundance estimate can be used instead of an accurate individual or biomass count. Hiscock (1979) discusses the validity of the qualitative or semi-quantitative approach and, quoting Moore (1974), points out that in some circumstances the qualitative assessment of a large species assemblage may be more useful in the long term than quantitative measurement. Such assessments can be done quickly and are particularly useful for rock surfaces where the biota are often crustose and hence difficult to sample or quantify. The assessments can be made according to both physical and biological criteria (Hiscock & Mitchell, 1980).

The Zurich–Montpelier phytosociological methods based upon scales of abundance/dominance and sociability were originally adapted for underwater use by French diving scientists (Fig. 5.10) working in the Mediterranean (Molinier & Picard, 1953; Pérès & Picard, 1964), and more recently on coral reefs (Pichon, 1978). However, as Drew (1971) points out, although such methods are extremely valuable for differentiation and classification of communities and for long-term surveillance, they give little indication of quantities of biota present.

Quantification itself can present a problem not only in terms of collection time but also in terms of what constitutes the quantity. Many organisms, for instance, are calcareous or have other mineral hard parts which greatly affect simple weight determinations. Drew (1971) recommends ash-free dry weights or calorific determination for such organisms. Because of the competition for space in epibenthic biota Denisov (1972) proposed the index $\sqrt{AB}$, where A is the number and B the biomass of a particular species, as a means of accounting for area in a particular biotope, i.e. where biomass is equivalent, the species of greater individual abundance is more dominant on a surface.

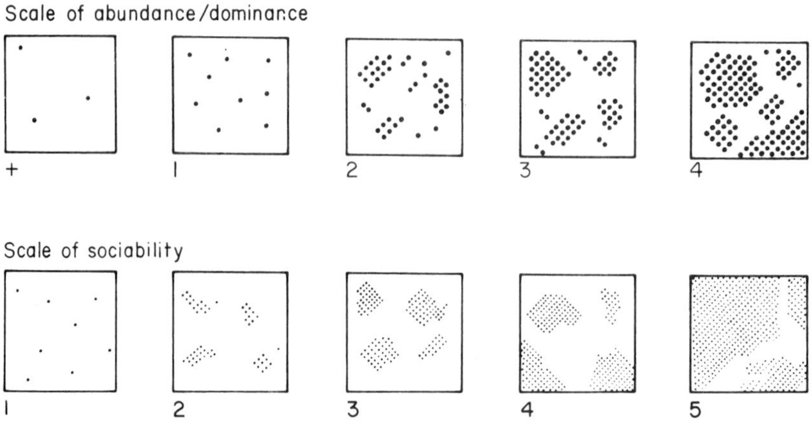

Abundance/dominance	Sociability
+ Sparse/very sparse, little cover	
1. Plentiful but little cover	1. Growing singly
2. Very numerous or covering > $\frac{1}{20}$ of area	2. Grouped or tufted
3. Any number of individuals covering $\frac{1}{4}$ to $\frac{1}{2}$ area	3. Small patches, cushions
4. Any number of individuals covering $\frac{1}{2}$ to $\frac{3}{4}$ area	4. Small colonies, extensive patches, carpets
5. Covering more than $\frac{3}{4}$ area	5. Great crowds or pure stands

Suffixes
< Reduced vitality
E Epiphytic
SE Upper storey
SS Lower storey

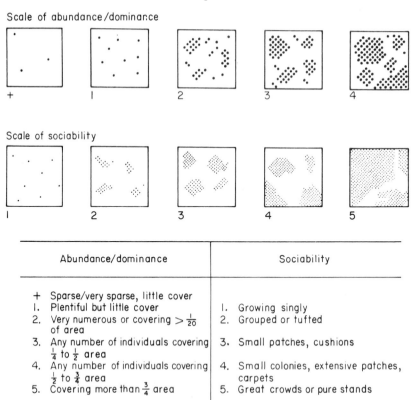

Fig. 5.10. Phytosociological scales devised according to Zurich–Montpelier methods by Pérès & Picard (1964). Suffixes are also applied to the scales. (From Drew, 1971.)

The statistical routines most frequently applied to information collected by diving are concerned not only with establishing the density of the organisms and community competition but also with similarity between and dispersion within habitats. Most statistical methods require random sampling procedures for maximum validity but, as emphasized by Loya (1978), this is often neither possible nor practicable within the limitations of the diving technique nor is it necessarily a sensible strategy on the highly zoned and patchy reef ecosystem. Similar arguments can be applied to all sublittoral epilithic habitats. Loya (1978) suggests that a stratified random sampling

Table 5.1. Rankings of diving ecological survey methods for coral-reef organisms (adapted from Weinberg, 1981). The lowest number indicates the best method.

Method	Survey criteria		
	No. species	Relative cover	Density
Individual counting and cover estimate (quadrat)	2	1	1
Line transect—species count and intercept length	6	5	4
Point intercept—surface (quadrat)	2	6	N.A.
Point intercept—line (transect)	5	7	N.A.
Point centre quarter method—plotless	6	3	2
Photographic record	1	2	N.A.
In situ mapping	4	4	3

N.A.: not applicable.

scheme be adopted, where the area of study is sub-divided, e.g. by depth or gross habitat type, and an appropriate number of samples taken according to subdivision, size and statistical procedure.

Gage & Coghill (1977) have presented an argument for using a regular rather than a random sample pattern for *in situ* investigation of the infauna of soft substrata. Like Angel & Angel (1967) they adopted the contiguous quadrat method developed by plant ecologists (Greig-Smith, 1965; Kershaw, 1973). Both used a series of contiguous cores along a transect line laid over the sediment surface and both were able to show aggregations of species on a scale which was not detectable by conventional ship-board larger scale samplers.

Most of the conventional indices of similarity and diversity have been used to describe a variety of sublittoral communities through diving. Boudouresque (1971) and Thomassin (1978) summarize the methods, while specific examples of their application are to be found in Denisov (1972), Weinberg (1978) and Gulliksen *et al.* (1980).

Weinberg (1981) has recently presented an important study which compares seven different methods of surveying a coral reef. The sampling methods were all scaled down to the same time on site underwater, and were all on the same $100\,m^2$ area of reef at 20 m depth which had previously been completely mapped visually and photographically. The rankings, in terms of effectiveness, for the various methods for three survey criteria are shown in Table 5.1. Surprisingly, the line transect length intercept method used by many reef workers was not very effective. Weinberg discards both point intercept methods, is sceptical about the line transect and plotless methods, and concludes that the quadrat-based individual counting and cover estimate (ICCE), photographic recording and *in situ* mapping methods are best. He opts for the ICCE method as it is simplest to use in the field and easiest to evaluate: photography, while providing a permanent record, can miss covered species and is laborious to evaluate (but see Bohnsack, 1979; Laxton & Stablum, 1974). *In situ* mapping gives a permanent record of limited accuracy. While it must be emphasized that this study is strictly associated with coral assemblages at one depth its objectives, design and conclusions are most relevant to anyone planning a detailed benthic diving survey.

Site identification and location

Precise relocation of a diving site in turbid water is best achieved by means of a marker buoy attached to a heavy sinker. Coarser scale location usually uses small boat navigation methods which rely on sitings or prominent shore-line features. Murray (1966) has provided a helpful example of the use of a sextant while other descriptions of methods are available in diving and navigation handbooks (e.g. British Sub-Aqua Club, 1978; Chapter 2 in this Handbook). Where precision is needed, microwave position finders such as the Decca Trisponding Position Fixing System, which give location to about 1 m, can be used.

Position finding underwater can be difficult, although diver-operable pinger locators are available (NOAA, 1979). In clear water it is possible to modify terrestrial survey methods using alidade and plane table (Milne, 1972, 1973), and even a theodolite (Farrington-Wharton, 1976), whereas in limited visibility distance is best measured with acoustic measuring equipment. Kelland (1976) describes the method of conducting an accurate planimetric survey with a diver-held acoustic rangemeter and transponder array.

Underwater experiments

Many aspects of the biology of benthic organisms favour *in situ* studies on ecological relationships, physiology and behaviour. Ott (1973) and Riedl (1980) have extensively reviewed such work; the aim here will be to draw attention to particular aspects of experimental method and objective. Many procedures, particularly those which do not require great dexterity, can be undertaken by divers and, with suitable waterproofing, many instruments such as light meters (Drew, 1971, 1973; Dustan, 1982) and respirometers (Svoboda, 1978) can be used on site underwater. The syringe, which is a basic tool in underwater experiments, is essential for both provision and collection of discrete fluid samples.

Physiological ecology

Some of the earliest experimental *in situ* work was carried out by Goreau (1959), who used radioactive ^{45}Ca to study calcification and growth of corals, and ^{14}C to measure primary production by zooxanthellae (Goreau & Goreau, 1959). Barnes & Taylor (1973) provide a more recent study, while McCloskey *et al.* (1978) present a methodological critique of reef coral photosynthesis and respiration studies.

Drew (1971, 1973) has developed the *in situ* ^{14}C uptake method for studies on macrophyte primary production in a range of environments (see also pp. 269–70; Chapter 8). He has been particularly concerned with depth and zonation and, as well as carrying out translocation experiments with the same species of alga, has also simulated light attenuation effects down to 130 m by rate of production measurements within a cave (Larkum *et al.*, 1967; Drew, 1969). Drew & Jupp (1976) modified the technique to study sea grass (*Posidonia oceanica*) photosynthesis.

Estimates of biomass and productivity of sublittoral macrophytic algae have been carried out in several instances (see Jupp & Drew, 1974) but, until the technique developed by Mann (1972b) became known, all used the 'peak cropping method' (Bellamy *et al.*, 1973), in which annual production was estimated from the biomass of laminae of algae of known age cropped at known intervals. This method does not account for the continual terminal erosion of those laminae during the growing season, and, taking account of this, Mann (1972b) estimated that *Laminaria longicruris* and *Agarum cribrosum* in Nova Scotian waters have an annual biomass turnover of 4–10 times. The method is based on *in situ* measurement of lamina growth by noting the rate of movement of a hole punched through the blade of the alga above the meristem (Fig. 5.11). Mann *et al.* (1979) have modified this method for production studies on the morphologically complex *Ecklonia maxima* where, in part of the mature plant, the growth region (meristem) is not localized. Mann & Mann (1981) have recently re-assessed this method, and techniques such as these have been used in the investigation of the production, trophic pathways and energy flow in kelp-ecosystems (Field *et al.*, 1977), and, more recently, in growth and production of sea grasses (Ott, 1980).

Respiration measurements are an important component of energy determination. Svoboda (1978) has described a method for *in situ* oxygen uptake determination of coral colonies using sealed chambers with implanted oxygen electrodes. Similar methods have been employed for determining oxygen demands of soft bottom biota, e.g. Davies (1975), Boynton *et al.* (1981), and Loeb (1981), where small areas of the surface of bottom sediments have been isolated in sealed chambers fitted with sampling ports, electrodes and circulation pumps. Houlihan *et al.* (1980) adopted a different approach to

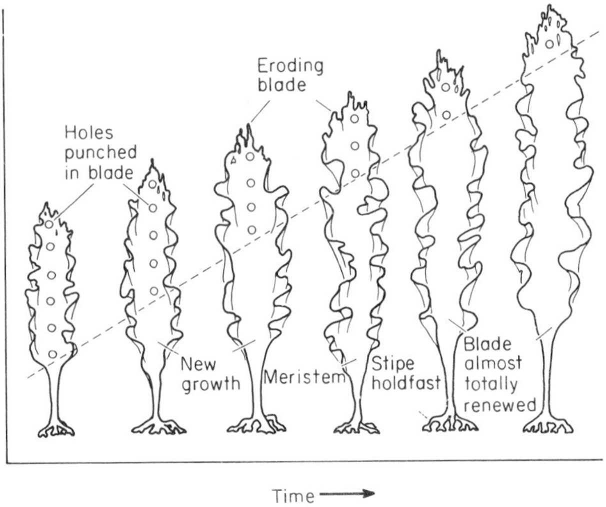

Fig. 5.11. Growth in *Laminaria longicruris* as shown by the distal movement of holes punched through the lamina above the meristem. (From Mann *et al.*, 1979.)

respiration in their use of syringe samples of pre- and post-branchial blood taken by divers from free-living crabs. pO_2, pCO_2, pH and blood protein levels were measured in a mobile shore laboratory within minutes of collection by divers. An objective of this research was to make comparisons with previous laboratory studies.

Many underwater experiments have involved removal, exclusion, introduction or translocation of particular organisms. The effect of sea urchin grazing on algal colonization and succession has been studied by exclusion of urchins from a particular strip of sea bed throughout a three year period (Jones & Kain, 1967) and by observation of macrophyte growth on boulders enclosed in echinoid-proof cages (Paine & Vadas, 1969). Boudouresque (1973) similarly followed recolonization after artificial clearance of a rock surface, while the use of introduced settlement panels is a well known technique (e.g. Pearse & Chess, 1971; Panzini & Pronzato, 1981). Such experiments are of use in testing anti-fouling materials and, by selected siting, can also be used in a natural ecological context (Riedl, 1966). Sarnthein & Richter (1974) and Richter & Sarnthein (1977) deployed trays of different sediments (Fig. 5.12) to test settlement into soft substrata. Larger scale artificial reefs (Stone & Buchanan, 1970; Lewis & Nichols, 1979) have also been constructed.

Zeitzschel & Davies (1978) reviewed the use of larger benthic isolation chambers in the study of processes at the benthic sediment interface but Arntz (1977) contributes a cautionary note about events which can go amiss with such experiments. Caging off a portion of benthic environment can lead to

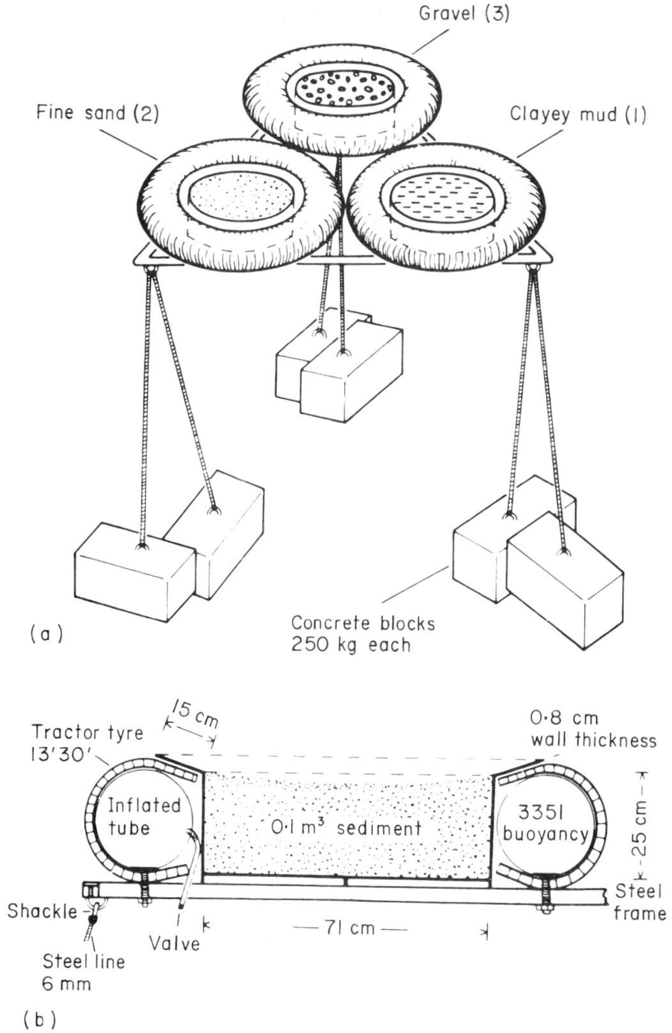

Fig. 5.12. Platform supporting three containers of different sediments used in faunal settlement experiments. Note the use of tyre inner tubes for buoyancy. (a) General view (not to scale); (b) cross-section sketch (to scale). (From Sarnthein & Richter, 1974.)

problems associated with the creation of a completely new microhabitat, including invasion of unsuspected predators (e.g. starfish), provision of refuge, and accumulation of detrital material. The most ambitious benthic isolation system yet developed has been the Kiel 'Plankton Tower' (von Bodungen *et al.*, 1976) which isolated both the sea bed and its superimposed water column.

The effects of depth and other physical variables can be tested by translocation experiments (Drew & Larkum, 1967). North (1964) has moved *Macrocystis pyrifera* plants, while the effects of local conditions on the orientation and growth of gorgonians and milleporids has been studied in relocation experiments by Velmirov (1973, 1974).

Behaviour

Many aspects of behaviour have been observed *in situ* by divers, although it should be remembered that diving equipment is noisy and the diver's presence can affect the behavioural patterns of many organisms. It is not intended to review methods in ethological studies since many of the techniques involved have to be specifically tailored for a particular investigation.

Fricke (1973b) has written a comprehensive account of behavioural studies on a coral reef which clearly outlines the importance and versatility of *in situ* observations. Feeding behaviour has featured in many underwater studies: a useful experimental investigation of fish feeding within the benthic ecosystem has been carried out by Kingett & Choat (1981). Further to the grazing effects of echinoids already mentioned, Krumbein & Van der Pers (1974) have studied sea urchins feeding on rock-burrowing polychaetes and have estimated the combined erosional effects of both organisms while Bernstein *et al.* (1981) have related grazing impact and behavioural response of sea urchins to predators. Rates of food transport have been measured directly with coloured food material in crinoids (Zmarzly & Holland, 1981), while Shepherd (1973) noted differences in the food choice and feeding behaviour of several abalone species through direct observation. Distributional aspects of predator–prey relationships between species have been evaluated by Schmid & Schaerer (1981) for two *Astropecten* species, and by Kastiendiek (1982) for the pennatulid *Renilla kollikeri*.

Much behavioural work involves estimation of movement: particular individuals can be recognized if large and conspicuous (e.g. Altman, 1967), but identification tags are often necessary. In some cases it is possible to use features of the organism itself, for example by slipping a coded sleeve over the spine of a sea urchin (Gamble, 1966) or starfish (Crump, 1976) or utilizing the pore holes of the abalone (Tutschulte, 1968; Shepherd, 1973). Other *in situ* tagging methods (NOAA, 1979) include injecting dyes, threading labels with nylon monofilament or wire, sticking with rapid setting epoxy resin and the attachment of miniature ultrasonic transmittors. Chapman *et al.* (1975) developed the latter method for *Nephrops norvegicus* since *in situ* attachment avoided the damage to this crustacean's vision which would occur if they were brought to the surface. See also the review by Jones (1979).

The movement pattern of slow moving animals can be plotted by triangulation from fixed points (Gamble, 1966) or by observation in relation to major features (Crump, 1976). However, continual records can be made with ultrasonic tags and an array of hydrophones (Chapman *et al.*, 1975). These systems are expensive, and a somewhat cheaper continuous nocturnal tracking method based on photographs recording a dim light attached to the shell of an abalone has been developed by Tutschulte (1968). It should be remembered that many benthic organisms are more active in the dark.

The efficiency of fishing gear has been examined extensively using diving techniques (e.g. Hemmings, 1973; High & Ellis, 1973). Originally, divers clung onto moving gear but recently Main & Sangster (1978b, 1979) have pioneered the use of a two-man towed wet submersible that permits both observation at greater fishing speeds and the use of underwater TV systems (see also Chapter 4). Caddy (1968) and Chapman *et al.* (1979) have made field observations on swimming behaviour of pectinid bivalves in response to dredging.

Underwater habitats and submersibles

It is technologically feasible to extend the limitations of the conventional air-breathing diver (see Trillo, 1976–1980). Commercial developments, concerned particularly with continental shelf margin oil extraction, are continually demanding deeper diving systems and, since 1960, underwater habitats have been set up for use for prolonged periods and at considerable depth.

Habitats

Habitats which range from inexpensive 1–2 man 24 hour work stations to very costly 4–6 man, deep, heliox-atmosphered systems (NOAA, 1979) enjoyed a rapid popularity in the period 1962–1970 but are, at present, going through an evaluation phase with only one or two, such as the US 'Hydrolab', currently functioning. High *et al.* (1973) present an evaluation of three major US projects and Collette (1972) in his summary of the 'Tektite' programme points out the great logistic advantage of saturation techniques for a repeated and intense diving study—e.g. 86 hours diving from a two week stay in the habitat were estimated to be equivalent to two months' continuous effort on conventional non-decompression surface schedules. Examples of biological work carried out from habitats are the fish trap behaviour studies of High & Ellis (1973), cleaner shrimp behaviour (Mahnken, 1972), and predation by *Asterias rubens* (Anger *et al.*, 1977).

Submersibles

Although some habitats with heliox atmospheres have been situated below

100 m, all are restricted to relatively shallow shelf seas. Diving to deeper depths is only possible in submersible vehicles where the crew are maintained at 1 ata in a pressurized compartment. Many such vessels are now available (Trillo, 1976–1980), some of which are being used for benthic research (Heirtzler & Grassle, 1976).

There are also many unmanned submersibles, controlled from surface vessels, which carry a variety of recording equipment and remotely operated tools. Geyer (1977), however, argues strongly for the need for manned submersibles in benthic research, pointing out the advantage of having 'real-time' observation and decision making capacity on site. Discussions of specific uses of submersibles are to be found in Geyer (1977), including study of fauna and sediments on the continental shelf (Wilson, 1977), coral reef and fishing bank reconnaissance (Bright & Rezak, 1977) and surveillance of ocean waste disposal (Palmer, 1977). Lissner (1979) has reviewed the use of submersibles in marine biological surveys.

Examples of specific biological studies from submersibles in shelf areas include those of Caddy (1970, 1976) on scallop beds, Wilson *et al.* (1977) on brittle star populations and Meinesz *et al.* (1981), who mapped sea grass beds. Grassle *et al.* (1975) have investigated the distribution patterns of large benthic organisms in the deep sea while, more recently, much multi-disciplinary work is being carried out in active geothermal regions (e.g. Corliss *et al.*, 1979: Spiess *et al.*, 1980; Baross *et al.*, 1982), where exciting discoveries are being made both about crustal tectonics and unique geothermally-based benthic food chains.

References

Aleem, A.A., 1956. A quantitative study of the benthic communities inhabiting the kelp beds off the California coast, with a self-contained diving apparatus. pp. 149–152 in T. Braarud & N.A. Sørensen (eds) *Second International Seaweed Symposium*. Pergamon Press, New York, 220 pp.

Altman, J.S., 1967. The behaviour of *Octopus vulgaris* Lam. in its natural habitat: a pilot study. *Underwater Association Report 1966–67*, 77–84.

Angel, H.H. & Angel, M.V., 1967. Distribution pattern analysis of a benthic community. *Helgoländer wissenshaftliche Meeresuntersuchungen*, **15**, 445–454.

Anger, K., Rogal, U., Schriever, G. & Valentin, C., 1977. *In situ* investigation on the echinoderm *Asterias rubens* as a predator of soft bottom communities in the western Baltic Sea. *Helgoländer wissenshaftliche Meeresuntersuchungen*, **29**, 439–459.

Arntz, W.E., 1977. Results and problems of an 'unsuccessful' benthos cage predator experiment (Western Baltic), pp. 31–44 in B.F. Keegan, P.Ó Céidigh & P.J.S. Boaden (eds) *Biology of Benthic Organisms. Eleventh European Symposium on Marine Biology, Galway, October 1976*. Pergamon Press, Oxford, 630 pp.

Atkinson, R.J.A., 1974. Spatial separation of *Nephrops* burrows. *Estuarine and Coastal Marine Science*, **2**, 171–176.

Baddeley, A.D., Defigueredo, J.W., Curtis, J.W.M. & Williams, A.N., 1968. Nitrogen narcosis and performance underwater. *Ergonomics*, **11**, 157–164.

Bailey, J.H., Nelson-Smith, A. & Knight-Jones, E.W., 1967. Some methods for transects across steep rocks and channels. *Underwater Association Report, 1966–67*, 107–111.

Barnes, D.J. & Taylor, D.L., 1973. *In situ* studies of calcification and photosynthetic carbon fixation in the coral *Montastrea annularis. Helgoländer wissenschaftliche Meeresuntersuchungen*, **24**, 284–291.

Barnett, P.R.O. & Hardy, B.L.S., 1967. A diver operated quantitative sampler for sand macrofaunas. *Helgoländer wissenschaftliche Meeresuntersuchungen*, **15**, 390–398.

Baross, J.A., Lilley, M.D. & Gordon, L.I., 1982. Is the CH_4, H_2 and CO venting from submarine hydrothermal systems produced by thermophilic bacteria? *Nature, London*, **298**, 366–368.

Bellamy, D.J., John, D.M. & Whittick, A., 1968. The 'kelp forest ecosystem' as a 'phytometer' in the study of pollution of the inshore environment. *Underwater Association Report 1968*, 79–82.

Bellamy, D.J. & Whittick, A., 1968. The kelp project. *Triton*, **13**, 16–17.

Bellamy, D.J., Whittick, A., John, D.M. & Jones, D.J., 1973. A method for the determination of seaweed production based on biomass estimates. *Monographs in Oceanographic Methodology, UNESCO, Paris*, **3**, 27–33.

Bernstein, B.B., Williams, B.E. & Mann, K.H., 1981. The role of behavioral responses to predators in modifying urchins (*Strongylocentrotus droebachiensis*) destructive grazing and seasonal foraging patterns. *Marine Biology*, **63**, 39–49.

Bleakney, J.S., 1969. A simplified vacuum apparatus for collecting small nudibranchis. *Veliger*, **12**, 142–143.

Bohnsack, J.S., 1979. Photographic quantitative sampling of hard-bottom benthic communities. *Bulletin of Marine Science*, **29**, 242–252.

Boudouresque, C-F., 1971. Méthodes d'étude qualitative et quantitative du benthos (en particulier du phytobenthos). *Téthys*, **3**, 79–104.

Boudouresque, C-F., 1973. Etude *in situ* de la réinstallation d'un peuplement sciaphile de mode battu après sa destruction experimentale, en Méditerranée. *Helgoländer wissenschaftliche Meeresuntersuchungen*, **24**, 202–218.

Boudouresque, C-F., 1974. Aire minima et peuplements algaux marins. *Bulletin Societé phycologique de France*, **19**, 141–157.

Boynton, W.R., Kemp, W.M., Osborne, C.G., Kaumeyer, K.R. & Jenkins, M.C., 1981. Influence of water circulation rate on *in situ* measurements of benthic community respiration. *Marine Biology*, **65**, 185–190.

Brett, C.E., 1964. A portable hydraulic diver operated dredge-sieve for sampling subtidal macrofauna. *Journal of Marine Research*, **22**, 205–209.

Bright, T.J. & Rezak, R., 1977. Reconnaissance of reefs and fishing banks of the Texas continental shelf. Chapter 6, pp. 113–150, in R.A. Geyer (ed.) *Submersibles and their use in Oceanography and Ocean Engineering*. Elsevier, Amsterdam, 383 pp.

British Sub-Aqua Club, 1978. *Diving Manual*, 10th ed. (revised). British Sub-Aqua Club, London, 573 pp.

Byers, G.J., 1977. A mini cassette recorder for use by divers underwater. *Progress in Underwater Science*, **2**, 131–134.

Caddy, J.F., 1968. Underwater observations on scallop (*Placopecten magellanicus*). Behaviour and drag efficiency. *Journal of the Fisheries Research Board of Canada*, **25**, 2123–2141.

Caddy, J.F., 1970. A method for surveying scallop populations from a submersible. *Journal of the Fisheries Research Board of Canada*, **27**, 535–549.

Caddy, J.F., 1976. Practical considerations for quantitative estimation of benthos from a submersible. pp. 285–298 in E.A. Drew, J.N. Lythgoe & J.D. Woods (eds) *Underwater Research*. Academic Press, London, 430 pp.

Chapman, C.J. & Howard, F.G., 1979. Field observations on the emergence rhythm of the Norway lobster, *Nephrops norvegicus*, using different methods. *Marine Biology*, **51**, 157–165.

Chapman, C.J., Johnstone, A.D.F., Dunn, J.R. & Creasey, D.J., 1974. Reactions of fish to sound generated by divers' open-circuit underwater breathing apparatus. *Marine Biology*, **27**, 357–366.

Chapman, C.J., Johnstone, A.D.F. & Rice, A.L., 1975. The behaviour and ecology of the Norway lobster, *Nephrops norvegicus* (L). pp. 59–74 in H. Barnes (ed.) *Ninth European Marine Biology Symposium*. Aberdeen University Press, Aberdeen, 760 pp.

Chapman, C.J., Main, J., Howell, T. & Sangster, G.I., 1979. The swimming speed and endurance of the queen scallop *Chlamys opercularis* in relation to trawling. *Progress in Underwater Science*, **4**, 57–72.

Chapman, C.J. & Rice, A.C., 1971. Some direct observations on the ecology and behaviour of the Norway lobster, *Nephrops norvegicus*. *Marine Biology*, **10**, 321–329.

Christie, N.D., 1976. The efficiency and effectiveness of a diver-operated suction sampler on a homogeneous macrofauna. *Estuarine and Coastal Marine Science*, **4**, 687–693.

Christie, N.D. & Allen, J.C., 1972. A self-contained diver-operated quantitative sampler for investigating the macrofauna of soft substrates. *Transactions of the Royal Society of South Africa*, **40**, 299–307.

Church, J. & Church, C., 1972. *Beginning Underwater Photography*, 2nd ed. J. and C. Church, Gilroy, California.

Clark, K.B., 1971. The construction of a collecting device for small aquatic organisms and a method for rapid weighing of small invertebrates. *Veliger*, **13**, 364–367.

Collette, B.B., 1972. Conclusions. pp. 171–174 in B.B. Collette & S.A. Earle (eds) *Results of the Tektite Program: Ecology of Coral Reef Fishes*. Natural History Museum, Los Angeles County. Science Bulletin, **14**, 179 pp.

Corliss, J.B., Dymond, J., Gordon, L.I., Edmund, J.M., Von Herzen, R.P., Ballard, R.D., Green, K., Williams, D., Bainbridge, A., Crane, K. & Van Andel, T.H., 1979. Submarine thermal springs on the Galápagos Rift. *Science, New York*, **203**, 1073–1083.

Cottam, G. & Curtis, J.T., 1956. The use of distance measures in phytosociological sampling. *Ecology*, **37**, 451–460.

Crump, R.G., 1976. Nocturnal behaviour in aggregations of *Acanthaster planci* in the Sudanese Red Sea. pp. 313–318 in E.A. Drew, J.N. Lythgoe & J.D. Woods (eds) *Underwater Research*. Academic Press, London, 430 pp.

Davies, J.M., 1975. Energy flow through the benthos in a Scottish sea loch. *Marine Biology*, **31**, 353–362.

Denisov, N.Y., 1972. Some aspects of the use of divers to investigate bottom communities. *Okeanologia, Moscow*, **12**, 884–891 (in Russian): *Oceanology*, **12**, 738–744 (English translation).

Drew, E.A., 1969. Photosynthesis and growth of attached marine algae down to 130 metres in the Mediterranean. pp. 151–159 in R. Margalef (ed.) *Proceedings of the Sixth International Seaweed Symposium, Santiago de Compostela, September 9–13, 1968*. Dirección General de Pésca Maritima, Madrid, 782 pp.

Drew, E.A., 1971. Botany. Chapter 6, pp. 175–233 in J.D. Woods & J.N. Lythgoe (eds) *Underwater Science. An introduction to experiments by divers*. Oxford University Press, London, 330 pp.

Drew, E.A., 1973. Primary production of large marine algae measured *in situ* using uptake of ^{14}C. *Monographs in oceanographic methodology, UNESCO, Paris*, **3**, 22–26.

Drew, E.A. & Jupp, B.J., 1976. Some aspects of the growth of *Posidonia oceanica* in Malta. pp. 357–367 in E.A. Drew, J.N. Lythgoe & J.D. Woods (eds) *Underwater Research*. Academic Press, London, 430 pp.

Drew, E.A. & Larkum, A.W.D., 1967. Photosynthesis and growth of *Udotea*, a green alga from deep water. *Underwater Association Report 1966–67*, 65–71.

Dustan, P., 1982. Depth-dependent photoadaptation by zooxanthellae of the reef coral *Montastrea annularis*. *Marine Biology*, **68**, 253–264.

Earll, R., 1977. A methodology for primary surveys of the shallow sublittoral zone. *Progress in Underwater Science*, **2**, 47–63.

Earll, R., 1979. Underwater bird watchers. *New Scientist*, **83**, 362–364.

Earll, R. & Erwin, D., 1979. The species recording scheme—results of the 1977 season. *Progress in Underwater Science*, **4**, 105–120.

Emig, C.C. & Lienhart, R., 1967. Un nouveau moyen de récolte pour les substrates meubles infralittoraux: l'aspirateur sous-marin. *Recuil des Travaux de la Station Marine d'Endoume*, **58**, 115–120.

Erwin, D.G., 1977a. A cheap SCUBA technique for epifaunal surveying using a small boat. *Progress in Underwater Science*, **2**, 125–129.

Erwin, D.G., 1977b. A diving survey of Strangford Lough: the benthic communities and their relation to substrate—a preliminary account. pp. 215–224 in B.F. Keegan, P.Ó Céidigh & P.J.S. Boaden (eds) *Biology of Benthic Organisms, Eleventh European Symposium on Marine Biology, Galway, October 1976*. Pergamon Press, Oxford, 630 pp.

Fager, E.W., Flechsig, A.O., Ford, R.F., Clutter, R.I. & Ghelardi, R.J., 1966. Equipment for use in ecological studies using SCUBA. *Limnology and Oceanography*, **11**, 503–509.

Farrington-Wharton, R., 1976. Towards the development of a practical underwater theodolite. pp. 267–276 in E.A. Drew, J.N. Lythgoe & J.D. Woods (eds) *Underwater Research*. Academic Press, London, 430 pp.

Farrow, G.E., 1975. Techniques for the study of fossil and recent traces. pp. 537–554 in R.W. Frey (ed.) *The Study of Trace Fossils*. Springer-Verlag, New York.

Field, J.G., Jarman, N.G., Dieckmann, G.S., Griffiths, C.C., Velimirov, B. & Zoutendyk, P., 1977. Sun, Waves, Seaweed and Lobsters: The dynamics of a West Coast Kelp-bed. *South African Journal of Science*, **73**, 7–10.

Finnish IBP-PM Group, 1969. Quantitative sampling equipment for the littoral benthos. *Internationale Revue der Gesamten Hydrobiologie*, **54**, 185–193.

Flemming, N.C. & Stride, A.H., 1967. Basal sand and gravel patches with separate indications of tidal current and storm-wave paths, near Plymouth. *Journal of the Marine Biological Association of the United Kingdom*, **47**, 433–444.

Forster, G.R., 1959. The ecology of *Echinus esculentus* L. Quantitative distribution and rate of feeding. *Journal of the Marine Biological Association of the United Kingdom*, **38**, 361–367.

Fricke, H.W., 1971. Fische als Feinde tropischer Seeigal. *Marine Biology*, **9**, 328–338.

Fricke, H.W., 1973a. Der Einfluss des Lichtes auf Körperfärbung und Dämmerungsverhatten des Korallen-fisches *Chaetodon melanotus*. *Marine Biology*, **22**, 251–262.

Fricke, H.W., 1973b. Behaviour as part of ecological adaptation—*In situ* studies in the coral reef. *Helgoländer wissenschaftliche Meeresuntersuchungen*, **24**, 120–144.

Gage, J.D., 1975. A comparison of the deep sea epibenthic sledge and anchor-box dredge samplers with the van Veen grab and hand coring by divers. *Deep-Sea Research*, **22**, 693–702.

Gage, J.D. & Coghill, G.G., 1977. Studies on the dispersion patterns of Scottish sea loch benthos from contiguous core transects. pp. 319–338 in B.C. Coull (ed.) *Ecology of Marine Benthos. The Belle W. Baruch Library in Marine Science*, No. **6**. University of South Carolina Press, Columbia.

Gamble, J.C., 1966. Some observations on the behaviour of two regular echinoids. *Proceedings of the Symposium of the Underwater Association, Malta 1965*, 47–50.

George, J.D., 1980. Photography as a marine biological research tool. Chapter 3, pp. 45–115 in J.H. Price, D.E.G. Irvine & W.F. Farnham (eds) *The Shore Environment, Vol. 1: Methods*, Systematics Association Special Vol. 17(a). Academic Press, London, 321 + 41 pp.

Geyer, R.A. (ed.), 1977. *Submersibles and their use in Oceanography and Ocean Engineering*. Elsevier, Amsterdam, 383 pp.

Glover, T., Harwood, G.E. & Lythgoe, J.W., 1977. *A Manual of Underwater Photography*. Academic Press, London, 219 pp.

Godden, D., 1975. Cold, wet and hostile. *New Behaviour*, **1**, 422–425.

Godden, D. & Baddeley, A.D., 1975. Context-dependent memory in two natural environments on land and underwater. *British Journal of Psychology*, **66**, 325–331.

Goodall, D.W., 1952. Quantitative aspects of plant distribution. *Biological Reviews of the Cambridge Philosophical Society*, **27**, 194–245.

Goreau, T.F., 1959. The physiology of skeleton formation in corals. I. A method for measuring the rate of calcium deposition by corals under different conditions. *Biological Bulletin, Marine Biological Laboratory Woods Hole*, **116**, 59–75.

Goreau, T.F. & Goreau, N.I., 1959. The physiology of skeleton formation in corals. II. Calcium deposition by hermatypic corals under various conditions in the reef. *Biological Bulletin, Marine Biological Laboratory Woods Hole*, **117**, 239–250.

Grassle, J.F., Sanders, H.L., Hessler, R.R., Rowe, G.T. & McLellan, T., 1975. Pattern and zonation: a study of the bathyal megafauna using the research submersible *Alvin*. *Deep-Sea Research*, **22**, 457–481.

Greig-Smith, P., 1965. *Quantitative Plant Ecology*, 2nd ed. Butterworth, London, 242 pp.

Gulliksen, B., 1980. The macrobenthic rocky bottom fauna of Borgenfjorden, North-Tröndelag, Norway. *Sarsia*, **65**, 115–138.

Gulliksen, B. & Derås, K.M., 1975. A diver-operated suction sampler for fauna on rocky bottoms. *Oikos*, **26**, 246–249.

Gulliksen, B., Haug, T. & Sandnes, O.K., 1980. Benthic macrofauna on new and old lava grounds at Jan Mayen. *Sarsia*, **65**, 137–148.

Guonot, M., 1969. *Méthodes d'étude quantitative de la Végetation.* Masson, Paris, 314 pp.

Hass, H., 1949. Beitrag zur Kenntnis der Reteporiden. *Zoologica Stuttgart*, **37**, 1–138.

Health and Safety, 1981. *Diving Operation at Work Regulations 1981.* UK Govt. Statutory Instrument 1981/399.

Heirtzler, J.R. and Grassle, J.F., 1976. Deep-sea research by manned submersibles. *Science, New York*, **194**, 294–299.

Hemmings, C.C., 1973. Direct observation on the behaviour of fish in relation to fishing gear. *Helgoländer wissenschaftliche Meeresuntersuchungen*, **24**, 348–360.

High, W.L. and Ellis, I.E., 1973. Underwater observations of fish behavior in traps. *Helgoländer wissenschaftliche Meeresuntersuchungen*, **24**, 341–347.

High, W.L., Ellis, I.E., Schroeder, W.W. & Loverich, G., 1973. Evaluation of the undersea habitats—Tektite II, Hydro-Lab and Edalhab—for scientific saturation diving programs. *Helgoländer wissenschaftliche Meeresuntersuchungen*, **24**, 16–44.

Hiscock, K., 1979. Systematic surveys and monitoring in nearshore sublittoral areas using diving. pp. 55–74 in D. Nichols (ed.) *Monitoring the Marine Environment*, Symposia of the Institute of Biology, **24**. Institute of Biology, London, 205 pp.

Hiscock, K. (in prep.). Surveys of sublittoral benthos using diving. in J.M. Baker, R. Mitchell & W.J. Wolff (eds) *Biological Surveys of Estuarine and Coastal Waters.* Cambridge University Press, London.

Hiscock, K. & Hoare, R., 1973. A portable suction sampler for rock epibiota. *Helgoländer wissenschaftliche Meeresuntersuchungen*, **25**, 35–38.

Hiscock, K. & Hoare, R., 1975. The ecology of sublittoral communities at Abereiddy Quarry, Pembrokeshire. *Journal of the Marine Biological Association of the United Kingdom*, **55**, 833–864.

Hiscock, K. & Mitchell, R., 1980. The description and classification of sublittoral epibenthic ecosystems. Chapter 1, pp. 323–370 in J.M. Price, D.E.G. Irvine & W.F. Farnham (eds) *The Shore Environment, Vol. 2: Ecosystems*, Systematics Association Special Volume, 17(b). Academic Press, London, 945 + 100 pp.

Houlihan, D.F., Duthie, G., Smith, P. & Talbot, C., 1980. *In situ* sampling of crab blood by SCUBA divers. *Journal of Experimental Marine Biology and Ecology*, **45**, 219–228.

Howard, A.E., 1980. Substrate and tidal limitations on the distribution and behaviour of the lobster and edible crab. *Progress in Underwater Science*, **5**, 165–169.

Johnston, C.S., Morrison, I.A. & MacLachlan, K., 1969. A photographic method for recording the underwater distribution of marine benthic organisms. *Journal of Ecology*, **57**, 453–459.

Jones, N.S. & Kain, J.M., 1967. Subtidal colonisation following the removal of *Echinus*. *Helgoländer wissenschaftliche Meeresuntersuchungen*, **15**, 460–466.

Jones, R. (ed.), 1979. Materials and methods used in marking experiments in fishery research. *FAO Fisheries Technical Paper*, **190**, 134 pp

Jupp, B.P. & Drew, E.A., 1974. Studies on the growth of *Laminaria hyperborea* (Gunn.) Fosl. I. Biomass and productivity. *Journal of Experimental Marine Biology and Ecology*, **15**, 185–196.

Kain, J.M., 1960. Direct observations on some Manx sublittoral algae. *Journal of the Marine Biological Association of the United Kingdom*, **39**, 609–630.

Kastendiek, J., 1982. Factors determining the distribution of the sea pansy, *Renilla kollikeri*, in a subtidal sand-bottom habitat. *Oecologia*, **52**, 340–347.

Keegan, B.F. & Könnecker, G., 1973. *In situ* quantitative sampling of benthic organisms. *Helgoländer wissenschaftliche Meeresuntersuchungen*, **24**, 256–263.

Kelland, N.C., 1976. A method for carrying out accurate planimetric surveys underwater. *The Hydrographic Journal*, **2(4)**, 17–32.

Kershaw, K.A., 1973. *Quantitative and Dynamic Plant Ecology*, 2nd ed. Edward Arnold, London, 318 pp.

Kingett, P.D. & Choat, J.H., 1981. Analysis of density and distribution patterns in *Chrysophrys auratus* (Pisces: Sparidae) within a reef environment: an experimental approach. *Marine Ecology—Progress Series*, **5**, 283–290.

Kirchner, W.B., 1974. A SCUBA operated corer for determining vertical distribution in the benthos. *Progressive Fish Culturist*, **36**, 27–28.

Kitching, J.A., Macan, T.T. & Gilson, H.C., 1934. Studies in sublittoral ecology. I. A submarine gully in Wembury Bay, South Devon. *Journal of the Marine Biological Association of the United Kingdom*, **19**, 677–705.

Krumbein, W.E. & van der Pers, J.N.C., 1974. Diving observations on bio-deterioration by sea-urchins in the rocky sublittoral of Helgoland. *Helgoländer wissenschaftliche Meeresuntersuchungen*, **26**, 1–17.

Lambert, G., Alexander, A.J. & Alletson, J., 1972. Method used in a preliminary study of the ecology of a marine reef off Durban Beach. *South African Journal of Science*, **68**, 221–224.

Lang, J., 1973. Interspecific aggression by scleractinian corals. 2. Why the race is not only to the swift. *Bulletin of Marine Science*, **23**, 260–279.

Larkum, A.W.D., Drew, E.A. & Crossett, R.N., 1967. The vertical distribution of attached marine algae in Malta. *Journal of Ecology*, **55**, 361–371.

Larsson, B.A.S., 1968. SCUBA—studies on vertical distribution of Swedish rocky bottom echinoderms. A methodological study. *Ophelia*, **5**, 137–156.

La Touche, R.W., 1978. The feeding behaviour of the featherstar, *Antedon bifida* (Echinodermata: Crinoidea). *Journal of the Marine Biological Association of the United Kingdom*, **58**, 877–890.

Laxton, J.M. & Stablum, W.J., 1974. Sample design for quantitative estimation of sedentary organisms of coral reefs. *Biological Journal of the Linnean Society*, **6**, 1–18.

Levin, V.S. & Shenderov, E.L., 1975. Some problems of macrobenthos census methods with the use of diving equipment. *Biologiya Morya*, **1**, 64–70 (in Russian) *Soviet Journal of Marine Biology*, **1**, 135–143 (English translation).

Lewis, G.A. & Nichols, D., 1979. Colonisation of an artificial reef by the sea-urchin *Echinus esculentus*. *Progress in Underwater Science*, **4**, 189–95.

Lissner, A.L., 1979. Use of submersibles in marine biological surveys. pp. 788–789 in *Oceans '79*. Institute of Electrical and Electronics Engineers Inc., New York.

Loeb, S.C., 1981. An *in situ* method for measuring the primary productivity and standing crop of the epilithic periphyton community in benthic systems. *Limnology and Oceanography*, **26**, 394–399.

Loya, Y., 1972. Community structure and species diversity of hermatypic corals at Eilat, Red Sea. *Marine Biology*, **13**, 100–123.

Loya, Y., 1978. Plotless and transect methods. *Monographs on oceanographic methodology, UNESCO, Paris*, **5**, 197–217.

Loya, Y. & Slobodkin, L.R., 1971. The coral reefs of Eilat (Gulf of Eilat, Red Sea). *Symposia of the Zoological Society of London*, **28**, 117–139.

Lundälv, T., 1971. Quantitative studies on rocky bottom biocoenoses by underwater photogrammetry. A methodological study. *Thalassa Jugoslavia*, **7**, 201–208.

Mahnken, C., 1972. Observations on cleaner shrimps of the genus *Periclimenes*. pp. 71–83 in B.B. Collette & S.A. Earle (eds) *Results of the Tektite Program: Ecology of Coral Reef Fishes*. Natural History Museum Los Angeles County, Science Bulletin, **14**, 179 pp.

Main, J. & Sangster, G.I., 1978a. The value of direct observation techniques by divers in fishing gear research. *Scottish Fisheries Research Report*, **12**, 15 pp.

Main, J. & Sangster, G.I., 1978b. A new method for observing fishing gear using a towed wet submersible. *Progress in Underwater Science*, **3**, 259–267.

Main, J. & Sangster, G.I., 1979. A study of bottom trawling gear on both sand and hard ground. *Scottish Fisheries Research Report*, **14**, 15 pp.

Mann, K.H., 1972a. Ecological energetics of the seaweed zone in a marine bay on the Atlantic Coast of Canada. I. Zonation and biomass of seaweeds. *Marine Biology*, **12**, 1–10.

Mann, K. H., 1972b. Ecological energetics of the sea-weed zone in a marine bay on the Atlantic Coast of Canada. II. Productivity of the seaweeds. *Marine Biology*, **14**, 199–209.

Mann, K.H., Jarman, N. & Dieckmann, G., 1979. Development of a method for measuring the productivity of the kelp *Ecklonia maxima* (Osbeck). *Transactions of the Royal Society of South Africa*, **44**, 27–41.

Mann, K.H. & Mann, C., 1981. Problems of converting linear growth increments of kelps to estimates of biomass production, pp. 699–704 in T. Levring (ed.) *Tenth International Seaweed Symposium*. Walter de Gruyter, New York, 780 pp.

Mariscal, R.N., 1970. The nature of the symbiosis between Indo-Pacific anemone fishes and sea anemones. *Marine Biology*, **6**, 58–65.

Massé, H., 1967. Emploi d'une suceuse hydraulique transformée pour les prélèvements dans les substrats meubles infralittoraux. *Helgoländer wissenschaftliche Meeres-untersuchungen*, **15**, 500–505.

Massé, H., 1970. La suceuse hydraulique, bilan de quatre années d'emploi sa manipulation, ses avantages et inconvénients. Peuplements benthiques. *Téthys*, **2**, 547–556.

Massé, H., Plante, R. & Reys, J.P., 1977. Etude comparative de l'efficacité de deux bennes et d'une suceuse en fonction de la nature du fond. pp. 433–443 in B.F. Keegan, P. Ó Céidigh & P.J.S. Boaden (eds) *Biology of Benthic Organisms. Eleventh European Symposium on Marine Biology, Galway, October, 1976*. Pergamon Press, Oxford, 630 pp.

McCloskey, L.R., Wethey, D.S. & Porter, J.W., 1978. Measurement and interpretation of photosynthesis and respiration in reef corals. *Monographs on oceanographic methodology, UNESCO, Paris*, **5**, 379–396.

McIntyre, A.D., 1971. Deficiency of gravity corers for sampling meiobenthos and sediment. *Nature, London*, **231**, 260.

McKenzie, J.D. & Moore, P.G., 1981. The microdistribution of animals associated with the bulbous holdfasts of *Saccorhiza polyschides* (Phaeophyta). *Ophelia*, **20**, 201–213.

Meinesz, A., Cuvelier, M. & Laurent, R., 1981. Méthodes récentes de cartographie et surveillance des herbiers de Phanérogames marines: Leur application sur les côtes françaises de la Méditerranée. *Vie et Milieu*, **31**, 27–34.

Miles, S. & Mackay, D.E., 1976. *Underwater Medicine*, 4th ed. Adlard Coles, London, 370 pp.

Milne, P.H., 1972. *In situ* underwater surveying by plane table and alidade. *Underwater Science and Technology Journal*, **4**, 59–63.

Milne, P.H., 1973. *In situ* seabed surveying by divers. pp. 221–228 in N.C. Flemming (ed.) *Science Diving International.* Proceedings of the Third Symposium of the Scientific Committee of the Confederation Mondiale des Activités Subaquatiques. British Sub-Aqua Club, London.

Ministry of Defence (Navy), 1972. *Diving Manual BR 2806*. H.M.S.O., London.

Molinier, R. & Picard, J., 1953. Recherches analytiques sur les peuplements littoraux Mediterraneens se developpant sur substrat solide. *Recueil des Travaux de la Station Marine d'Endoume*, **9(4)**, 1–18.

Moore, E.J., 1978. Underwater photogrammetry. *Progress in Underwater Science*, **3**, 101–110.

Moore, P.G., 1974. The kelp fauna of northeast Britain. III. Qualitative and quantitative ordination, and the utility of a multivariate approach. *Journal of Experimental Marine Biology and Ecology*, **16**, 257–300.

Muckelroy, K., 1978. *Maritime Archaeology*. Cambridge University Press, Cambridge, 270 pp.

Murray, J.W., 1966. A study of the seasonal changes of water mass of Christchurch Harbour, England. *Journal of the Marine Biological Association of the United Kingdom*, **46**, 561–578.

Myers, A.C., 1979. Summer and winter burrows of a mantis shrimp, *Squilla empusa*, in Narragansett Bay, Rhode Island (USA). *Estuarine and Coastal Marine Science*, **8**, 87–98.

Nash, R.D.M., 1980. *The behavioural ecology of small demersal fish associated with soft sediments*. Ph.D. Thesis, University of Glasgow, 173 pp.

Neushul, M., 1965. SCUBA diving studies of the vertical distribution of benthic marine plants. *Botanica Gothoburgensia*, **3**, 161–176.

Nichols, D., 1978. A nationwide survey of the British sea-urchin, *Echinus esculentus*. *Progress in Underwater Science*, **4**, 161–187.

Nichols, D., 1981. The role of volunteer divers in a biological project. *Journal of the Society for Underwater Technology*, **6**, 17–20.

NOAA, 1979. *NOAA Diving Manual*. National Oceanic and Atmospheric Administration Manned Undersea Science and Technology Office. U.S. Government Printing Office, Washington, D.C.

North, W.J., 1964. Experimental transplantation of the giant kelp, *Macrocystis pyrifera*. pp. 248–255 in A.D. de Virville & J. Feldmann (eds) *Proceedings of the Fourth International Seaweed Symposium, Biarritz—September 1961*. Pergamon Press, Oxford, 467 pp.

Okamoto, M., Sato, O., Kuroki, T. & Murai, T., 1981. The effect of divers on the behavior of fishes. *Bulletin of the Japanese Society of Scientific Fisheries*, **47**, 1567–1573 (English abstract).

Ott, J.A., 1973. Concepts of underwater experimentation. *Helgoländer wissenschaftliche Meeresuntersuchungen*, **24**, 54–77.

Ott, J.A., 1980. Growth and production in *Posidonia oceanica* (L.) Defile. *Marine Ecology*, **1**, 47–64.

Paine, R.T. & Vadas, R.L., 1969. The effects of grazing by sea urchins, *Strongylocentrotus* spp., on benthic algal populations. *Limnology and Oceanography*, **14**, 710–719.

Palmer, H.D., 1977. The use of manned submersibles in the study of ocean waste disposal. Chapter 13, pp. 317–334 in R.A. Geyer (ed.) *Submersibles and their use in Oceanography and Ocean Engineering*. Elsevier, Amsterdam, 383 pp.

Panzini, M. & Pronzato, R., 1981. Étude des spongiares de substrats artificiels immergés durant quatre ans. *Vie et Milieu*, **31**, 77–82.

Pearse, J.B. & Chess, J.R., 1971. Comparative investigations of the development of epibenthic communities from Gloucester, Massachusetts to St Thomas, Virgin Islands. pp. 55–62 in D.J. Crisp (ed.) *Fourth European Marine Biology Symposium*. Cambridge University Press, London, 599 pp.

Pearson, R.G., 1981. Recovery and recolonization of coral reefs. *Marine Ecology—Progress Series*, **4**, 105–122.

Pemberton, G.S., Risk, M.J. & Buckley, D.E., 1976. Supershrimp: deep bioturbation in the Strait of Canso, Nova Scotia. *Science, New York*, **192**, 790–791.

Pérès, J.M. & Picard, J., 1964. Nouveau manuel de bionomie benthique de la Mer Mediterranée. *Recueil de Travaux de le Station Marine d'Endoume*, **47**, 5–137.

Pichon, M., 1978. Quantitative benthic ecology of Tuléar reefs. *Monographs on Oceanographic Methodology, UNESCO, Paris*, **5**, 163–174.

Poulet, G., 1962. *Connaissance et Technique de la Plongée*. Editions Denöel, Paris.

Reiswig, H.M., 1971. *In situ* pumping activities of tropical Demospongiae. *Marine Biology*, **9**, 38–50.

Rice, A.L. & Chapman, C.J., 1971. Observations on the burrows and burrowing behaviour of two mud-dwelling decapod crustaceans, *Nephrops norvegicus* and *Goneplax rhomboides*. *Marine Biology*, **10**, 330–342.

Rice, A.L. & Johnstone, A.D.F., 1972. The burrowing behaviour of the gobiid fish *Lesueurigobius friesii* (Collet). *Zeitschrift fur Tierpsychologie*, **30**, 431–438.

Richter, W. & Sarnthein, M., 1977. Molluscan colonization of different sediments on submerged platforms in the western Baltic Sea. pp. 531–539 in B.F. Keegan, P. Ó Céidigh & P.J.S. Boaden (eds) *Biology of Benthic Organisms. Eleventh European Symposium on Marine Biology, Galway, October 1976*. Pergamon Press, Oxford, 630 pp.

Riedl, R., 1966. *Biologie der Meereshöhlen*. Paul Parey, Hamburg, 636 pp.

Riedl, R., 1976. The role of sea cave investigation in marine sciences. *Pubblicazione delle Stazione Zoologica di Napoli*, **40**, 492–501

Riedl, R., 1980. Marine ecology—a century of changes. *Marine Ecology*, **1**, 3–46.

Riley, J.D. & Holford, B.H., 1965. A sublittoral survey of Port Erin Bay, particularly as an environment for young plaice. *Report of the Marine Biological Station of Port Erin (1964)*, **77**, 49–53.

Rørslett, B., Green, N.W. & Kvalvågnaes, K., 1978. Stereophotography as a tool in aquatic biology. *Aquatic Botany*, **4**, 73–81.

Russell, B.C., Talbot, F.H., Anderson, G.R.W. & Goldman, B., 1978. Collection and sampling of reef fishes. *Monographs on Oceanographic Methodology, UNESCO, Paris*, **5**, 329–345.

Sarnthein, M. & Richter, W., 1974. Submarine experiments on benthic colonization of sediments in the Western Baltic Sea. I. Technical layout. *Marine Biology*, **28**, 159–164.

Schmid, P.H. & Schaerer, R., 1981. Predator–prey interaction between two competing sea star species of the genus *Astropecten*. *Marine Ecology*, **2**, 207–214.

Schmid, W.D., 1965. Distribution of aquatic vegetation as measured by line intercept with SCUBA. *Ecology*, **46**, 816–823.

Schulke, F., 1978. *Underwater Photography for Everyone*. Prentice-Hall, Engelwood Cliffs, New Jersey, U.S.A., 220 pp.

Shepherd, S.A., 1973. Studies on southern Australian abalone (genus *Haliotis*). I. Ecology of five sympatric species. *Australian Journal of Marine and Freshwater Research*, **24**, 217–257.

Shilling, C.W., Werts, M.F. & Schandelmeier, N.R. (eds), 1976. *The Underwater Handbook. A Guide to Physiology and Performance for the Engineer*. Plenum Press, New York, 912 pp.

Shinn, E.A., 1968. Burrowing in recent lime sediments of Florida and the Bahamas. *Journal of Palaeontology*, **42**, 879–894.

Sigl, W., Von Rad, U., Oeltzschner, H., Braune, K. & Fabricius, F., 1969. Diving sled: a tool to increase the efficiency of underwater mapping by SCUBA divers. *Marine Geology*, **7**, 357–363.

Spiess, F.N., MacDonald, K.C., Atwater, T., Ballard, R., Carranza, A., Cordoba, D., Cox, D., Diaz Garcia, V.M., Francheteau, J., Guerrero, J., Hawkins, J., Haymon, R., Hessler, R., Juteau, T., Kastner, M., Larson, R., Luyendyk, B., MacDougall, J.D., Miller, S., Normark, W., Orcutt, J & Rangin, C., 1980. East Pacific Rise: hot springs and geophysical experiments. *Science, New York*, **207**, 1421–1433.

Stoddart, D.R. & Johannes, R.E. (eds), 1978. Coral reefs: research methods. *Monographs on Oceanographic Methodology, UNESCO, Paris*, **5**, 581 pp.

Stone, R.B. & Buchanan, C.C., 1970. Old tires make new fishing reefs. *Underwater Naturalist*, **6(4)**, 23–28.

Strauss, R.H. (ed.), 1976. *Diving Medicine*. Grune and Stratton, New York, 420 pp.

Svoboda, A., 1978. *In situ* monitoring of oxygen production and respiration in Cnidaria with and without zooxanthellae. pp. 75–82 in D.S. McLusky & A.J. Berry (eds) *Physiology and Behaviour of Marine Organisms. Proceedings of the Twelfth European Symposium on Marine Biology, Stirling, Scotland, September 1977*. Pergamon Press, Oxford, 388 pp.

Tanner, C., Hawkes, M.W., Lebednik, P.A. & Duffield, E., 1977. A hand-operated suction sampler for the collection of subtidal organisms. *Journal of the Fisheries Research Board of Canada*, **34**, 1031–1034.

Taylor, H., 1977. *Underwater with the Nikonos and Nikon Systems*. American Photographic Book Publishing Co. Inc., New York, 160 pp.

Terrell, M., 1965. *The Principles of Diving*. Stanley Paul, London, 240 pp.

Thomassin, B., 1978. Soft-bottom communities. *Monographs on Oceanographic Methodology, UNESCO, Paris*, **5**, 263–298.

Trillo, R.L., 1976–1980. *Jane's Ocean Technology, 1976, 1977, 1978 and 1979.* Macdonald and Jane's Publishers, London.

True, M.A., 1970. Étude quantitative de quatre peuplements sciaphiles sur substrate rocheux dans la région marseillaise. *Bulletin de l'Institut Océanographique*, **69, No. 1401**, 48 pp.

Tutschulte, T.C., 1968. Monitoring the nocturnal movement of abalones. *Underwater Naturalist. Bulletin of the American Littoral Society*, **5(3)**, 12–15.

Underwater Association for Scientific Research Ltd, 1979. *Code of Practice for Scientific Diving*, 3rd ed. Natural Environment Research Council, Swindon, 82 pp.

U.S. Naval Ship Systems Command Supervisor of Diving, 1973. *U.S. Navy Diving Manual.* NAVSEA 0994-LP-001-9010. U.S. Govt. Printing Office, Washington, D.C.

Velmirov, B., 1973. Orientation in the sea fan *Eunicella cavolinii* related to water movement. *Helgoländer wissenschaftliche Meeresuntersuchungen*, **24**, 163–173.

Velmirov, B., 1974. Orientiertes Wachstum bei *Millepora dichotoma* (Hydrozoa). *Helgoländer wissenschaftliche Meeresuntersuchungen*, **26**, 18–26.

Von Bodungen, B., Von Brockel, K., Smetacek, V. & Zeitzschel, B., 1976. The plankton tower. I. A structure to study water/sediment interactions in enclosed water columns. *Marine Biology*, **34**, 369–372.

Walker, B., 1967. A diver-operated pneumatic core sampler. *Limnology and Oceanography*, **12**, 144–146.

Warner, G. & Woodley, J.D., 1975. Suspension-feeding in the brittle-star *Ophiothrix fragilis. Journal of the Marine Biological Association of the United Kingdom*, **55**, 199–210.

Weinberg, S., 1978. The minimal area problem in invertebrate communities of Mediterranean rocky substrata. *Marine Biology*, **49**, 33–40.

Weinberg, S., 1981. A comparison of coral reef survey methods. *Bijdragen tot de Dierkunde*, **51**, 199–218.

Wilson, J.B., 1977. The role of manned submersibles in sedimentological and faunal investigations on the United Kingdom continental shelf. Chapter 7, pp. 151–167 in R.A. Geyer (ed.) *Submersibles and their Use in Oceanography and Ocean Engineering*. Elsevier, Amsterdam, 383 pp.

Wilson, J.B., Holme, N.A. & Barrett, R.L., 1977. Population dispersal in the brittle-star *Ophiocomina nigra* (Abildgaard) (Echinodermata: Ophiuroidea). *Journal of the Marine Biological Association of the United Kingdom*, **57**, 405–439.

Woods, J.D. & Lythgoe, J.N. (eds), 1971. *Underwater Science. An Introduction to Experiments by Divers.* Oxford University Press, London, 330 pp.

Zeitzschel, B. & Davies, J.M., 1978. Benthic growth chambers. *Rapports et Procès-Verbaux des Réunions Conseil International pour l'Exploration de la Mer*, **173**, 31–42.

Zmarzly, D.L. & Holland, N.D., 1981. Rates of food transport down the ambulacral grooves and through the gut of *Comanthus bennetti* (Echinodermata: Crinoidea) observed *in situ. Marine Ecology—Progress Series*, **6**, 229–230.

Chapter 6. Macrofauna Techniques

A. ELEFTHERIOU AND N. A. HOLME

Intertidal observation and collection

Study of the intertidal fauna and flora is in some ways easier than that of subtidal areas, but since the habitat is subjected to both aerial and aquatic climates, environmental factors influencing distribution are more complex.

When visiting the shore it should be remembered that low tide is a quiescent period for many animals which are active only when covered by the sea. Also, certain predators invade the shore at different periods of the tidal cycle: fish and crustaceans at high tide, and man, birds, insects, and sometimes rats or cattle, at low tide.

On *sandy* or *muddy shores* estimates of the standing crop of macrofauna and flora are made by driving a square sheet metal frame of appropriate area (e.g. $0{\cdot}1\,m^2$ or $0{\cdot}25\,m^2$) into the substratum, the sediment within the frame being excavated to the desired depth. Plastic or metal tubes, which remove an undisturbed core of sediment, are also widely used: a large version ($0{\cdot}1\,m^2$) which is slid into the sediment to remove a sample to a depth of 10 cm has been successfully employed on some beaches (Grange & Anderson, 1976).

Preliminary excavations should be made to find a suitable sampling depth. On some shores the majority of species and individuals occur in the top 15 cm, in others it may be necessary to excavate to >30 cm.

A plentiful supply of water is required for sieving; sometimes a nearby stream may be utilized, but if this is fresh water it may damage the more delicate organisms. If no stream is available and the distance to low water mark is great it may be possible to arrange sampling for a time when the water's edge is not too far from the sampling position. Subsequent sorting is made easier if all traces of mud can be washed out of the sample when sieving.

The question of sieve mesh is discussed on pp. 194–6. On muddy or fine sandy shores it may be possible to use an aperture as small as 0·5 mm, but on coarse sand or gravel shores a coarser mesh of 1·0 or 1·4 mm may be necessary. Since the majority of small individuals occur in the top 5–10 cm it may be possible to use a fine sieve for the surface layers, the deeper layers being passed through a coarser mesh (Pamatmat, 1968). In this way much unnecessary labour may be avoided. Organisms which would pass the smallest mesh employed can be collected by taking smaller volumes of deposit with a Perspex or glass coring tube, and adopting the methods described for meiofauna in Chapter 7.

At high tide fish and crustaceans invade the intertidal zone. These may be sampled qualitatively by shrimping net, beam trawl (Edwards & Steele, 1968), or by a small sledge such as that described by Pullen *et al.* (1968). The Riley push-net (Fig. 6.1) may be used in shallow water, either just below low tide mark, or on the flooded beach. If operated at a standard speed and for a fixed time it can produce comparative data on small active animals such as juvenile flatfish. Some small sand-burrowing crustaceans swim freely in the overlying water at high tide, returning to their zones in the sand as the tide retreats (Watkin, 1941). These animals can be sampled by nets such as those described by Colman & Segrove (1955), and Macer (1967).

It is usual to survey an intertidal flat by traverses running from high to low

Fig. 6.1. Riley push-net in use. (Photograph J.D. Riley.)

water mark, with sampling stations at regular intervals. Where possible, two or more samples should be taken at each station, as a measure of the variability of the populations. For intensive and repetitive investigations the shore can be divided by intersecting transects to provide samples at regularly-spaced intervals. Since the surface of an intertidal flat is often more or less uniform in appearance, the question of possible bias in selecting the position of the traverses and of individual stations does not usually arise, although the presence of irregularities on the beach should be carefully observed, since these are often associated with turbulent conditions at particular tidal levels. On smaller areas, such as pocket beaches, the profile may be much more obviously irregular, and the various features—ridges and runnels, streams, pools of fresh or salt water, should be taken into account when sampling. Examples of surveys are given by McIntyre & Eleftheriou (1968), and Eleftheriou & McIntyre (1976).

On *rocky shores* sampling is carried out by means of a square frame of heavy gauge wire laid on the substratum, the animals within the frame being counted, weighed, or estimated in terms of percentage cover of the surface. Sometimes estimates may be made *in situ*, otherwise growths must be scraped off for subsequent examination in the laboratory. For larger organisms, a frame of 1 or 0·25 m^2 is suitable, but for smaller organisms or where the rock surface is irregular frames of 0·1 m^2 (316 × 316 mm) should be used. A flexible frame might be appropriate on some rock surfaces. For the sampling of filamentous algae and their associated fauna in shallow water, a square-quadrat box-sampler with a sampling area of 1109 cm^2 is described by the

Finnish IBP-PM Group (1969), and other methods used by divers, as described in Chapter 5, may sometimes be appropriate.

For counting small organisms such as barnacles, squares of $0{\cdot}01\,m^2$ (100×100 mm) are suitable. For such counts a piece of thick (6 mm) Perspex, exactly 100 mm square and etched with a grid of 10 mm squares is convenient, allowing smaller areas to be counted and also minimizing the possibility of missing individuals or counting them twice. Besides organisms living on the rock surface, estimates should be made of crevice-living and boring species, and of those sheltering among weeds. The importance of photography (Chapter 4) for non-destructive recording of seashore species and habitats must be emphasized.

Stations should be spaced out at regular distances (or height intervals) along traverses from high to low water mark. The lack of uniformity of most rocky shore habitats will often make it necessary to make a number of estimates in different types of habitat (e.g. rock pool, rocks exposed/sheltered from waves/sun, crevices, under stones) at each station. Since such habitats cannot normally be selected by predetermined measurements, the question of bias will arise. Bias cannot be entirely eliminated when making such estimates, and indeed on a rocky shore having an almost infinite variety of microhabitats, it is questionable whether it is possible to take a sample which is in any degree representative of a wider area.

Space does not permit detailed consideration of the literature on intertidal methods, but a useful discussion of methods for both sediment and rocky shores is given by Gonor & Kemp (1978), and in the separate contributions in Price *et al.* (1980). Methods for coral reefs are described in Stoddart & Johannes (1978)—see also Chapter 5.

Habitat cards for description of rocky and sediment shores are described by Holme & Nichols (1980), and the subject of species and habitat recording is further considered on pp. 199–201.

Position fixing and levelling on the shore

The position of stations or transects on the shore may be fixed by standard surveying techniques as outlined, for example, by Southward (1965), and by Jones (1980). A cheap and simple level which may be used by one person is described by Kain (1958), and other levelling methods are described by Emery (1961), Stephen (1977), and Nelson-Smith (1979). Fuller treatment of levelling techniques is given by the Admirality (1948), Ingham (1975), and Pugh (1975). Other relevant references are Kissam (1956), Morgans (1965) and Zinn (1969).

For repeat sampling, positions on a rocky shore may be marked with paint, marks chiselled into the rock, expanding bolts inserted into crevices or holes drilled in the rock, or by small concrete blocks cast *in situ*. Boalch *et al.* (1974) used an underwater-setting resin compound for fixing metal bolts into holes drilled into rocks on the shore.

On sediment shores positions may be marked by posts driven deeply into the sediment. However, such posts may cause some alteration to the environment and may also attract the attention of people gathering shellfish so that the sediment around the posts receives more trampling than elsewhere. In such instances it is advisable to position reference marks on rocks or other permanent structures, from which locations are determined by tape measure, and/or transit lines.

For intertidal organisms, the duration of exposure at each low tide is important. This may be roughly determined from the zonation of plants and animals on the shore, and by observations of the length of time for which the selected sites are exposed over a number of low tides. Where tidal data are available the positions may be levelled to a bench-mark or other mark of known height; it will then be possible to calculate exposure from data given in the tide tables.

In estuaries or in parts of the world where accurate tidal data are not available a series of observations on a graduated tide pole should be carried out; positions on the shore can then be levelled in relation to this. Allowance should be made for spring and neap tides, the effects of wind and barometric pressure, and for river outflow in estuaries. It should not be assumed that the tide follows a symmetrical harmonic curve, nor that tidal heights are necessarily the same on the two tides of one day (Doodson & Warburg, 1941; Lewis, 1964). On wave-exposed coasts levels are elevated through spray and swash, so that plants and animals will tend to occur much higher than on sheltered shores.

Remote collection

There are a number of reviews in which the equipment and techniques used for sampling the benthos are described (e.g. Gunter, 1957; Thorson, 1957; Holme, 1964; Hopkins, 1964; Longhurst, 1964; Reys, 1964; Bouma, 1969, Kajak, 1971; Lamotte & Bourlière, 1971; Menzies *et al.*, 1973; Eagle *et al.*, 1978), and a bibliography of benthic samplers is given by Elliott & Tullet (1978).

While choice of equipment depends largely on local conditions—size of ship, power and capabilities of lifting gear, whether sampling in exposed or sheltered conditions, depth of water, bottom deposit, and type of sample required—the multiplicity of samplers which have been described is evidence not only of such factors, but also of a widespread dissatisfaction with existing methods of collection.

Because the many samplers which are available have been described and discussed in reviews such as those listed above, this chapter will not attempt to cover the whole field nor to give a historical review, but rather to guide the reader towards the most suitable instruments for his own particular purpose: to this end a summary of the attributes of selected samplers is given in Table 6.3 (pp. 188–9).

There are some grounds—notably those of rocks and boulders—which cannot be adequately sampled by any of the instruments listed here. At best, they can be sampled only by dredge, which may prove inadequate even as a qualitative sampler. Such habitats are often better investigated by diver observation from a submersible (Chapter 5), or by photography and television (Chapter 4).

Trawls

Beam, Agassiz and otter trawls may be used for qualitative sampling of the epifauna. These nets are designed to skim over the surface of the bottom, and because of the large area covered, are useful for collecting scarcer members of the epifauna, and species of fish, cephalopods and crustaceans associated with the bottom. The efficiency of such gear, in terms of numbers of animals captured in relation to those in the area swept by the net, is generally low, and is selective for particular species. Attempts at quantifying results from trawl catches are considered on p. 156.

The *beam trawl* is still used commercially for fishing for shrimps, prawns and flatfish. The mouth of the net is held open by a wooden beam of 2–10 m length, with metal runners at either end. The net is a fairly long bag, of mesh about 12·5 mm knot-to-knot, the lower leading edge of which is attached to a weighted chain, forming a ground rope which curves back behind the top of the net attached to the beam so that fish disturbed by the ground rope cannot escape upwards.

The *Agassiz* or *Blake trawl* (Fig. 6.2) is virtually a double-sided beam trawl which was designed for deep-sea collecting, where it is not possible to control which way up the trawl lands on the bottom (Agassiz, 1888). Compared to the beam trawl it suffers from the disadvantage that the ground and head ropes, being interchangeable in function, are necessarily of the same length. Consequently few fish are caught by the Agassiz trawl, which is not, therefore, used commercially.

Both beam and Agassiz trawls can be towed on a pair of bridles attached to a single tow-rope or wire. Unfortunately, the cross-bars or beams are liable to be damaged if they meet an obstruction, and a weak link should be used (pp. 151–3), especially when working unexplored grounds or in deep water. Carey & Heyamoto (1972) describe a beam trawl with a flexible connection

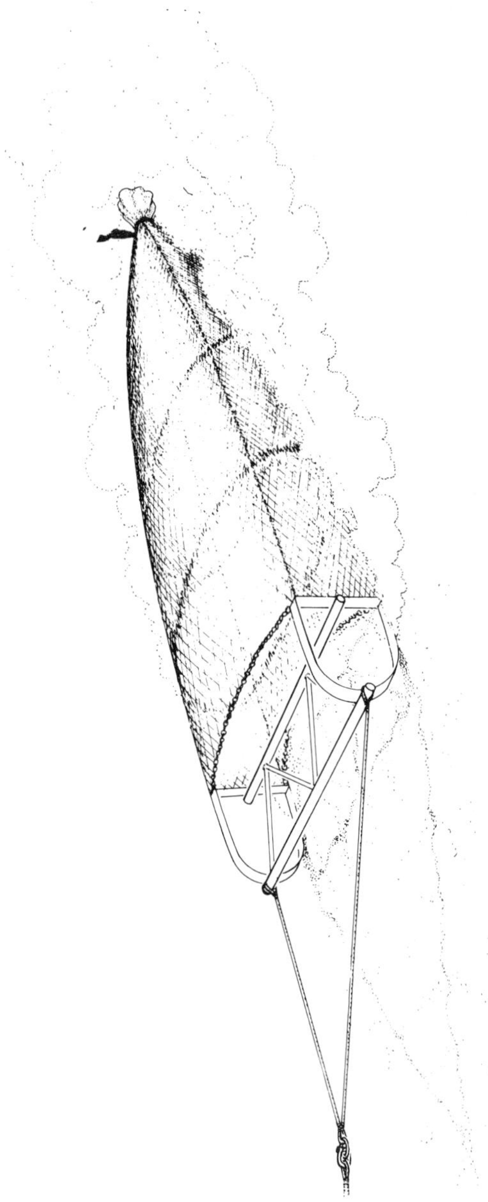

Fig. 6.2. Agassiz trawl.

between beam and runners, allowing momentary collapse of the net when an obstruction is encountered.

Besides animals retained in the cod-end of a trawl, many small organisms may be found attached to the net meshes, and trawls usefully supplement the information obtained by dredge and grab when making benthic surveys.

Otter trawls used for commercial fishing also capture members of the invertebrate epifauna, but because of the rather large meshes only the larger animals are retained. The rigging, shooting and working of otter trawls is a specialized subject; instructions for making up a small trawl are given by Steven (1952), and details of various types of fishing net are given by Andreev (1962); von Brandt (1972); Davis (1958); Garner (1962, 1973, 1977); Kristjonsson (1959), and Strange (1977). FAO (1972) have published a catalogue of fishing gear designs, and further references are given by von Brandt (1978).

Bottom sledges

Many types of sledge have been designed for sampling the epifauna and larger members of the plankton immediately above the bottom. Some are little more than plankton nets on runners (e.g. Myers, 1942), but Ockelmann's detritus-sledge (Fig. 7.3) has a tickler-chain to stir up newly-settled benthic invertebrates (Ockelmann, 1964). Others have opening and closing mechanisms, and sometimes a meter to measure the quantity of water filtered (Bossanyi, 1951; Wickstead, 1953; Beyer, 1958 (illustrated in Holme, 1964) Frolander & Pratt, 1962; Macer, 1967). A sledge net for hand towing in the intertidal zone is described by Colman & Segrove (1955).

An improved version of Macer's sled-mounted 'supra-benthic' sampler (Macer-GIROQ) has been developed by Brunel *et al.* (1978), who have used it successfully to 200 m depth (Fig. 6.3).

Epibenthic sledges have a heavy frame enclosing the net, and are particularly useful for deep-sea sampling. Hessler & Sanders' (1967) sled (Fig. 6.4) has a net 80 cm wide and 30 cm high mounted within steel runners which prevent it sinking too deeply into deep-sea muds, in spite of a total weight of 160 kg. Aldred *et al.* (1976) describe an epibenthic sledge with an opening and closing mechanism, measuring wheel (Rice *et al.*, 1982) and forward-pointing camera (Fig. 6.5). This sledge is designed to land on the bottom, and operate, on one side only.

There are a number of instruments which are intermediate between the Agassiz trawl and dredge in that they consist of a net attached to a rectangular steel frame fitted with runners. These include the *Small Biology Trawl* of Menzies (1962) (Fig. 6.6), used for deep-sea sampling. A feature of this trawl is that the net is held within a rigid frame, preventing it becoming wrapped

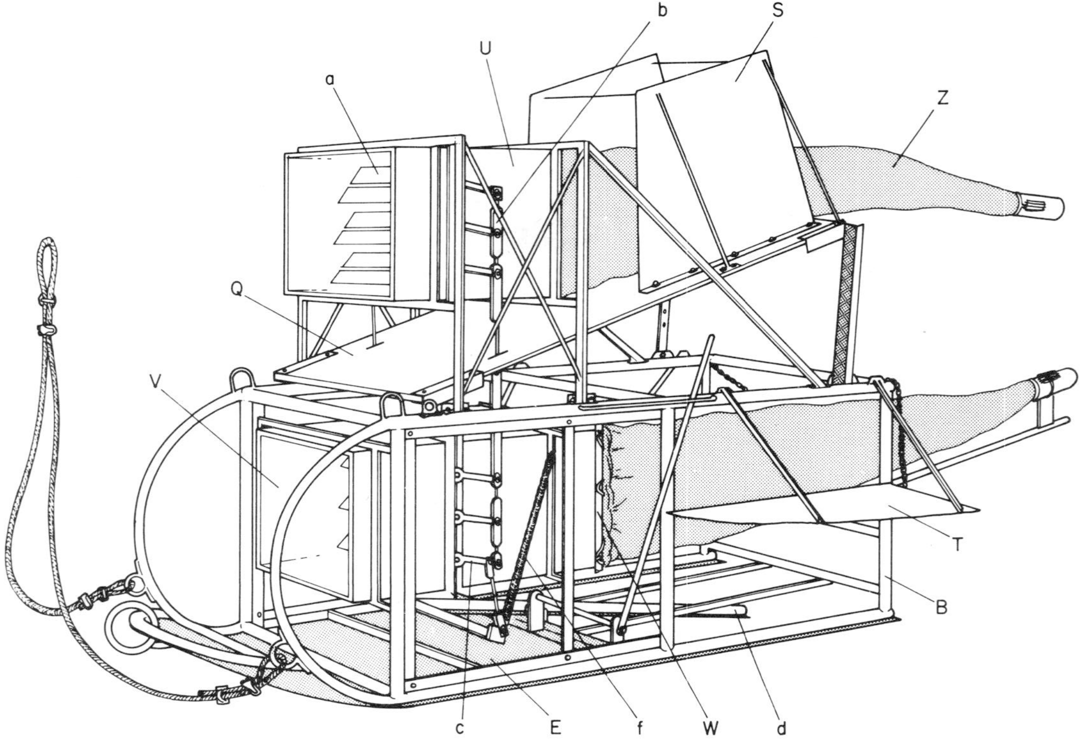

Fig. 6.3. Macer–GIROQ sledge. B, tubular chassis; E, sheet-metal gliding plate, turned upward at front; Q, adjustable wooden depressor; S, vertical fin; T, horizontal fin; U, wooden box at front of upper net; V, wooden box at front of lower net; W, metal strip for attachment of net; Z, zooplankton net; a, shutter closing mechanism; b, adjustable control link; c, crank lever; d, lever operating closing mechanism; f, closing spring. (After Brunel *et al.*, 1978.)

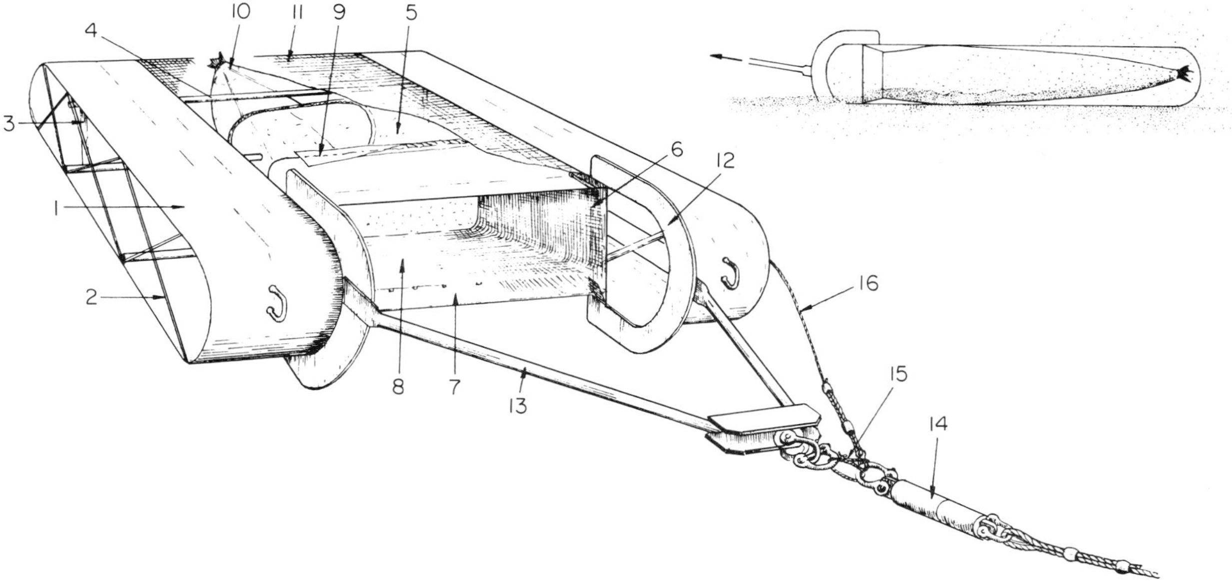

Fig. 6.4. Epibenthic sled. In this sketch much of the top protective wire screen and part of the anterior tubular cross-piece are cut away to show additional details. 1, runners; 2,3, strengthening members inside runners; 4, tubular cross piece; 5, collecting net (nylon);6, side-plate at mouth of net; 7, biting edge at top and bottom of net, adjustable for height; 8, canvas collar at front of net, which is tied by canvas flaps (9) to the tubular crosspieces and struts; 10, net tied at posterior end; 11, heavy wire screen to protect net; 12, flange preventing mud entering net from the side; 13, towing yoke; 14, swivel; 15, weak link; 16, safety-line. The smaller drawing shows the mode of operation of the sledge. (Redrawn from Hessler & Sanders, 1967.)

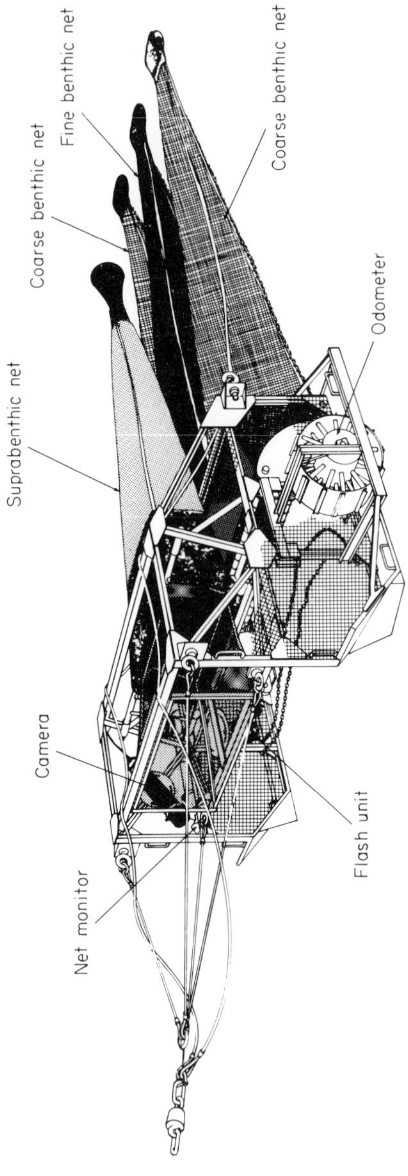

Fig. 6.5. IOS epibenthic sledge, shown in attitude adopted on the sea bed. (From Rice *et al.*, 1982.)

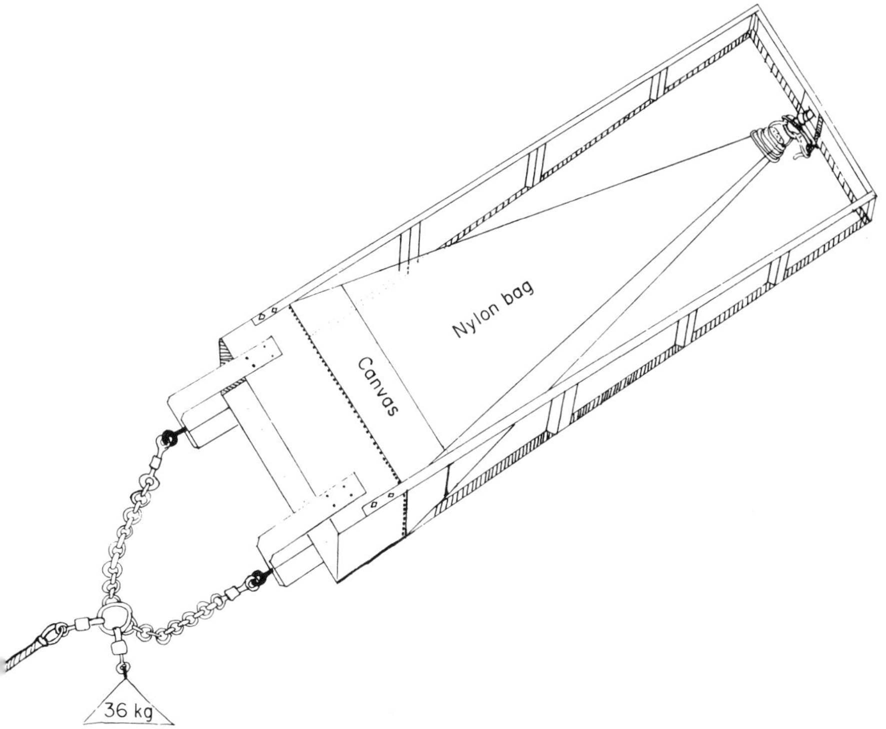

Fig. 6.6. Small Biology Trawl of Menzies (1962). The frame length is 3 m and width 1 m.

around the mouth of the trawl during lowering. A small dredge with runners used at Plymouth (Fig. 6.7) has been extensively employed for sampling the epifauna on the continental shelf.

Dredges

Dredges have a heavy metal frame and are designed for breaking off pieces of rock, scraping organisms off hard surfaces, or for limited penetration and collection of sediments.

The *naturalists'* or *rectangular dredge* (Fig. 6.8) is a useful instrument for exploratory purposes as it can obtain samples on a variety of grounds. One of the dredge arms is attached directly to the tow-rope, the other being joined to it by a few turns of twine, which act as a weak link to release the dredge should it come fast on the sea bed. It is important not to use too many turns of twine, since, particularly with synthetic twine, it may then be too strong to part should the need arise. More sophisticated links, involving a metal shear pin,

Fig. 6.7. Dredge with runners, as used by the Marine Biological Association, Plymouth. The width is 1 m. Note canvas chafing strips on either side of net.

are available commercially. An alternative arrangement is to attach a chain or wire from the towing point to some point towards the back of the dredge, for retrieval when the weak link parts (Figs 6.4; 6.9). A swivel should always be inserted between towing warp and the dredge (or any other sampling gear), and this must be of a ball- or roller-bearing type, turning freely under load, when the towing warp is of wire.

The dredge net is usually about half as deep as it is wide, the mesh varying

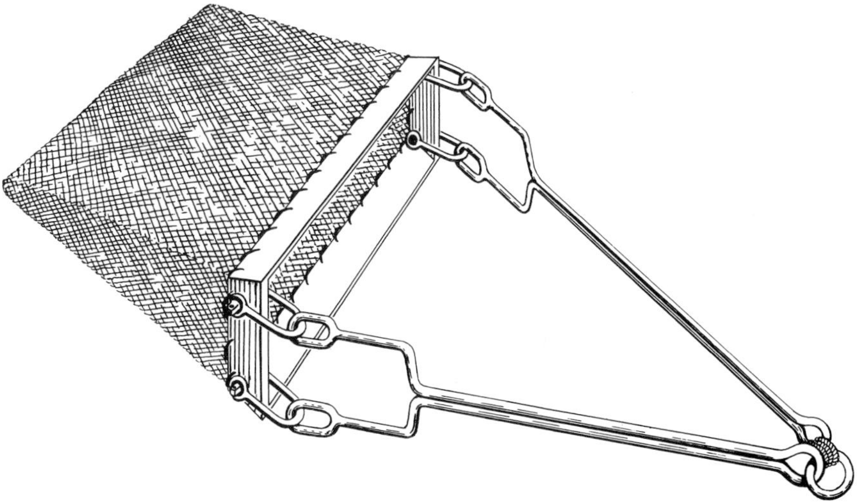

Fig. 6.8. Naturalists' or rectangular dredge. Note weak link of twine joining one arm to ring.

according to circumstances. Machine-made netting of mesh 10–12 mm knot-to-knot is generally suitable, but material used for shrimp trawls is usually too light in construction. For collection of sediment the bag can be lined with an inner bag of sacking, stramin, or burlap. Impervious material such as canvas should not be used, as water must drain away when the dredge comes on board. Where a deeper net bag is required (e.g. to avoid washing out of the sample when hauling from deep water), the net should be an open-ended sleeve, tied at the bottom with a rope which is untied to release the sample. Too

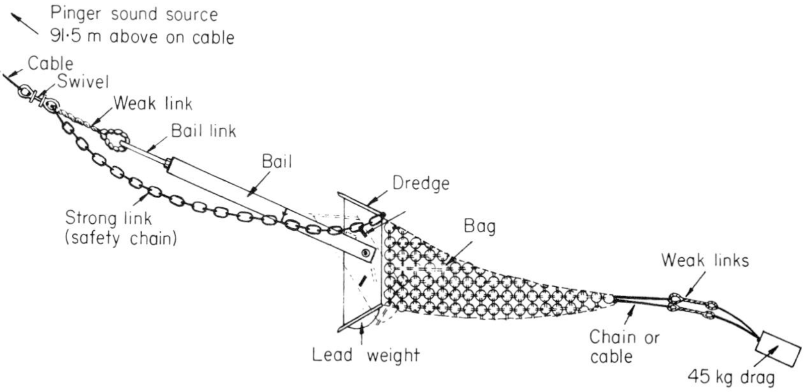

Fig. 6.9. Rock dredge, showing arrangement of weak link, safety chain and swivel. The bag is of interlaced metal rings. (Redrawn from Nalwalk *et al.*, 1962.)

Table 6.1. A selection of ropes and wires for different purposes. Steel ropes with a central fibre core. Specifications of individual brands should be checked from manufacturers' data.

Material	Diameter (mm)	Strength (kg/mm^2)	Construction	Breaking strength (kg)	Use
Manila (natural)	16	—	3 strand	2000	Hand or capstan-hauled ropes
Ulstron (synthetic)	12	—	3 strand	1930	Hand or capstan-hauled ropes
Ulstron (synthetic)	16	—	3 strand	3302	Hand or capstan-hauled ropes
Polyethylene (synthetic)	12	—	3 strand	1524	Hand or capstan-hauled ropes
Polyethylene (synthetic)	16	—	3 strand	2794	Hand or capstan-hauled ropes
Polyester (synthetic)	12	—	3 strand	2270	Hand or capstan-hauled ropes
Polyester (synthetic)	16	—	3 strand	4100	Hand or capstan-hauled ropes
Galvanized steel	6	180	6 × 19	1990	Hydrographic use. Light grabs and corers
	8	145	6 × 19	2850	Grabs. Light dredges
	11	145	6 × 19	5390	Large grabs. Dredging on shelf
	11	180	6 × 19	6690	Grabs, dredges, geological corers. Shelf and slope
	12	145	6 × 19	6420	Otter-trawls, dredging. Size according to that of ship
	16	145	6 × 19	11 400	
	20	145	6 × 19	17 800	
	24	145	6 × 19	25 700	

deep a net, however, may raise an unnecessarily large sample and burst when a heavy load is brought up. Chafing of the net while being towed along the bottom can be minimized by fitting sheets of rubber, hide or canvas (Fig. 6.7) on either side of the net bag.

When operating from a small boat where the dredge is hand-hauled a rectangular dredge frame of length 300–380 mm is large enough—this should be towed on a rope of at least 10 mm, and preferably 12 mm, diameter for ease of hand-hauling (Table 6.1). For larger boats with a power hoist a dredge frame of 450–600 mm is suitable, and 750–1300 mm is used for trawler-sized vessels. On ships equipped with winches, wire ropes will be used for working the dredge, the diameter of the wire being related to the size of the ship and the depth of water, bearing in mind that, in spite of weak links, very severe strains can be exerted on the towing warp from time to time. On the continental shelf it is usual to tow with warp equal to $2\frac{1}{2}$–3 times the depth of water, but in the deep sea a factor $< 1\frac{1}{2}$ times may be sufficient, particularly where an acoustic pinger is used to show when the dredge is on the bottom (pp. 176–7). When dredging, the ship should drift or steam slowly (1–2 knots), a check being made on the performance of the dredge by the angle of the warp and any jerks or vibrations, strain being recorded if possible on a strain gauge. The dredge should normally be towed on the bottom for 5–10 minutes, but on some grounds it may fill up at once and can be hauled almost immediately. Special techniques for deep-sea dredging are given on pp. 175–9.

Many types of dredge have been produced for different purposes. For geological sampling there are sturdy rock dredges, often with teeth (Nalwalk *et al.*, 1962—see Fig. 6.9; Boillot, 1964), and Clarke (1972) has described a heavy dredge for sampling under difficult conditions in mixed boulder and mud substrata. Rock dredges typically have bags of metal rings or wire grommets, similar to those used on oyster and scallop dredges. These may be lined with a finer mesh of synthetic netting if desired. Where digging into the sediment is required, dredges which are bowed, oval or circular in shape are more effective than a straight edge, but the penetrating powers of most are limited, except in soft mud, so that they typically sample only the shallower-burrowing members of the infauna. Dredges with teeth are more effective for certain purposes (Baird, 1955, 1959), and the use of inclined steel diving plates to make the dredge sink more quickly and to help maintain contact with the bottom is worth considering.

Although seldom providing satisfactory quantitative samples, trawls and dredges are indispensable sampling instruments and should always be carried on board ship. Dredges, in particular, are invaluable for preliminary surveys to discover the nature of the bottom and its fauna, and may be the only instruments which can be used under adverse sea conditions; being mechanically unsophisticated they are a standby when more complex equipment has broken down, and on some grounds they may be the only means of obtaining a sample and would then be the prime means of investigation.

Semi-quantitative estimates with trawl and dredge

When trawling it may be possible to standardize conditions and duration of tow, so as to obtain estimates of population density which are of value for comparative purposes, and such hauls are commonly used for estimating demersal fish populations. An account of statistical methods used in fisheries biology is given by Gulland (1966), Nikolskii (1969) and Ricker (1975). However, it is not easy to estimate the exact time for which gear is on the bottom: both trawls and dredges will continue to drag along the bottom after hauling has commenced. Carey & Heyamoto (1972) used a time-depth recorder to monitor the behaviour of a trawl on the bottom, and methods involving acoustic signals are described by Laubier *et al.* (1971) (Fig. 6.24), and also by Rice *et al.* (1982). It is well known that trawls and dredges sample only a fraction of the fauna lying on the surface of the sea bed (e.g. Mason *et al.*, 1979) and very few of the burrowing animals, so that results merely represent minimum densities on the ground.

A number of workers have fitted measuring wheels on trawl or dredge frames in order to measure distance covered (e.g. Belyaev & Sokolova, 1960; Riedl, 1961; Wolff, 1961; Gilat, 1964; Richards & Riley, 1967; Bieri & Tokioka, 1968; Carey & Heyamoto, 1972; Pearcy, 1972; Carney & Carey, 1980; Rice *et al.*, 1982; see also the television sledge of Holme & Barrett (Chapter 4). The performance and success of such wheels seems to be variable, and is complicated by the fact that some sampling instruments tend to progress by leaps and bounds (Baird, 1955; Menzies, 1972). In addition the wheel may jam or misfunction for all or part of the tow, giving a false reading. The fitting of an odometer wheel each side of the frame would seem to overcome this problem, but Carney & Carey (1980) found considerable variation between readings from wheels fixed either side of a beam trawl. They considered that slippage might result in under-reading of distance by as much as 40 %. If a continuous record of rotation can be obtained, so that it is known that the wheel is rotating throughout the tow, a reasonably satisfactory measure of distance should be obtained (Holme & Barrett, 1977), although Rice *et al.* (1982) have some doubts over the accuracy of a continuously recording wheel attached to their epibenthic sledge.

Where the dredge bag is lined with closely-woven material to retain the sediment, semi-quantitative data in terms of numbers of animals per unit volume of sediment can be obtained. The results are naturally dependent on depth of digging, but under difficult conditions where grab samples cannot be taken this may be the best that can be achieved.

Anchor dredges

Forster's anchor dredge (Forster, 1953) is an invaluable instrument for semi-quantitative sampling of sands and other firmly-packed deposits. This dredge has an inclined plate intended to dig in deeply at one place, and it is not towed, as are other dredges. It is shot by allowing the ship to drift while warp equal to five times the depth is gradually paid out; the warp is then made fast, and the strain exerted as the ship is brought to a standstill drives the dredge into the sand to a depth of up to 25 cm. This dredge is best used from a small launch fitted with a power hoist, since when used from a larger ship there is a tendency for it to be jerked out of the sediment. This also occurs if insufficient length of warp is paid out. The sample theoretically taken by Forster's anchor-dredge is wedge-shaped in section and approximately the same area as the digging plate, so that it may be used for semi-quantitative studies. A double-sided anchor dredge capable of working either way up is shown in Fig. 6.10.

Thomas (1960) describes a modified anchor dredge with adjustable digging plate and with a self-sifting mesh net. Because this dredge is intended to be towed through the sediment the results are non-quantitative, but it appears to be a useful collecting instrument for deeper-burrowing animals.

Sanders *et al.* (1965) describe a rather different type of anchor dredge (Fig. 6.11) for use on the shelf and the deep sea. This is double-sided with two angled digging plates between which is a wide horizontal plate which limits penetration to 11 cm. Since the dredge is very heavy it is assumed to sink consistently to this depth in the sediment, and, therefore, to sample to a constant depth. Once the bag is full further material is rejected out of the mouth of the dredge. Because of the observed consistency of operation, at least on silt-clay grounds, this dredge has been employed for quantitative studies. A lighter model is described in Sanders (1956).

A modified version of the Sanders anchor-dredge—the *Anchor-Box Dredge*—was developed by Carey & Hancock (1965). This samples an area of 1·3 m^2, to a depth of 10 cm, and has been used with success to depths of 2800 m, at which depth it was worked with a warp to depth ratio of 1·39:1.

Grabs

Quantitative samples of animals inhabiting sediments are usually taken by grab. The grab, which is lowered vertically from a stationary ship, captures slow-moving and sedentary members of the epifauna, and infauna to the depth excavated.

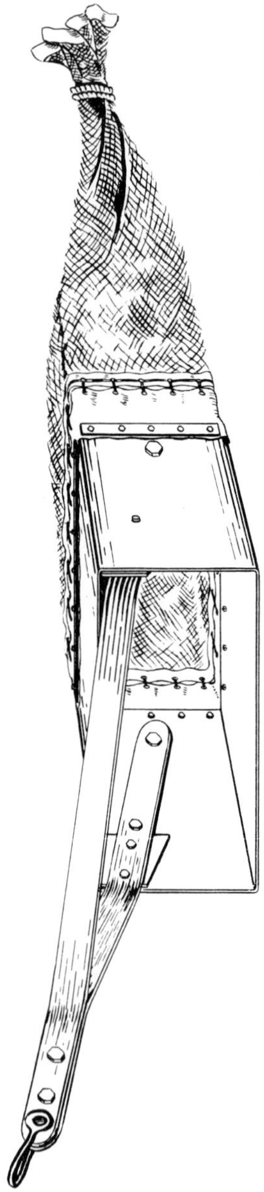

Fig. 6.10. Double-sided anchor dredge as used by Holme (1961). The wishbone towing arms are free to swivel or can be locked to one side if required.

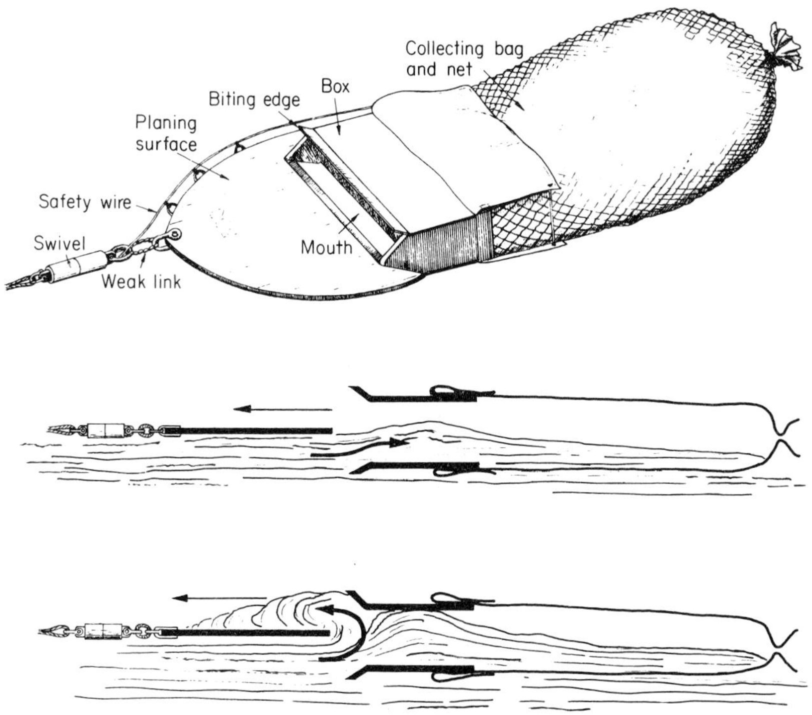

Fig. 6.11. Deep-sea anchor dredge. Above, general view; centre, movement of sediment into dredge before clogging; below, movement of sediment after clogging. (Redrawn from Sanders *et al.*, 1965.)

There has been much discussion on the depths to which animals burrow into the sea floor (MacGinitie, 1935, 1939; Thorson, 1957; Holme, 1964). The majority inhabit the top 5–10 cm, but some burrow more deeply (Barnett & Hardy, 1967; Kaplan *et al.*, 1974; Thayer *et al.*, 1975). Exceptionally, some crustaceans have been found to burrow to depths ≥ 3 m (Pemberton *et al.*, 1976; Myers, 1979). Few grabs are designed to dig deeper than 15 cm, and in practice many dig to less than 10 cm in firmly-packed deposits (Table 6.3). Thus a grab may be an adequate sampling instrument for some grounds (Ankar, 1977b), while on others it may leave significant elements of the fauna unsampled below the depth of bite.

If a grab is used as the prime means of investigation it would be advisable, at the pilot survey stage, to make comparative hauls with a deeper-digging instrument such as the Forster anchor dredge (p. 157), box corer (p. 167), or suction sampler (p. 169) in order to check whether there are deeper-burrowing individuals out of reach of the grab. If this were so one of these methods should be used from time to time to supplement the information obtained by grab. The sampling efficiency of grabs is further discussed on pp. 182–4.

For sampling the macrofauna, grabs covering a surface area of 0·1 or 0·2 m^2 are commonly employed, several samples being taken to aggregate to 0·5 or 1·0 m^2 per station. Samples of this total size are usually considered adequate for quantitative determinations of the commoner species, measurements of biomass, etc. (see pp. 6–17), but do not adequately sample scarcer animals, which are often members of the epifauna. Moreover, some fast-moving species escape the grab altogether. It is, therefore, advisable to supplement grab estimates of the epifauna by hauls with an Agassiz or beam trawl, or by underwater photography, television or diving.

The *Petersen grab* used by C.G.J. Petersen for investigations in the Danish fiords at the beginning of the century, is the prototype from which many modifications and improvements have been made (Petersen & Boysen Jensen, 1911). It consists of two buckets (Fig. 6.12) hinged together, which are held in an open position during lowering. When on the bottom the lowering rope slackens, allowing a release hook to operate so that on hauling up the two

Fig. 6.12. Petersen grab approaching the sea bed. After the release hook has actuated, an upward pull exerted on the central chain closes the two halves of the grab. (After Hardy, 1959.)

buckets close together before the grab leaves the bottom. The disadvantages of the Petersen grab for sampling in other than soft muds and in sheltered waters have often been discussed (e.g. Davis, 1925; Thorson, 1957; Holme, 1964). These relate to premature operation of the release during descent due to momentary slackening of the rope as the ship rolls, failure to penetrate sufficiently deeply into the sediment, losses due to the jaws not closing completely and inadequate sampling due to an oblique upward pull when closing due to drift of the ship on station.

The *Campbell grab* (Hartman, 1955) is similar to the Petersen grab, but its greater efficiency is due to its larger size (0·55 m^2 sampling area, contrasted with 0·1 or 0·2 m^2 for the Petersen Grab), and greatly increased weight (410 kg).

The *Okean grab* (Lisitsin & Udintsev, 1955). The Petersen grab has a gauze-covered window at the top of each bucket to allow water to escape while the grab is closing. This offers some resistance to the swift lowering which is highly desirable for deep-sea sampling. In the Okean grab (Fig. 6.13) the tops of the buckets form hinged doors which are held open during the descent, and

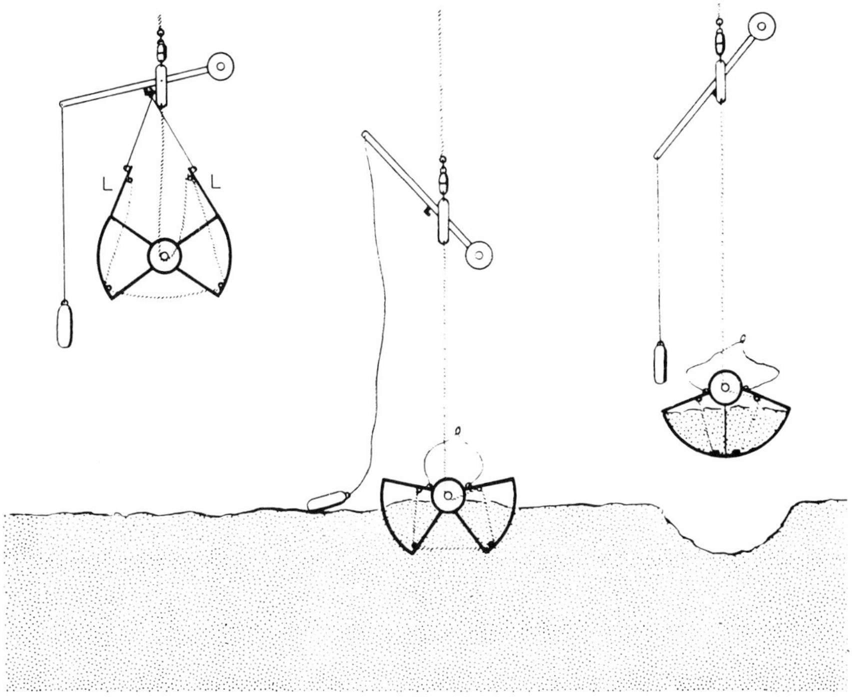

Fig. 6.13. Operation of the Okean grab. Note the counterweight release and the lids (L) of the two buckets, which are open during the descent. (Redrawn from Lisitsin & Udintsev, 1955.)

which close when the grab reaches bottom. Very rapid rates of lowering are possible with the Okean grab, which has, in addition, a counterweight mechanism to prevent tripping in mid-water. A comparison of the sampling efficiency of the Okean and van Veen grabs, which showed that the latter sampled more effectively, is given by Ankar *et al.* (1978).

The *van Veen grab* (van Veen, 1933) improves on the Petersen grab in having long arms attached to each bucket, so giving better leverage for closing (Fig. 6.14). The arms also tend to prevent the grab being jerked off the bottom

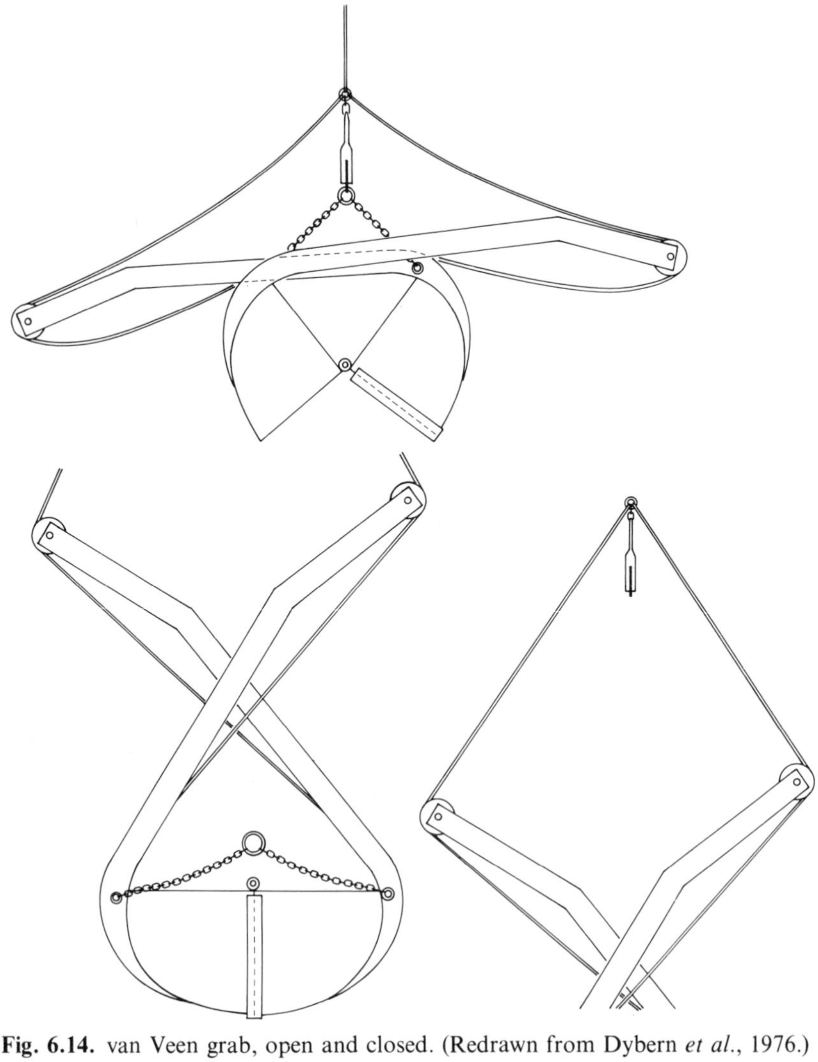

Fig. 6.14. van Veen grab, open and closed. (Redrawn from Dybern *et al.*, 1976.)

should the ship roll as the grab is closing. On the other hand, the arms may pull the grab to one side if, through drift of the ship, the upward pull for closing is oblique. The sampling efficiency of this grab in different sediments has been tested by Christie (1975), Ankar (1977a), and Ankar *et al.* (1979). The van Veen grab has been adopted as the standard sampler for benthic investigations in the Baltic Sea (Dybern *et al.*, 1976), but some improvements to the mechanism have been proposed by Sjölund & Purasjoki (1979). Under open-sea conditions the counterweight release mechanism described by Lassig (1965) can be employed.

The *Ponar grab* (Powers & Robertson, 1967) is a limnological grab closed by a scissor action of the arms attached to the buckets. It appears to be an efficient sampler under sheltered conditions, but the release mechanism is not suited for open-sea use.

The *Hunter grab* (Hunter & Simpson, 1976) is a robust and compact grab with jaws which are extended to form levers, with the upper surfaces providing lids which allow a free flow of water through the sampler during the descent.

The *Smith–McIntyre grab* (Smith & McIntyre, 1954) was designed for sampling under the difficult conditions often encountered when working from a small boat in the open sea. This grab has hinged buckets mounted within a stabilizing framework (Fig. 6.15), and powerful springs to assist penetration of the sediment. Trigger plates on either side of the frame ensure that the grab is resting flat on the bottom before the springs are released. Closing of the grab is completed as hauling commences by cables linked to arms attached to each bucket. The Smith–McIntyre grab covers an area of $0{\cdot}1\ m^2$, and on firm sands penetrates to about the same depth as the $0{\cdot}1\ m^2$ van Veen grab, but the greater reliability of its release makes it preferable for open-sea use.

A number of workers have adopted the Smith–McIntyre grab as their standard sampling instrument, but some consider it complicated and sometimes dangerous to use. The *Day grab* (Day, in preparation) (Fig. 6.16) represents an attempt to simplify the design of this type of instrument. It incorporates a frame to keep the grab level on the sea bed, and two trigger plates for actuating the release, but there are no springs to force the hinged buckets into the bottom. The Day grab seems to sample as efficiently as the Smith–McIntyre grab (Tyler & Shackley, 1978), and is preferred by some workers because of its greater simplicity.

The *orange-peel grab* has been much used in the United States and also in France and Australia. It has four curved jaws closing to encircle a hemisphere of sediment. Penetration of the sediment is aided by the fact that the jaws narrow to a point, but there is likely to be loss of material between the jaws, and also from the top, unless a canvas cover, as described by Reish (1959b), is fitted. A spring-loaded version of this grab, mounted on a frame, has been described by Briba & Reys (1966). The orange-peel grab is available in a range

Fig. 6.15. Smith–McIntyre grab. Above, in open position ready for lowering; below, in closed position. Note the trigger plates on either side, both of which must rest on the bottom before the release is actuated. The threaded studs with butterfly nuts are for attachment of lead weights. (Photograph A.D. McIntyre.)

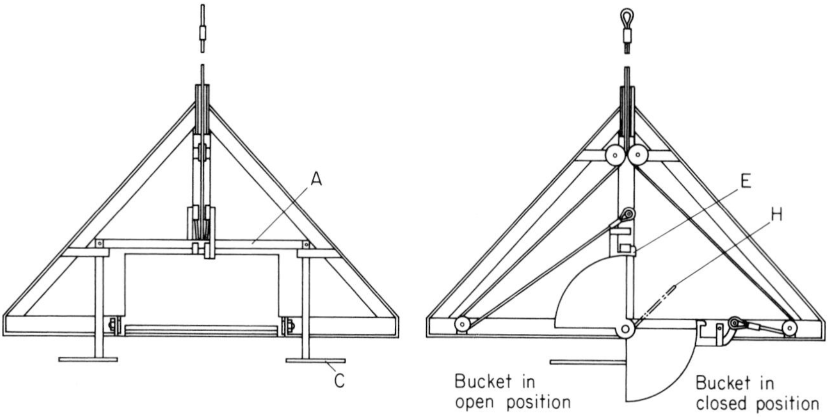

Fig. 6.16. Day grab. Left, end view, open for lowering; right, side view, one bucket open, the other closed. On reaching the sea bed the two pressure plates (C) are pushed upward, releasing the transverse beam (A) so that the hooks (E) holding the buckets open are released. The buckets are closed by tension on the two cables, the hinged flap (H) allowing water to escape during the descent but acting as a cover during hauling. (From Day, in preparation)

of sizes. Thorson (1957) does not consider it a satisfactory quantitative sampler.

The *Baird grab* (Baird, 1958) was designed for sampling the epifauna of oyster beds, but has also been used in the open sea. It has two inclined digging plates which are pulled together by springs and levers (illustrated in Holme, 1964). It covers an area of $0{\cdot}5\,m^2$ and has been found to dig into sediments quite well, having applications for sampling the infauna where a sample of large area is required. Since the surface of the sample is not covered there may be some washing out during hauling.

The *Hamon grab* (Fig. 6.17) samples an area of about $0{\cdot}29\,m^2$ by means of a rectangular scoop rotating through 90°. It appears to be particularly effective in coarse, loose, sediments, but, because of its mode of action, may not sample entirely quantitatively. Dauvin (1979) reports that in spite of its considerable weight (350 kg) and size (height 2 m) it is easy to handle on board ship, and is capable of sampling the deeper infauna of sediments, below 10 cm.

The *Holme grab* (Holme, 1949) samples by means of a single semi-circular scoop rotating through 180°. This design minimizes loss of material while hauling, and a later model (Holme, 1953) with two independently-operating scoops (Fig. 6.18) reduces any tendency to sideways movement during digging.

The *Shipek sediment sampler* has a single semi-circular scoop actuated by powerful springs. Covering an area of only $0{\cdot}04\,m^2$, this instrument is rather

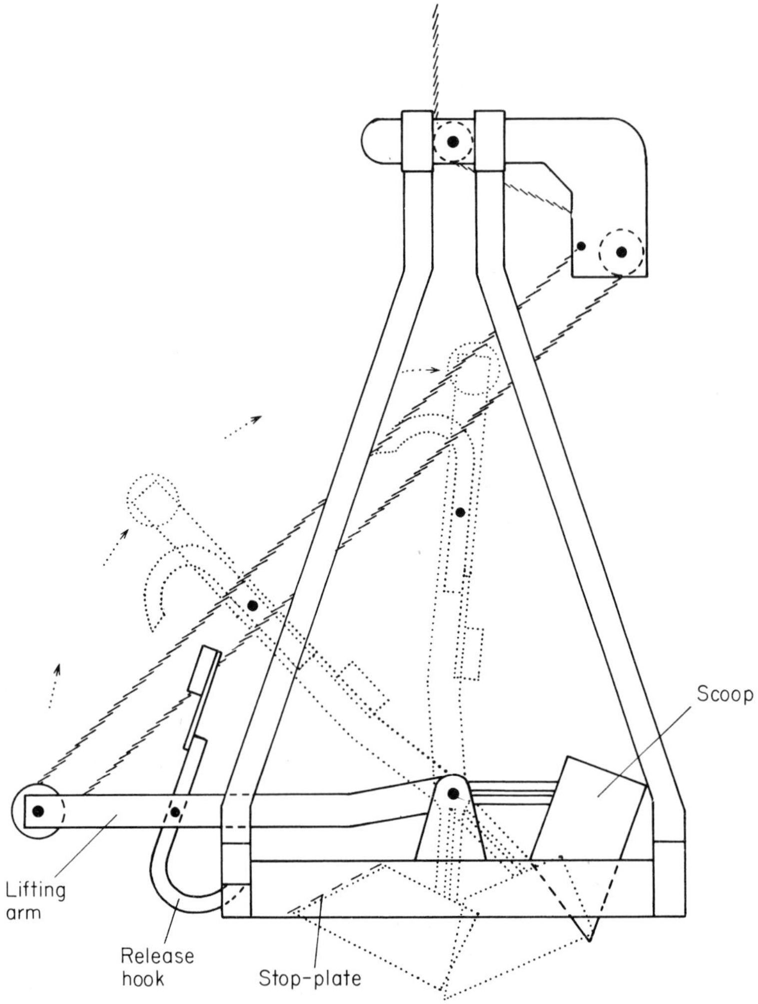

Fig. 6.17. Hamon grab, showing mode of action. The lifting arm rotates through 90° to drive the scoop through the sediment, closing against the stop-plate. (After Dauvin, 1979.)

small for macrofauna investigations, and its main use is by geologists to obtain a small sample of bottom sediment.

Many small grabs such as the *Ekman* and *Birge–Ekman grab* (see bibliography in Elliott & Tullett, 1978) are suitable for use from a small boat on soft sediment, but since they cover an area of only about 0·04 m^2 are not very suitable for macrofauna sampling. However a modified Birge–Ekman grab has been designed by Rowe & Clifford (1973) for use by SCUBA divers or from deep submergence vessels.

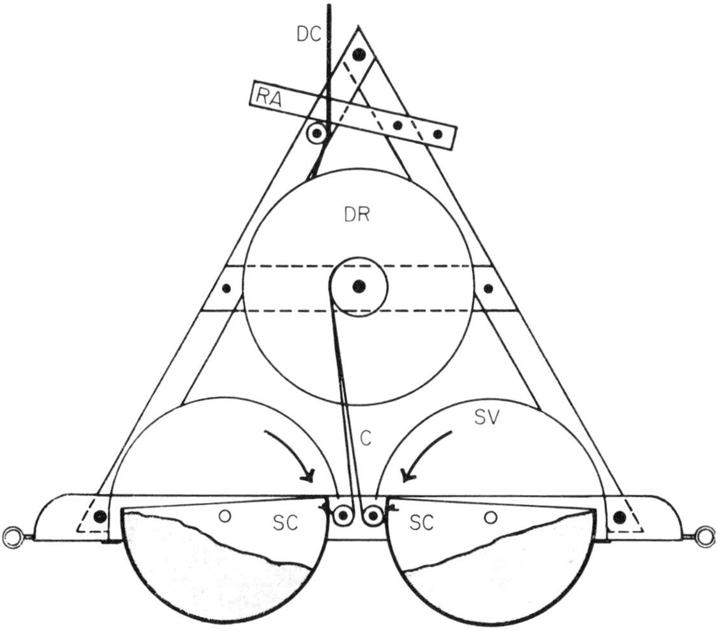

Fig. 6.18. Holme's double-scoop sampler (Holme, 1953), with scoops closed. This model takes two samples, each of $0{\cdot}05\,m^2$, but a later model, with scoops twice the width, appears similar in section. DC, cable from ship; DR, cable drum; C, cables actuating scoops; RA, release arm; SC, scoop; SV, scoop-cover.

Box samplers and corers

The *Reineck box sampler* (Reineck, 1958, 1963) consists of a rectangular corer supported in a pipe frame with a hinged cutting arm which is pulled down to close the bottom of the tube (Fig. 6.19). This instrument samples an area 20×30 cm to a depth of 45 cm and weighs 750 kg, weighted for use. A similar instrument is described by Bouma & Marshall (1964). There have been a number of other modified versions of the original box samplers: Hessler & Jumars (1974) describe a 'spade corer' covering an area of $0{\cdot}25\,m^2$, further modified by Jumars (1975) in having the sample subdivided into 25 contiguous subcores. Farris & Crezée (1976) improved the sealing of the corer so that it gave better retention of coarse sand cores, together with the overlying water. Another type of box corer, sampling 30×30 cm, is described by Jonasson & Olausson (1966). The IOS box corer described by Peters *et al.* (1980) has direct lowering control from the ship for penetration of the sediment, allowing elimination of the main framework and of a piston assembly previously needed to control penetration.

Box corers provide a means of obtaining deep and relatively undistorted

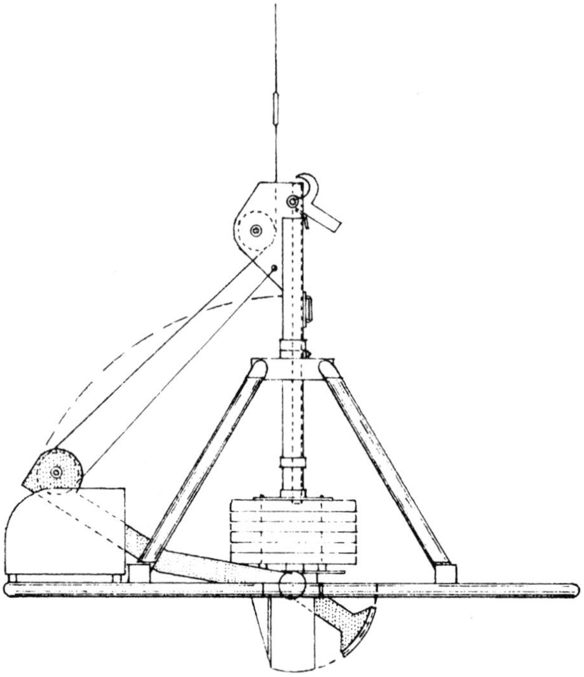

Fig. 6.19. Reineck box sampler. The rectangular coring tube is closed by a knife edge actuated by pulling up on the lever on the left. An attachment can be fitted to show the inclination and compass orientation of the core. (Redrawn from Reineck, 1963.)

samples suitable for evaluation of macrofauna from a variety of sediments and are suitable for deep-sea use. The surface of the bottom, and supernatant water, appear to be taken without undue disturbance, as evidenced by the persistence of animal tracks and burrows and sessile forms in their living positions in the samples (Hessler & Jumars, 1974).

Owing to their large size and great weight box corers are difficult to work, needing a large vessel and calm conditions for safe deployment. The above authors state that 'Launch and recovery are always challenging and potentially dangerous.... Finally, box corers are very expensive, although they are rarely lost and seldom damaged.'

The *LUBS sampler* (Menzies & Rowe, 1968), is a gravity corer fitted with a canvas or nylon bag which is drawn tight to close the lower end of the core to retain the sample. It was designed for quantitative sampling of soft bottom communities in the deep sea. There are three sizes, sampling areas from 617 to 2490 cm^2, and in oceanic oozes penetration is ≥ 20 cm. However, penetration of sand and other compact sediments is poor. The design allows recovery and inspection of undisturbed samples, each being retained in its own container to facilitate subsequent study.

There are a number of other corers which can be used in certain situations. These include the *Haps corer* (Kanneworff & Nicolaisen, 1973) and a number of narrower diameter corers more appropriate for meiofauna or geological studies—these are referred to in Chapter 7.

Large diameter (> 10 cm) gravity, piston or vibrocorers designed for geological work (McManus, 1965; Reineck, 1967; Burke, 1968) are, on the whole, untried for biological work. However, their deep penetration, relatively large sampling area and effective core catchers are features which could make them suitable for macrofauna sampling.

Suction samplers

A number of samplers employ suction, either to force a coring tube down into the substratum or to draw the sediment and its fauna up into a tube leading to some form of self-sieving collector.

The *Knudsen sampler* (Knudsen, 1927) was the first sampler to use suction to take a core of a size suitable for sampling the macrofauna. A pump attached to the top of a wide coring tube (36 cm diameter × 30 cm long) is actuated by unwinding cable from a drum when the sampler is on the sea bed. On lifting, the coring tube is inverted as it comes out of the bottom, so retaining the sample.

The Knudsen sampler meets many of the specifications for the perfect sampler, but may anchor itself in the sea bed so firmly that the cable parts when attempts are made to break it out of the bottom. It can only be used under calm conditions, but any tendency to fall over is minimized in Barnett's (1969) modification in which the sampler is supported by a pipe frame (Fig. 6.20). The same principle of suction sampling was adopted by Kaplan *et al.* (1974) who describe a manual and an automatic sampler for use to water depths of 3·5 m, and by Thayer *et al.* (1975), who describe a more complex device with a similar depth capability.

A number of samplers use pumped water to suck up samples of sediment and fauna. For example, the *Benthic Suction Sampler* of True *et al.* (1968) employs a jet of water acting through a venturi to suck a coring tube of 0·1 m^2 area into the bottom, the sediment and fauna being drawn up into a wire mesh collecting basket. The instrument is powered by a submerged electric motor or by pressure hose from the ship, and has been successfully operated in deep water from a submersible. Other remotely controlled suction samplers are described by Emig & Lienhart (1971), and Emig (1977) (Fig. 6.21), while diver-controlled suction samplers are described in Chapter 5. van Arkel & Mulder (1975) describe a hand-held corer, working on a counterflush system, which can be used from a small boat in shallow water (Fig. 6.22). A modification of this system, which enables the corer to suck itself into the sediment, is described by Mulder & van Arkel (1980). An instrument for deep sampling to 65–70 cm in intertidal sands is described by Grussendorf (1981), and another shallow water sampling system is described by Larsen (1974).

Fig. 6.20. Knudsen sampler fitted with framework to keep it upright on the sea bed (Barnett, 1969). The sampler is operated by unwinding cable from the drum at the top. This operates a pump which sucks the wide coring tube down into the sediment. When hoisted off the bottom, strain on the wishbone arm causes the sampling tube to invert as it comes out of the sediment. (Photograph E. Elliott.)

Drake & Elliott (1982) have carried out comparative studies on three air-lift samplers for use in rivers.

There are a number of suction samplers which are towed slowly along the bottom, sediment and fauna being sucked into a collecting bag. These include the vacuum sled of Allen & Hudson (1970), used for making quantitative estimates of young pink shrimps buried in the sand. In addition, there are suction dredgers used for commercial harvesting of shellfish (e.g. Pickett, 1973) (Fig. 6.23).

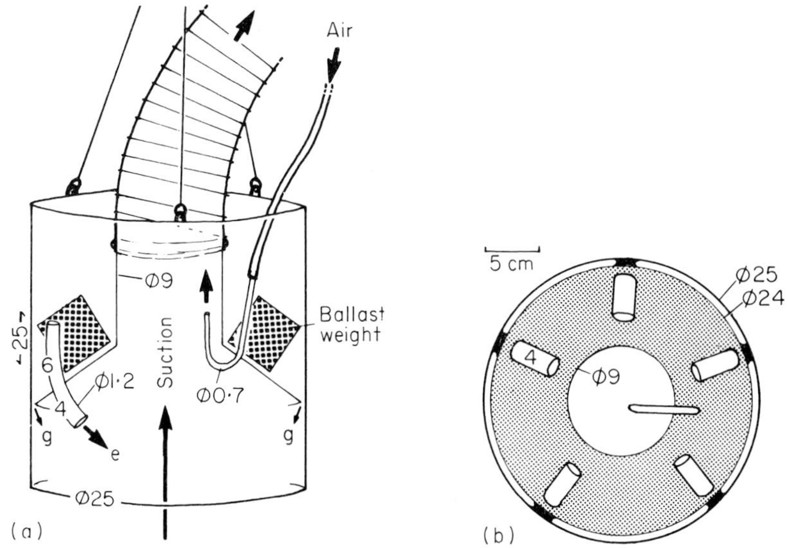

Fig. 6.21. Suction sampler of Emig (1977). (a) In profile; (b) from below. Air introduced through the small tube produces suction in the central tube, through which sediment and fauna is drawn up. The five compensating tubes (e) and the gap between cylinder and cone (g) are provided not only to enable the tube to dig into the substratum but also to help bring the sediment into suspension so that it is more easily collected. Diameters of the various tubes, in centimetres, are shown.

Other methods of sampling

Some species are not readily taken by conventional sampling gear such as dredges, trawls or grabs, either because they occur too sparsely to be represented in samples covering a limited area or because they live in habitats inadequately sampled by the instrument employed. Alternative methods, which do not necessarily sample other members of the fauna, are available for such species. Techniques for underwater photography and television, of special value in estimating scarce members of the epifauna, are described in Chapter 4.

After storms, burrowing animals are often washed in from shallow water on to the beach, and this may be the best means of obtaining some deep-burrowing species not readily taken by ship-borne samplers. Empty mollusc shells and other remains cast up on the beach are usually some guide to the nature of the shallow-water offshore fauna. Occurrence of burrowing species on the sediment surface following dinoflagellate blooms has been noted by Dyer *et al.* (1983) and others.

Fish stomachs often contain deep-burrowing or active members of the benthos seldom taken by sampling instruments, but as the fish are likely to have been feeding selectively little idea of the abundance of the prey species

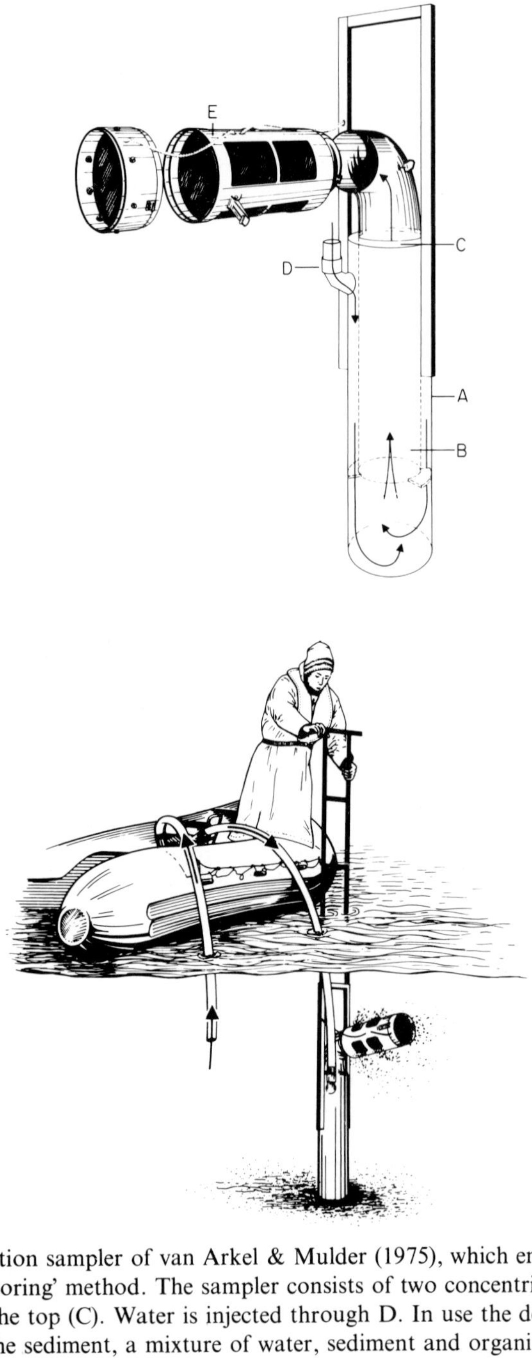

Fig. 6.22. Suction sampler of van Arkel & Mulder (1975), which employs the 'counterflush coring' method. The sampler consists of two concentric pipes (A and B), united at the top (C). Water is injected through D. In use the device is pushed steadily into the sediment, a mixture of water, sediment and organisms passing up pipe B to the cylindrical sieve (E).

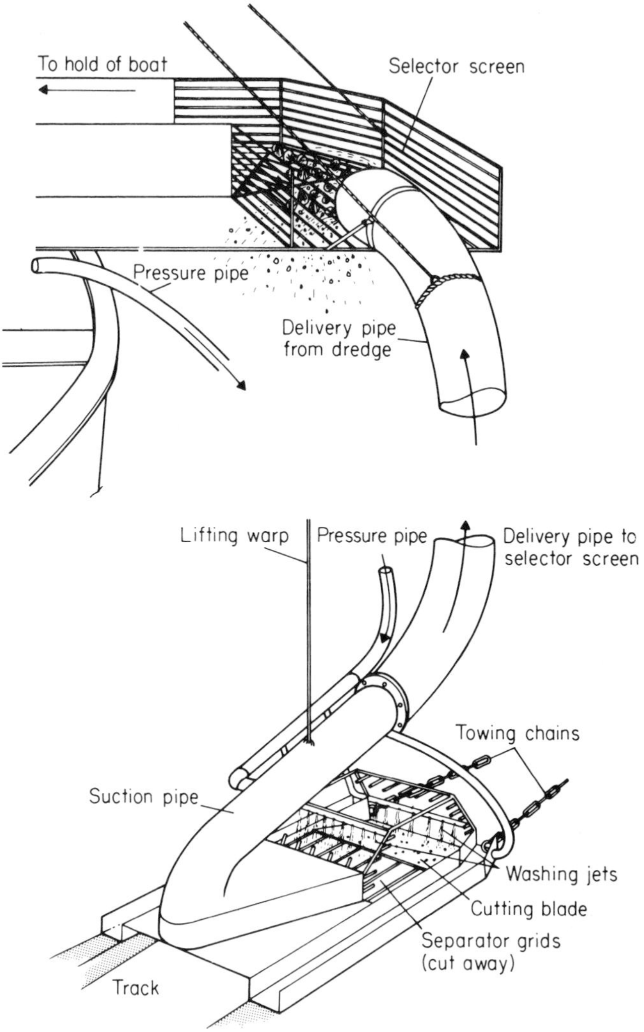

Fig. 6.23. Hydraulic dredge used for harvesting cockles in the Thames Estuary. The lower part is on the sea bed, the upper part on the boat. (From Pickett, 1973.)

can be obtained by this means. Similarly, the stomachs of birds feeding in estuaries and salt marshes may show the presence of otherwise unreported species.

Species of fish and crustaceans which hide away in rock crevices may be taken by using one of the chemical fish collectors which drive them out of their hiding places. Some such as Rotenone are toxic, but the Quinaldine compounds (Gibson, 1967), have an anaesthetic action and usually cause no permanent damage. The range of chemical techniques available, together with other methods such as use of explosives, are reviewed by Lagler (1978), and by Russell *et al.* (1978) for coral reefs. An electrofishing technique, claimed to be less selective for sampling rock lobsters in that 'soft' individuals which have moulted are also taken, is described by Phillips & Scolaro (1980).

Tagging and marking techniques have been used for studies of growth, migration and population size of fish and invertebrates. A review of techniques is given by Jones (1979); see also pp. 126–7 in Chapter 5.

Free-living invertebrates may be captured in baited traps or with light lures. Different types of trap are described in the commercial fisheries literature, including Davis (1958), von Brandt (1972), FAO (1972), Kawamura & Bagarinao (1980) and Motoh (1980). Traps for invertebrates and small fish, some with light, are described by Zismann (1969), Beamish (1972), Espinosa & Clark (1972); Ervin & Haines (1972), Thomas & Jelley (1972), and Haahtela (1978). Traps for the deep sea are described by Albert, Prince of Monaco (1932), Isaacs & Schick (1960), Paul (1973) and Shulenberger & Hessler (1974), and the use of bait to attract deep-sea animals into the field of view of a time-lapse camera is described by Isaacs & Schwartzlose (1975). The occurrence of fish and amphipods around baited traps has been recorded photographically by Dayton & Hessler (1972), Dahl *et al.* (1976), Hessler *et al.* (1978) and Thurston (1979).

There are many active animals, both fish and invertebrates, which are poorly sampled because they actively avoid trawls and other towed gear. There is still a need for the development of methods for their assessment, although photographic methods are giving good results for some species (e.g. Kanneworff, 1979). An attempt to sample such populations by a cage lowered on to the bottom is described by Van Cleve *et al.* (1966), and, on a smaller scale, the use of throw traps for sampling small fish in shallow marshes is described by Kushlan (1981). References to small drop and pull-up traps are given by Aneer & Nellbring (1977).

Working sampling gear at sea

Continental shelf

In the earlier part of this chapter (p. 155) comments were made on the working of dredges and other towed gear. Grabs and other instruments operated vertically require a different technique. When using a grab of moderate weight (up to, say, 100–150 kg) it is important not to use too thick a wire for lowering: many grabs are not hydrodynamically shaped and sink rather slowly, so that if too heavy a wire is used it may form a loop below the grab, causing kinking or entanglement. For the same reason the grab should be lowered at steady speed with gentle braking (or reverse torque) on the winch. This also helps to prevent the wire slackening as the ship rolls, which can trip the release prematurely. A light grab can be worked on a 6 mm wire, but some workers prefer to use a rope, which overcomes some of the above difficulties.

As soon as bottom is reached, paying out should be stopped, and hauling should commence immediately. Any delay will increase wire angle if the ship is drifting, causing the instrument to be pulled out obliquely, so that it samples less effectively. It is, however, important to haul in slowly and steadily until the sampler has left bottom because, with most grabs, closing is completed as hauling commences, so that if the warp is suddenly pulled up the grab will tend to be jerked out of the bottom while it is still closing. In addition, great strains are produced in breaking out a sampler which has dug deeply into the sediment, and these may cause the gear to buckle or the warp to part.

Deep sea

Successful sampling in the deep sea with grab, dredge, or trawl at the end of several kilometres of wire requires special skills, and as Menzies (1964) has emphasized, the proportion of failures, particularly with towed gear, has been high. Sampling in the deep sea is a time-consuming operation, and presupposes a research ship of adequate size, equipped with a deep-sea winch, and sophisticated position-fixing and echo-sounding equipment.

The importance of selecting the optimum size of wire for working the gear has already been indicated, and in deep-sea work the correct choice may make all the difference between success and failure. Ocean-going research ships are often equipped with a wire of 10–12 mm diameter, sometimes tapered (p. 179), for dredging and coring, but such a wire is too heavy for many grabs, which are best worked on a 6 mm wire. The Okean grab has been successfully used on a 4·7 mm wire on Russian ships.

With a wire of small diameter it should be possible to detect when the grab has reached bottom, and a strain gauge linked to a chart recorder is an asset for this purpose. Since the lowering wire is never quite vertical, the length of wire paid out can only be an approximate guide as to when bottom will be reached, and for precise control an acoustic pinger, which shows the position of the gear above the bottom at all times, should be attached to the wire a short distance above the gear, as described on pp. 35–7.

Box corers and other very heavy equipment used vertically require special precautions to prevent premature release. For example, Hessler & Jumars (1974) used a pressure-powered safety device designed to actuate at 1000 m off bottom to prevent accidental release while launching and during the early part of the descent. They describe the method of operating a box corer (with a pinger attached to the wire), and point out that bottom contact as indicated by a drop in wire tension will precede indication of contact in the pinger record by several seconds, because of the time required for acoustic impulses to travel through the water.

Dredging and trawling in the deep sea present particular problems, and Menzies (1964) has considered these in detail, although at that time pingers were not in general use. The main problems with towed gear relate to the rate of paying out of the wire, the total length paid out, and the ship's course and speed throughout the operation. Failure to take a sample may be due either to paying out too much warp, or at too high a speed, so that the wire becomes entangled with itself or with the gear, or too little may be paid out so that bottom is never reached. For dredges and Agassiz trawls the gear can either be lowered from a stationary ship, or the ship can go ahead slowly while shooting is underway, which is essential for working otter trawls (Laubier *et al.*, 1971). A heavy weight attached to the wire some distance ahead of the dredge aids descent, and gives a more horizontal pull on the gear, so reducing the length of wire to be paid out. Little and Mullins (1964) have shown that diving plates increase the speed of descent of a beam trawl, and reduce the length of wire required.

As to speeds of lowering, on the Swedish deep-sea expedition an otter trawl was lowered at 0·7–0·8 m sec^{-1}, and hauled at 0·4 m sec^{-1} (Nybelin, 1951). More recent data on lowering speeds and technique are given by Rowe & Menzies (1967), Carey & Heyamoto (1972) and Laubier *et al.* (1971).

If the net is not streamed through the water during lowering there is a risk of it becoming wrapped around its metal frame, which will prevent proper sampling. A simple expedient is to attach a heavy weight to the end of the net (Fig. 6.9). However most deep-sea trawls are now enclosed within a rigid framework (Menzies, 1964) to overcome this problem.

Kullenberg (1951) made calculations of the length of wire of various diameters required for trawling at different depths. However because of the uncertainties inherent in estimating ship's speed, and the likelihood of cross-currents at different depths, his figures can be taken as only a rough guide. It is now generally agreed that the most satisfactory way of monitoring the behaviour of a trawl or dredge is by means of a pinger placed some distance up the wire ahead of the gear, combined with the use of a recording strain gauge. Under such conditions sound signals from the pinger are oblique (Fig. 6.24) and may need to be received by a non-directional hydrophone, but they can provide precise information on time of bottom contact and duration of tow. Examples of use of pingers on towed gear are given by Backus (1966), Rowe & Menzies (1967), Laubier *et al.* (1971) and Gaunt & Wilson (1975). When working in the deep sea, and particularly when pingers and strain gauges are in use, it should be possible to tow on a wire length equal to little more than the depth of water. Laughton (1967) gives ratios of wire out:depth of between 1·1 and 1·2, and further information is given by Rowe & Menzies (1967), Laubier *et al.* (1971), and Aldred *et al.* (1976).

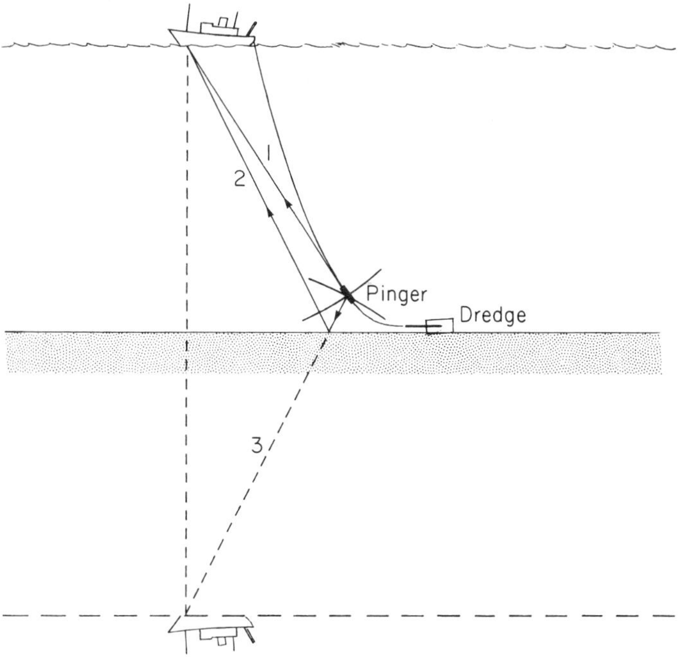

Fig. 6.24. Real (1, 2) and virtual (3) paths of sound from a pinger used to monitor dredging. (Redrawn from Laubier *et al.*, 1971.)

Even when the trawl or dredge has reached bottom it may not function satisfactorily. Menzies (1972; also Menzies *et al.*, 1973) describe results from mounting a forward-facing time-lapse camera in the mouth of a 1·52 m trawl. It was found that in a 90 minute tow the net was fishing normally for only about 4 minutes, and that most of the time was spent 'twisting, flipping and flopping over the sea bed, spilling its contents back on the sea floor, or being buried mouth first in the mud.' (Fig. 6.25.)

Aldred *et al.* (1976) and Rice *et al.* (1982) describe an epibenthic sledge (Fig. 6.5), with pinger attached to the frame, which monitors its performance on the bottom. This would seem to overcome many of the uncertainties attached to the achievement of satisfactory results with such gear.

Exploration of canyons on the continental slope requires special care. Where the bottom is steeply sloping it may not be possible to get a reliable depth sounding, because of echoes from the side walls of the canyon, and it may be difficult to place the gear in the required location on the sea bed. Unless the canyon is particularly well charted, and accurate position fixes can be obtained, working any type of gear may be hazardous. Because of the risk of loss it is inadvisable to use expensive instruments such as pingers, and indeed pinger signals may give a false impression by failing to give a return echo from vertical cliff-faces towards which the gear may be drifting. Cousteau's diving

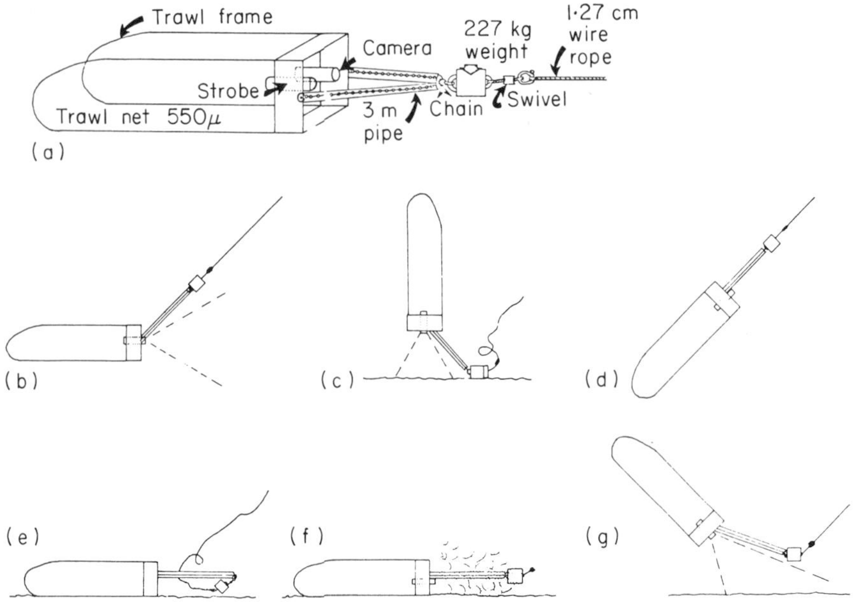

Fig. 6.25. Behaviour of a deep-sea trawl on the sea bed as indicated by a camera positioned in its mouth. (a) Diagram of 1·52 m beam trawl showing position of camera and strobe; (b) trawl being lowered, photographs show clear water and no bridle in picture; (c) trawl upset, camera faces bottom, of which it takes a vertical shot; (d) trawl on way up, bridle and clear water or mud falling off show in photograph; (e) trawl on bottom, weight drops with bridle towards trawl mouth; (f) trawling, mud clouds in photograph, bridle out of sight; (g) trawl oblique, camera takes oblique view of bottom. (From Menzies, 1972.)

saucer has been successfully used for exploration of submarine canyons (Shephard *et al.*, 1964; Shephard, 1965), and an ingenious device for photographing the side walls of a canyon is described by McAllister (1957). The latter would seem to be appropriate only for fairly shallow water and where the topography of the canyon is well known.

For general purpose collecting on the slope and in canyons a sturdy rock-collecting dredge (Fig. 6.9), with safety link can be used. This should be lowered vertically from the ship, which then drifts or steams slowly towards the canyon wall. Manoeuvering of the ship at slow speeds is more easily accomplished if the ship is steaming against the surface current. Exploration of canyons can only be carried out in good weather, and even then frequent losses of gear are to be expected.

One other aspect of deep sea sampling has been emphasized by Hessler & Sanders (1967) who point out that often much of the sample may be lost by winnowing (washing out) of the lighter components during hauling. Wherever practicable the sample should be covered when hauling. When sampling sediments it is best if the net bag is lined with closely woven material which will retain the sediment. This allows many of the smaller organisms which might otherwise be washed away to be brought to the surface still within the sediment, which not only gives mechanical protection but also buffers against sudden temperature changes as the sample is hauled through warm surface waters.

The relative sparsity of the deep-sea fauna and the risk that a grab will not close, so wasting much time, have encouraged the use of dredges, trawls and other towed gear such as the epibenthic sledge. These almost invariably obtain large samples, often rich both in species and individuals, and so provide more material to work on than could otherwise be obtained from the deep sea in the necessarily limited amount of ship's time available. For the use of submersibles for observation and sampling see pp. 127–8.

Note on warps

Emphasis has already been placed on the need for selecting warp of the appropriate diameter, and material, for a particular purpose.

When gear is to be hauled by hand or over a capstan head natural fibre or synthetic ropes should be used. Synthetic ropes, which are now in almost universal use, are very strong and the choice of a suitable size is determined more by ease of handling than by the breaking strain. The minimum diameter for hand-hauling is 10 mm, but 12 mm or larger is easier to hold. The only disadvantage of synthetic ropes as compared with wire are their greater bulk in storage and the possibility that they may get cut.

Most dredging, coring and grab sampling is carried out with galvanized steel warps. General purpose warps are of 6 × 19 construction, of steel with a nominal breaking strain of 145 kg mm^{-2}, but for deep-water work where maximum strength is required steel of 180 or 200 kg mm^{-2} is preferable, and the normal hemp core can be replaced by one of steel.

For hydrographic work 6 mm diameter wire is now used (in preference to 4 mm), and light corers and grabs can be used on the same wire. A selection of ropes and wires for different purposes is given in Table 6.1 (p. 154).

For deep-sea work a tapered warp is commonly used, as the wire nearest the surface has to support the weight of that paid out in addition to strains imposed by the gear. As an example, the R.V. *John Murray* was fitted with 9000 m of wire, with the following specification:

Galvanized flexible steel wire, nominal breaking strength 190–205 kg mm^{-2}. Fibre core.
1st section: 1500 m, 13 mm diameter, 6 × 19 construction, breaking strain 10262 kg.

2nd section: 3000 m, 12 mm diameter, 6 × 17 construction, breaking strain 9043 kg.
3rd section: 4300 m, 11 mm diameter, 6 × 12 construction, breaking strain 7824 kg.

Recovery of lost gear

With the increasing sophistication of underwater gear and the development of 'instrument packages' of electronic equipment, often purpose-made, the action to be taken should such valuable gear be lost or become fouled on the sea bed ought to be anticipated.

Loss of gear can be minimized by ensuring that all wires, shackles, swivels, etc. are in good condition and of appropriate size, and that weak links will fail when required to do so. Very often loss occurs through parting of the wire or associated components close to the underwater towing point either: (a) at the sea surface, through increasing strains as the gear and sample are brought out of the water, or the gear is accidentally hoisted right up into the towing block through failure to stop the winch; or (b) on the sea bed through entanglement with underwater obstructions, the side walls of canyons, etc.

Apart from precautions to minimize loss, the following guidelines are suggested to aid recovery:

(1) Name, address and telephone number should have been conspicuously marked on the gear.
(2) On towed instruments a light line (length equal to 2–3 times depth) with marker float is attached to the gear. This will greatly aid underwater location by a diver, particularly near the limit of diving depth.
(3) A similar pop-up marker, automatically released after a time delay, with light and radio beacon, may be used.
(4) The possibility of designing the equipment so that an instrument package can be released independently of the main framework should be considered. This might be brought up on a light line, or on a pop-up buoy system with preset time-delay.
(5) On vertically lowered gear it may be difficult to attach a marker line because of the likelihood of it twisting around the lowering wire. Under such circumstances, and whenever valuable equipment is involved, an acoustic transponder which can be interrogated by sonar from the ship will allow location after an interval of weeks or even months. Below diving depths the possibility of recovery will depend on the value of the lost gear, and whether a submersible can be deployed for salvage.

It is suggested that by common practice, if not in law, gear on the sea bed which has been clearly marked as indicated above is not 'lost', and could not be the subject of a salvage claim.

It is assumed that at the time of loss the ship's master would log the position, taking account of any transit lines to the shore. In the event of none of the suggested procedures being carried out an anchored marker float (which should have been made ready in advance) should be dropped over the side without delay. Grapnels and trawls have, on occasion, been used with success for recovery of lost gear.

Efficiency of benthos sampling gear

The efficiency with which a sampler operates is related both to its design and to its mode of operation. Sampling efficiency is a useful concept only when referring to quantitative or semi-quantitative gear.

Dredges and trawls

Most dredges, defined as collecting instruments which are towed along the bottom, are at best semi-quantitative. When used to collect fauna living on or just above the bottom, the efficiency of a dredge, judged by its ability to capture all the animals within its sweep, is usually low. The performance will vary with the configuration and nature of the bottom and since several types of ground may be encountered on any given tow, it is difficult to allow for such variations. Other complicating factors are the behaviour of the ship, the length of warp, and the speed of towing—increased speed above a low level usually reduces catches. Further, the type of warp used can affect performance. Wire, with a weight in water greater than the drag, can increase the effective weight of the dredge on the bottom, while rope, with the drag greater than the weight and a consequent backward and upward catenary, can produce a lift. The fitting of depressors or diving plates, tickler chains and the proper use of teeth on the leading edge can increase efficiency (Baird, 1959). Attempts have been made to employ odometers to measure the exact distance covered by the dredge on the bottom (see p. 156), which should help to quantify results.

The behaviour of the animals themselves is also of importance and if sampling is related to only one species, with consequent reduction in the range of habitat and behaviour encountered, it should be possible to make appropriate gear modifications to increase efficiency. Yet Dickie (1955) has shown that dredges specifically designed to capture scallops had an efficiency of only 5% on uneven inshore grounds and just over 12% on smoother offshore areas. Somewhat higher efficiencies were reported by Chapman *et al.* (1977), however. Again, juvenile stages of a flatfish have recently been the subject of population studies in Britain, and gear has been developed for their capture. Since they occur on relatively flat sandy grounds in shallow water where their behaviour can be observed by divers, it may be expected that high gear efficiencies should be possible. Riley & Corlett (1965) used a 4 m beam trawl and found it worked best when towed at a speed of 35 m per minute with three tickler chains attached. Efficiency ranged from 33 to 57 %, and even for fish in their first year of life it varied considerably at different times of the year, depending on the size and age of the fish. In other experiments Edwards & Steele (1968) found a catching efficiency of a 2 m beam trawl for plaice to range between 23 and 37 %. Efficiencies of this order, however, seem to be exceptional for dredges, and when total fauna is considered, a value nearer 10 % is probably more realistic (Richards & Riley, 1967).

Grabs

The concept of efficiency is more meaningful for grabs, which are lowered on (ideally) a vertical warp from a stationary ship, to take a deposit sample of a given surface area. In this context, the term efficiency tends to be used loosely to cover various purely functional aspects of an instrument's general performance and digging characteristics, as well as its ability to produce an acceptable picture of animal density and distribution. Although all these uses of the term are related, they should not be confused.

Performance

Considering first the functional aspect, this refers primarily to the ability of a grab to perform consistently and correctly, according to its design, in all conditions of deposit, depth, and weather. These features are perhaps better covered by the word 'reliability' rather than 'efficiency' and are best judged by the volume of deposit collected, so that an instrument which filled to capacity on every haul would be regarded as completely reliable within the limits of its design. The first requirement for reliability is that the grab must land on the bottom in a condition in which it will operate properly. Any grab which is activated by the slackening of the warp when the gear strikes bottom will tend to be set off in mid-water by the roll of the ship, and so may be difficult to use in bad weather or deep water. An instrument which is not stable when in an upright position on the bottom, or which must remain at rest for a time to collect the sample (e.g. the Knudsen sampler) may be upset by strong currents or by an oblique upward pull on the warp. Once correctly on the sea bed, a grab-type instrument covers a known surface area, and, assuming that it can be raised by a reasonably vertical warp and not pulled laterally off the bottom (the skill and experience of the operator is often important here, and comments on the use of grabs at sea are given on pp. 174–5, then the extent to which it attains its maximum depth and, therefore greatest volume of sample, depends largely on the weight of the instrument and the nature of the substratum. On soft mud most grabs will fill completely, but on the most difficult grounds of hard-packed sand conventional grabs will merely scrape the surface unless they are adequately weighted, and even those instruments which achieve some initial penetration by spring-loading (Smith & McIntyre, 1954; Briba & Reys, 1966) will be raised off the bottom if they are not heavy enough. If all of these factors are satisfactory the volume of deposit will serve as an index of the depth of penetration, but will not give an absolute measure unless the exact profile of the bite is known.

Given that a particular grab is fully reliable, as discussed above, one would further wish to know, still dealing with the functional aspect, how well its design allows it to collect the deposit below the surface area which it covers. This is the sense in which Birkett (1958) used the term 'efficiency', which he defined as the ratio between the volume of sediment collected and the theoretical volume, which is calculated by multiplying the area covered by the deepest penetration depth. This could perhaps be called the index of digging performance.

In recent years a generation of suction samplers has been introduced, working on the principle that the deposit can be raised by jets of air or water (p. 169). Such samplers tend to have a high digging performance (especially if operated by divers—see p. 108), since they can lift all of the deposit to a given penetration depth from within their area of operation. In contrast, it was considered until recently that the biting profile of grabs with horizontally placed spindles was more or less semi-circular and that the digging performance of such grabs was, therefore, low, since the deepest part of the bite sampled only a fraction of the surface area spanned by the open jaws. Observations by Gallardo (1965), Lie & Pamatmat (1965) and Ankar (1977a), however, indicate that the profile is more nearly rectangular, and that divergencies from a true rectangle can be explained in terms of the closing mechanisms of the various grabs. Thus the Petersen and van Veen, which have an upward leverage as the jaws close, tend to leave a hump of deposit in the middle of the sampling area, while the Smith–McIntyre, with a downward pull on the arms as the jaws close, digs deeper in the middle of the sampling area and thus takes a rather larger volume than the van Veen, for the same degree of penetration. These grabs thus appear to have higher digging performance than had been supposed, and on hard-packed sands this can be increased by the addition of extra weights.

Efficiency of capture

The second aspect of efficiency is more complex and relates to the ability of the gear to collect the fauna so as to give a reasonably accurate picture of its density and distribution. This may be called the efficiency of capture. Few studies have been made of how efficient a particular grab is in capturing all the fauna below a given surface area. This was attempted by Lie & Pamatmat (1965), who compared collections taken with a 0·1 m^2 van Veen grab at high tide with hand-dug samples of the same unit area at low tide. They showed that for the most abundant species there were significant differences between the two sets of samples in only 8 out of 37 cases.

Efficiency of capture may be defined as the ratio of the number of animals in the volume of deposit collected by the grab to the number present in the same volume *in situ*. By using as the denominator of this ratio the same volume as was collected rather than the theoretical volume based on a rectangular bite, the ratio excludes the functional efficiency of the grab and deals only with its ability to capture the fauna available to it.

One possible cause of low efficiency of capture is the downrush of water caused by the grab's descent, which may disturb the surface of the deposit and result in the loss of superficial fauna. Smith & McIntyre (1954) considered that the use of gauze windows on the upper surface of their grab reduced this, producing higher catches of small crustaceans on sandy grounds, and this is supported by experimental work by Wigley (1967) who studied the behaviour of grabs by motion pictures. On muddy grounds disturbance of the surface layers may present a serious problem, as has been shown by Andersin & Sandler (1981): comparison of the efficiency of two types of van Veen grab showed that large windows at the upper surface of the buckets minimized the shock wave, resulting in increased efficiency—by an average of 50 % for small amphipods and by 80 % for small polychaetes, as compared to a grab with small windows. It is probable, however, that most sampling instruments, even when landed carefully on the bottom, will not sample the fine surface layer adequately. Apart from this aspect, even a sampler which fulfills the criteria of reliability as described above, i.e. which consistently takes its maximum volume, may not be the most suitable for every task. For example, one with a low or medium maximum penetration will not sample adequately the deep burrowing animals, and an instrument which covers only a small surface area, no matter how efficient as a machine, may be quite unsuitable for sampling widely dispersed species.

In conclusion, selection of the most 'efficient' grab involves consideration of reliability, digging performance, and capture efficiency which are, in turn, influenced by such factors as the size of the sample required, type of deposit, depth of water, prevailing weather conditions, handling facilities available, and the experience of the operator.

Corers

From the purely functional aspect, coring devices have a higher digging performance and most tend to be relatively reliable in that a particular instrument of a given weight will usually provide consistently similar lengths of core, depending on the type of sediment sampled. The main difficulty may be loss of the sample during ascent, and on sandy grounds a core retainer is usually required.

As a means of providing quantitative data on the fauna, the main criterion of a corer's efficiency must be the accuracy with which the collected core represents the sediment column as it was *in situ*. Only the larger versions of the different types of corer used in meiofaunal and geological studies are suitable for the efficient sampling of macrofauna. With wide cores (> 10 cm diameter), core shortening due to wall friction, observed with the narrow diameter corers, is not noticeable and undisturbed cores can be obtained. The downwash of a corer, although less than that produced by a grab, will nevertheless lead to some loss of the superficial layer of most sediments.

In conclusion, the most efficient corer will be one which takes relatively undisturbed samples, penetrates deeply, and shows the smallest loss of surface layers. Box corers (p. 167), conforming to most of these characteristics have a distinct advantage over many other core samplers.

Comparative efficiency

Many workers have tested sampling devices and provided information as to their comparative efficiency. The criteria used in these comparisons are:

(1) The digging characteristics of the sampler (depth of penetration, volume of sediment, degree of disturbance).
(2) The efficiency of capture in order to give a representative picture of the density and distribution of the fauna.
(3) Technical characteristics of the samplers (ease of manipulation, weight, ease of access to the sample, mechanical reliability, etc.).

A selection of works dealing with the comparative efficiency of different samplers and sampling methods is given in Table 6.2, and a more complete bibliography of works on the comparative efficiency of samplers can be found in Elliott & Tullet (1978).

Choice of a sampler

The choice of a sampler must necessarily be a compromise, based on requirements of the survey, working conditions, and the availability of suitable gear. Table 6.3 represents an attempt to show the main characteristics of the samplers described in this chapter. Choice may often be restricted by weight, and the heavier equipment commonly used for deep-sea work (large grabs, epibenthic sledges, box corers) can only be worked from large research

Table 6.2. Guide to literature in which efficiency of different sampling gear and methods is compared. Cross-referencing is obtained by reading vertically and horizontally from the name of each type of gear. The categories are broad, and where different sizes or variants of the same sampler have been tested against each other the references are shown in brackets after the name. However, comparative hauls between different types of otter trawl are not included. Descriptions and references to the different types of sampling gear are in Chapter 6.

	Otter trawl	Agassiz trawl	Beam trawl	Dredge	Anchor dredge	Sledges [53]	Petersen grab	Campbell grab	Okean grab	van Veen grab [1, 3, 7]	Ponar grab	Hunter grab	Smith–McIntyre grab	Day grab	Orange-peel grab
Otter trawl															
Agassiz trawl															
Beam trawl															
Dredge			18												
Anchor dredge															
Sledges [53]					16,53										
Petersen grab			18	18											
Campbell grab						31									
Okean grab							24								
van Veen grab [1, 3, 7]		27			16	16	7,13, 17,26, 42,44		2						
Ponar grab							13,34, 39			13,51					
Hunter grab															
Smith–McIntyre grab					11		17,34, 41,50			6,17, 41,47, 51	34,51	23			
Day grab													43		
Orange-peel grab										35,51	22,51		8,29, 51		
Baird grab													19		

												54														Hamon grab
				53	53	20,21			35					35												Holme grab
						39			51	15,39, 51		51		51												Shipek grab
						4,13, 39			13,38	13,15, 22,39				22				15,39								Ekman grab [4, 15]
									5,51	51		40,51		51				51								Box/Spade corers
					31		31																			LUBS sampler
						25			45																	Knudsen sampler
				16	16	39			16,38, 52	15,39								15,39	15,33, 38,39							Corers
						21			9			29		28,29, 36			21									Suction samplers [28]
12,46	27		49		37		14,30, 32,48		27																	Photography
46																									46	Submersibles

The references are numbered as follows: 1. Andersin & Sandler (1981); 2. Ankar *et al.* (1978); 3. Ankar *et al.* (1979); 4. Bakanov (1977); 5. Beukema (1974); 6. Bhaud & Duchêne (1977); 7. Birkett (1958); 8. Bourcier *et al.* (1975); 9. Christie & Allen (1972); 10, Dickie (1955); 11. Dickinson & Carey (1975); 12. Dyer *et al.* (1982); 13. Elliott & Drake (1981); 14. Emery *et al.* (1965); 15. Flannagan (1970); 16. Gage (1975); 17. Gallardo (1965); 18. Gilat (1968); 19. Higgins (1972); 20. Holme (1949); 21. Holme (1953); 22. Hudson (1970); 23. Hunter & Simpson (1976); 24. Ivanov (1965); 25. Johansen (1927); 26. Kutty & Desai (1968); 27. McIntyre (1956); 28. Massé (1967); 29. Massé *et al.* (1977); 30. Menzies *et al.* (1963); 31. Menzies & Rowe (1968); 32. Owen *et al.* (1967); 33. Paterson & Fernando (1971); 34. Powers & Robertson (1967); 35. Reys (1964); 36. Reys & Salvat (1971); 37. Rice *et al.* (1982); 38. Rowe & Clifford (1973); 39. Sly (1969); 40. Smith & Howard (1972); 41. Smith & McIntyre (1954); 42. Thamdrup (1938); 43. Tyler & Shackley (1978); 44. Ursin (1954); 45. Ursin (1956); 46. Uzmann *et al.* (1977); 47. Wigley (1967); 48. Wigley & Emery (1967); 49. Wigley & Theroux (1970); 50. Wildish (1978); 51. Word (1976); 52. Jensen (1981); 53. Huberdeau & Brunel (1982); 54. Dauvin (1979).

Table 6.3. Suitability of different sampling gear for various applications.

Gear	Weight	Sample		Quanti-tative?	Depth of sample (firm sand)	Deposit			Sea depth			Difficult sea conditions	Ship size
		Width (m)	Area (m^2)			Rock/stones	Firm sand	Mud	Shallow	Shelf	Deep sea		
Beam trawl	L–M	2–10		SQ†	0	○	+	+	+	+		+	SML
Agassiz trawl	L–M	2–4			0	○	+	+	+	+	+	+	SML
Otter trawl	H	>10			0	○	+	+	+	+		+	SML
Macer–GIROQ sampler	H	0·5		Q	0	○	+	+	+	+			SML
Epibenthic sled (Hessler & Sanders)	H	0·8			0	○	+	+			+		ML
Epibenthic sledge (Aldred *et al.*)	H	2·3		SQ†	0	○	+	+			+		ML
Rectangular dredge	L	0·3–1·3			0	+	+	+	+	+	+	+	SML
Small Biology Trawl (Menzies)	L	1·0			0	○	+	+	+	+	+		ML
Anchor dredge (Forster)	L	0·5		SQ	3	○	+	+	+	+	○		SM
Anchor dredge (Thomas)	L	0·6			2	○	+	+	+	+			SM
Small anchor dredge (Sanders)	L	0·29		SQ	1	○	+	+	+	+	○		SM
Anchor dredge (Sanders *et al.*)	H	0·57		SQ	2	○	+	+		+	+		ML
Anchor-box dredge	H	0·5	1·33	SQ	1	○	+	+		+	+		ML
Petersen grab	L		0·1*	Q	1	○	○	+	+	+		○	SM
Campbell grab	H		0·55	Q	2	○	+	+		+	+		ML
Okean grab	L		0·08*	Q	1	○	+	+	+	+	+		SML
van Veen grab	L		0·1*	Q	1	○	+	+	+	+		○	SM
Ponar grab	L		0·055	Q	1	○	+	+	+	+	○		SM
Hunter grab	L		0·1	Q	1	○	+	+	+	+	○		SM
Smith–McIntyre grab	L		0·1	Q	1	○	+	+	+	+			SM
Day grab	L		0·1	Q	1	○	+	+	+	+			SM

Orange-peel grab	M		Various	Q	1	○	+	+	+	+	+		ML
Baird grab	L		0·5	Q	2	○	+	+	+		○	○	SM
Hamon grab	H		0·29	SQ	2	○	+	+	+	+			ML
Holme grab	M		2 × 0·05	Q	1	○	+	+	+	+			M
Shipek grab	L		0·04	Q	1	○	+	+	+	+	○		SM
Birge–Ekman grab	L		0·04	Q	M1	○		+	+	+			SM
Reineck box sampler	H		0·06*	Q	3	○	+	+	+	+	+		ML
LUBS sampler	M		0·06–0·25	Q	M2	○		+		+	+		ML
Haps corer	L		0·015	Q	M3	○		+	+	+			SM
Knudsen sampler	M		0·1	Q	3	○	+		+	+	○	○	SM
Suction sampler (True *et al.*)	L		0·1	Q	3	○	+		+	+	+‡	○	SM
Suction sampler (Kaplan *et al.*)	L		0·1	Q	3	○	+	+	+	○	○	○	S
Suction sampler (Thayer *et al.*)	L		0·07	Q	3	○	+	+	+	○	○	○	S
Flushing sampler (van Arkel)	L		0·02	Q	3	○	+	+	+	○	○	○	S
Diver-operated suction sampler (Barnett & Hardy)	L		0·1	Q	3	○	+	+	+	○	○	○	
Photography	L		<2	Q	0	+	+	+	+	+	+	○	ML
Television	L	1–2		Q	0	+	+	+	+	+		○	ML
Submersible observation	H	2–10		SQ†	0	+	+	+	+	+	+		
Traps	L	×	×		0	+	+	+	+	+	+		SML

General applications: +, suitable; blank, possible application; ○, unsuitable.

Weight (total with any additional weights included): L, <100 kg; M, 100–200 kg; H, >200 kg.

Sample width and area: * Other sizes available. ×, traps sample an indefinite area.

Quantitative?: Q, quantitative; SQ, semi-quantitative; SQ†, semi-quantitative if odometer wheel fitted.

Depth of sample (penetration of sampler into firm sand): 0, surface sample only; 1, 1–10 cm penetration; 2, 10–20 cm penetration; 3, >20 cm penetration; M, above penetration depths but in soft mud only.

Sea depth: shallow, diving depth (i.e. <30 m); shelf, 30–200 m; deep sea, >200 m (i.e. slope and abyss); ‡ from submersible.

Difficult sea conditions (most sampling gear cannot be used under severe conditions of swell, waves or currents): +, instruments likely to obtain a sample under such conditions; ○, these instruments can only be used under calm conditions and/or the absence of strong currents.

Ship size, S, launch with power hoist; M, trawler; L, large research vessel.

vessels which have the necessary lifting equipment. Many of the lighter samplers, on the other hand, have serious limitations imposed by their being less quantitative, having more limited penetration into firm sediments, or sampling too small an area.

Sampling of firmly-packed sediments presents particular problems, since many instruments fail to penetrate sufficiently deeply to sample all the fauna. If the objective is to sample to ≥ 20 cm, choice is limited to a few samplers, all of which have restrictions on their suitability. The Forster anchor dredge is a useful instrument for this purpose, but it samples most effectively when used in shallow water and from a small launch, and its samples must be considered as only semi-quantitative. Box corers such as the Reineck are only suitable for use from large ships, and the Haps corer samples too small an area ($0{\cdot}015\,m^2$) for general macrofauna studies. The Knudsen sampler takes a sample of adequate depth and surface area, but is unsuitable for open-sea conditions and for deep water. Other suction samplers, whether remote or diver-operated, are mostly restricted to shallow water, and generally sample less well in cohesive muds.

Where quantitative samples are required, using an instrument of only moderate weight, it must be accepted that penetration of firm sediments is unlikely to be adequate to capture the deeper burrowers. Choice is limited to a few grabs, notably the van Veen, Smith–McIntyre, and Day grabs. Most other grabs sample too small an area, do not dig sufficiently deeply, or have other drawbacks not necessarily indicated in the Table. The Baltic Marine Biologists have adopted the $0{\cdot}1\,m^2$ van Veen grab as the standard sampling instrument for macrofauna (Dybern *et al.*, 1976), but this is not the best choice under more open-sea conditions, for which purposes a number of workers have adopted the Smith–McIntyre grab. The more recently introduced Day grab is of simpler construction than the latter, and is without springs, making it safer in use. It appears to dig about as deeply as the Smith–McIntyre grab.

Treatment and sorting of samples

A benthos sample usually consists of a volume of sediment from which the animals must be extracted. Macrofauna samples may vary in size from a few to many litres, and the extraction process is often divided into two stages, the first being carried out in the field with a view to reducing the bulk of material to be taken back to the laboratory, where the second stage of separation takes place.

Initial treatment

At sea, the standard procedure is to receive the bottom sampler on deck on a wooden or metal-lined sieving table or hopper, which should be designed such

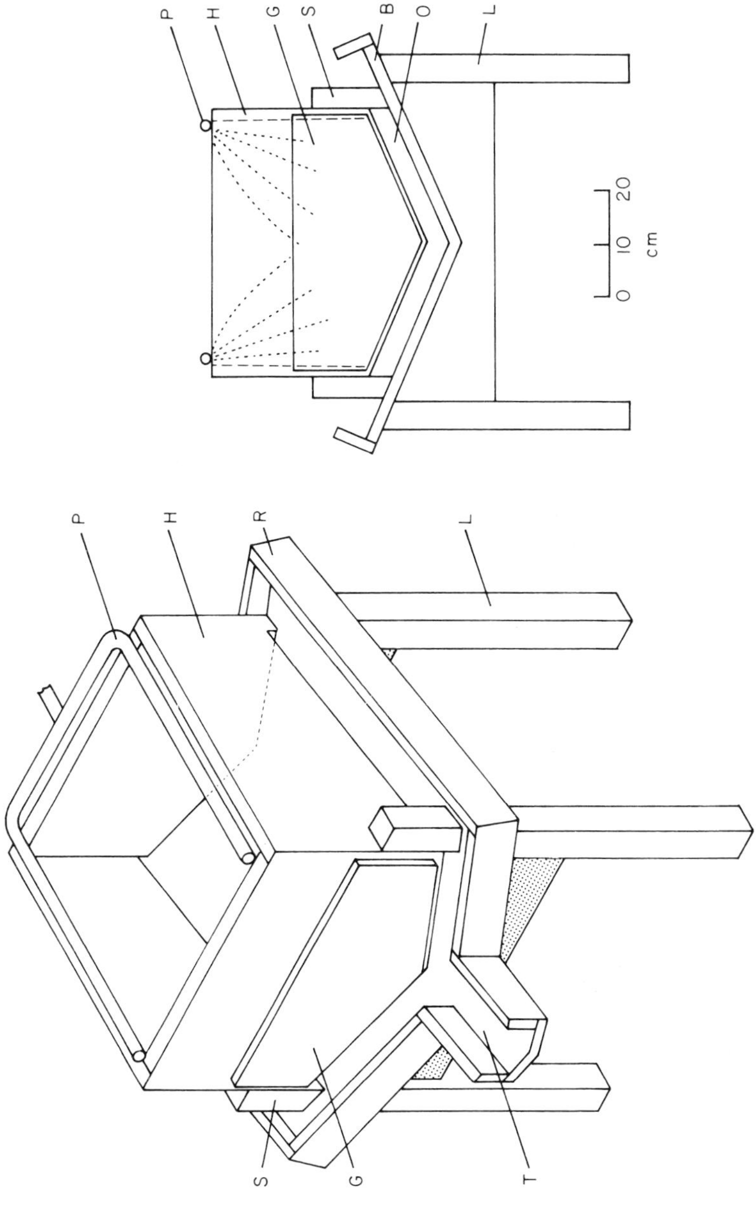

Fig. 6.26. Hopper for treatment of sediment samples (Holme, 1959). Left, general view; right, cross-section. P, pipes supplying jets along top of hopper; H, side-wall of hopper; R, retaining wall at side of base (B); T, spout; G, rising gate; S, short legs supporting hopper off base; L, legs; O, gap between hopper and base.

that the sampler can be emptied and the contents washed through a sieve without loss of material. A measure of the total volume of deposit collected is often required, since this helps in assessment of the performance of the gear. The volume may be measured either by using a dip stick before emptying the sampler, or by arranging that the contents pass into a graduated container before being sieved. Methods of dealing with the sample have been reviewed by Holme (1964), and vary from the simple arrangement of McNeely & Pereyra (1961) which consists of washing the sample through a nest of sieves with a hose, to the elaborate set-up described by Durham (in Hartman, 1955) in which a mechanical shaker agitates a graded series of screens under a set of sprinkler heads. A small hopper (Fig. 6.26) for general use has been described

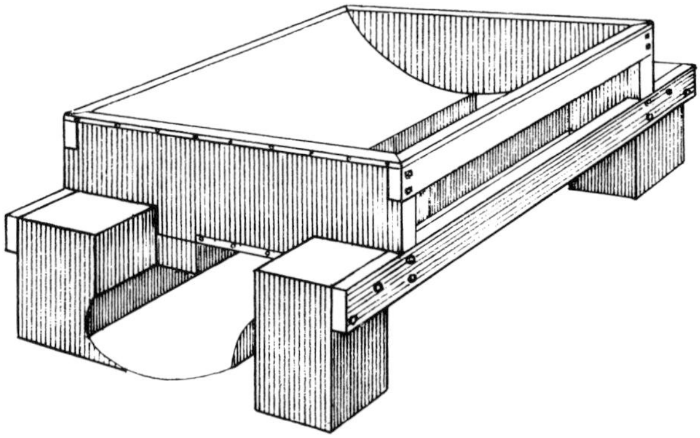

Fig. 6.27. Combined grab cradle and wash trough. (Redrawn from Carey & Paul, 1968.)

by Holme (1959), but there is probably no single set-up which would be satisfactory for the full range of sampling instruments and working conditions, and it may be desirable to modify an existing pattern appropriately to suit particular needs, or to construct a new unit such as the combined grab cradle and wash trough (Fig. 6.27) designed by Carey & Paul (1968) for the Smith–McIntyre grab.

Other mechanical means for washing large numbers of samples are given by Pedrick (1974), who describes a non-metallic device suitable for pollution studies. For the processing of large samples (over 20–30 l) an elaborate system has been used by M.H. Thurston & R.G. Aldred (Institute of Oceanographic Sciences) (Fig. 6.28). The sample is placed in a large tub and, after preliminary hand picking of the larger animals, it is passed through a series of sieve baskets (16, 2, and 0·5 mm mesh) where it is separated by washing and agitation into three separate size classes. Each size class is kept separately and is transferred to containers for fixation. The system, which is operated by one or two

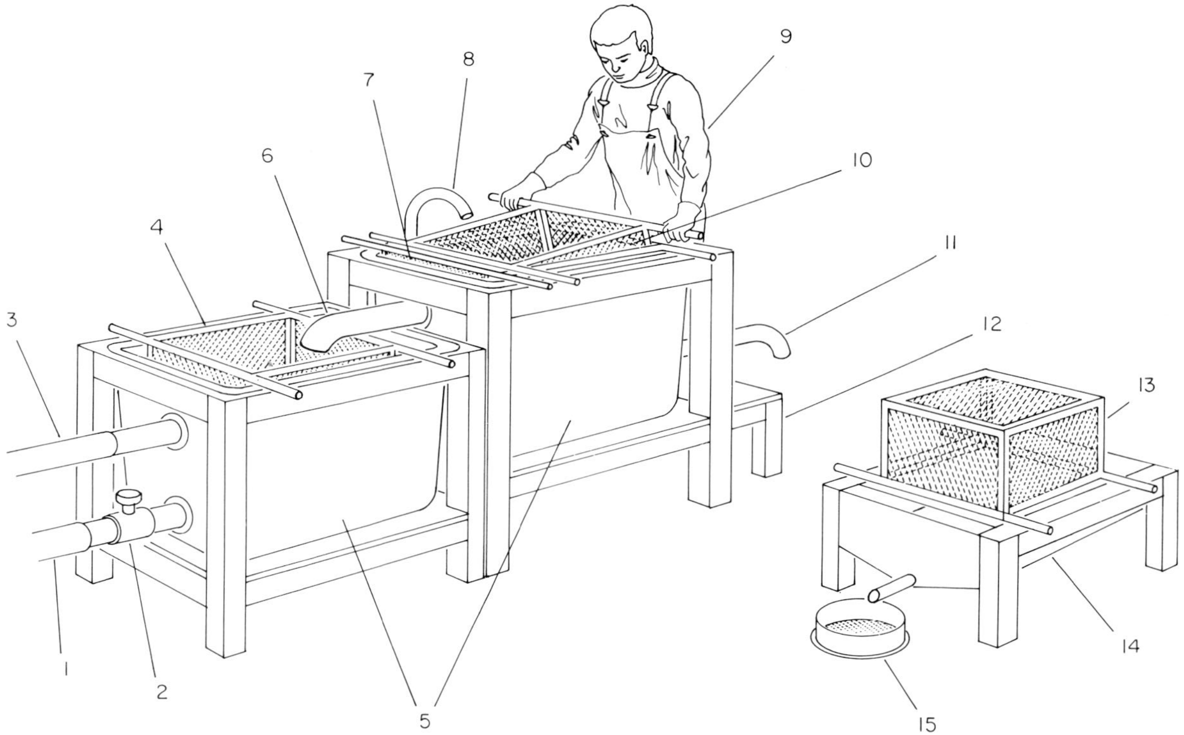

Fig. 6.28. IOS equipment for washing and sieving sediment samples. 1, hose for draining lower tank; 2, valve; 3, outflow hose for lower tank; 4, 500 μm mesh basket; 5, fibreglass tanks; 6, outflow for upper tank; 7, 2 mm mesh basket; 8, water supply; 9, operator agitating sieve basket; 10, 16 mm mesh basket; 11, outlet and valve for draining upper tank; 12, retractable step; 13, inverted sieve basket; 14, basket washing trough with V-section, sloping bottom; 15, sieve.

persons, enables large samples (200–300 l or more) to be processed in a reasonably short time.

The surface area of the sieve may depend to some extent on the design of the sieving table or hopper, but, unless the sample is to be washed through slowly in small sections, the sieve must be of a certain minimum size to allow an adequate sieving area and to prevent clogging by the sediment. For samples of more than a few litres a sieve of at least 30 × 30 cm is desirable. Washing can be done by single or multiple jets from a hose, but unless this is very gentle and, therefore, time-consuming, it can damage the animals. A gentler method, described by Sanders *et al.* (1965) utilizes a large vessel with spout about one-third of the way up. A continuous stream of water passes through the vessel, carrying the animals over into a sieve. If the animals are required in especially good condition sieving should be done by hand, the sieve being gently agitated in water so that flow takes place from below as well as from above.

The mesh size of the sieve is of critical importance and should be determined at an early stage of planning. Sieves with either round or square holes may be used, but a square mesh is to be preferred since it has a higher percentage open area, and this type is in general use for soil analysis so that a wide range of sizes to standard specifications is available. In practice, mesh sizes have varied from 2 to 0·5 mm, and even finer apertures have occasionally been used to capture juvenile stages or to make maximum use of deep-sea samples. The effects of different meshes on results have been referred to by McIntyre (1961) and Driscoll (1964). Jónasson (1955) showed that for one particular species a small decrease in mesh size, from 0·62 to 0·51 mm, resulted in a 47 % increase in numbers, and stressed that the use of too large a mesh could produce an erroneous picture of seasonal peaks in animal numbers. Lewis & Stoner (1981), testing the retention of 0·5 and 1·0 mm meshes, found that only 55–77 % of the total macrofauna was retained by the 1·0 mm mesh. Nalepa & Robertson (1981) found that use of a 0·595 mm mesh resulted in serious underestimates of the abundance of freshwater macrobenthos, although 97 % of the biomass was retained. They concluded that the optimum mesh size chosen should be small enough (e.g. 0·10 mm) to retain all or most of the individuals of the taxa studied. In a more detailed study Reish (1959a) passed five grab samples from a shallow water muddy ground through a series of 11 sieves with apertures ranging from 4·7 to 0·15 mm. His data have been recalculated in Table 6.4 to show cumulative percentages for the main species and groups. If only molluscs were required, a screen of 0·85 mm, which separated about 95 % of the individuals in his samples, would have been suitable, but a very much finer mesh was needed for nematodes and crustaceans. Analyses of polychaetes into species show the variation which can occur within a single taxonomic group: about 95 % of the *Lumbrinereis*

Table 6.4. Number of specimens retained on graded screens, and cumulative percentages. (Calculated from Reish, 1959a.)

	Mesh sizes (mm)											
	4·7	2·8	1·4	1·0	0·85	0·7	0·59	0·5	0·35	0·27	0·15	Total
Nematoda						1	2	6	26	456	90	581
%						0·2	0·5	1·5	6·0	72·0	100·0	
Nemertea	2	7	6	3	5	2		1				26
%	7·7	34·6	57·7	69·2	88·5	96·2	96·2	100·0				
Polychaeta:												
Lumbrineris spp.	8	10	33	9	3							63
%	12·7	28·6	81·0	95·2	100·0							
Dorvillea articulata			22	52	30	5	4	2	2	2		119
%			18·5	62·2	87·4	91·6	95·0	96·6	98·3	100·0		
Prionospio cirrifera	1	6	23	100	115	30	18	10		1		304
%	0·3	2·3	10·0	42·8	80·6	90·1	96·4	99·8	99·8	100·0		
Capitita ambiseta	3	6	29	104	109	23	21	13	2			310
%	1·0	2·9	12·3	45·8	81·0	88·4	95·2	99·4	100·0			
Cossura candida			1	11	129	100	157	265	88	105	10	866
%			0·1	1·4	16·3	27·8	46·0	76·6	86·7	98·8	100·0	
Other Polychaeta	10	23	27	38	31	10	11	7	4	5	2	168
%	6·0	19·6	35·7	58·3	76·8	82·7	89·3	93·4	95·8	98·8	100·0	
Crustacea	3				2		1	3	5	3		17
%	17·6	17·6	17·6	17·6	29·4	29·4	35·3	52·9	82·4	100·0		
Mollusca	5	4	7	5	2			1				24
%	20·5	37·5	66·7	87·5	95·8	95·8	95·8	100·0				
Pisces	1											1
%	100·0											
Total	33	56	148	322	426	171	214	308	127	572	102	2479
%	1·3	3·5	9·7	22·5	39·7	46·6	55·3	67·7	72·8	95·9	100·0	

was found on the 1·0 mm mesh, but to attain this level of separation for *Cossura candida* required a mesh of 0·27 mm. When the overall assemblage is considered it appears that a 0·27 mm mesh was needed to collect about 95 % of all individuals. On the other hand, if only biomass is required, Reish found that >90 % was retained on the 1·4 mm sieve. While these results clearly apply only to the particular ground studied, they emphasize the importance of correct selection of sieve mesh, according to the purpose of the survey.

The Baltic Marine Biologists have standardized the mesh used in their studies at 1·00 mm (Dybern *et al.*, 1976; Ankar *et al.*, 1979), with the recommendation that a 0·5 mm mesh should be used in addition whenever possible.

In general, it is suggested that a 0·5 mm sieve should be used for macrofauna separation, but since this may retain too large a volume of material on coarse grounds, a compromise may have to be made, the final mesh selected being related to the grade of deposit and the size of the organisms to be separated. The use of a mesh corresponding to one of those in International Standard (ISO) is recommended, and a choice should be made from one of the following apertures: 2·00; 1·40; 1·00; 0·71; or 0·50 mm.

In areas where coarse deposits necessitate the use of wide-meshed sieves, it is recommended that additional small samples be sieved through a finer mesh to assess the losses occurring through the coarser sieve.

Preservation

Having reduced the sample to a manageable size by initial sieving, the sorting of animals from the residue can proceed. If the final extraction is carried out soon after collection, and near the sampling site, sorting of living material may be possible, with the advantage that movement of the animals helps in their detection, especially if they are small. But it is frequently necessary, after initial sieving in the field, to preserve the collections for later sorting. In such circumstances it is important to ensure adequate labelling. It may be convenient to number the tops or sides of jars with a waterproof marker, but even if this is done, a properly annotated label strong enough to withstand water and preservatives should be placed inside the jar. A suitable paper is goatskin parchment, obtainable from Wiggins Teape (Mill Sales) Ltd, Gateway House, Basingstoke, Hampshire, England. Alternatively, sheet plastic with a matt surface which can be marked with pencil can be used, and this is particularly suitable for sediment samples.

Formalin is normally used for the initial preservation, and this can conveniently be diluted with sea water. While a 10 % solution of commercial formalin (equivalent to 4 % formaldehyde) is suitable for histological purposes, the strength may be reduced to between 2·5 and 5 % formalin for general storage, provided that the volume of preserving fluid is considerably greater than that of the specimens. In very large samples, perhaps containing much gravel, care should be taken to see that not only is there sufficient preservative, but that it is adequately mixed through the sample. Since formalin tends to become acid with storage and so cause damage to the specimens, a buffer such as borax or hexamine is often added to the formalin. These substances have been criticized for causing disintegration of labels, or for producing a precipitate, and for plankton samples sodium acetate is sometimes used. The addition of marble chips to the formalin is often used as minimum precaution to prevent the development of acidic conditions.

Although bulk treatment is satisfactory for general samples, particular animals required in good condition should be extracted and dealt with individually. It is an advantage to narcotize highly contractile animals before fixation, allowing subsequent preservation in an extended condition. An account of anaesthetic agents is given by Steedman (1976) and by Lincoln & Sheals (1979). Alcohol is often used for later storage of samples, but it is less satisfactory for initial field preservation because of its volatility, because mixing with sea water causes a precipitate, and because it may cause the separation of lamellibranchs from their shells. A mixture of 70 % ethanol–5 % glycerine is often used for permanent storage.

The use of 'Dowcil 100' in 10 % solution releases formaldehyde only in the presence of proteins and has many advantages over formalin because it does not give off irritant fumes. It is also preferable to alcohol because it is neither flammable nor volatile. Its rather high cost has prevented its wider usage so far.

Detailed information on the use of fixatives, preservatives and buffering agents is provided by the SCOR working group (Steedman, 1976), and by Lincoln & Sheals (1979).

Subsequent sorting

If the study is restricted to major species or to large individuals hand-sorting may be straightforward. This is best done in glass trays below which black or white material can be inserted to provide varying backgrounds suitable for distinguishing different types of animals. If every individual must be extracted this can be a time-consuming task, which may severely restrict the extent of sampling. It is often possible to divide a sample into fractions by agitating the light material into suspension and pouring it through a fine sieve. This separates small animals (such as crustaceans and polychaetes) together with fine debris, leaving large or heavy animals behind in the main sample.

Bulk staining of samples with vital stains (rose bengal, rhodamine B, eosin, etc.) to facilitate sorting has sometimes been used, and a counter-staining technique for samples containing large quantities of detritus is described by Williams & Williams (1974), where the primary stain, rose bengal or Lugol's iodine, is counterstained with chlorazol black E to provide a high colour contrast between the animals and the detritus in the samples. Hamilton (1969) used fluorescence for faster sorting of freshwater organisms from sediment and detritus: organisms stained with a dye (rhodamine B) fluoresce when examined under longwave ultra violet light. Of all the stains, rose bengal ($4\,g\,l^{-1}$ of 36% formaldehyde) is the most widely used, although there is some opposition to its widespread application as it may obscure diagnostic features used in species identification. See also methods for meiofauna in Chapter 7.

Two methods sometimes used to ease the work of sorting macrofauna are flotation and elutriation. Flotation is based on differences in specific weight between the organisms and the sediment—application of a medium of a suitably high density causes the animals to float free of heavier debris. Liquids such as carbon tetrachloride (Birkett, 1957; Dillon, 1964; Whitehouse &

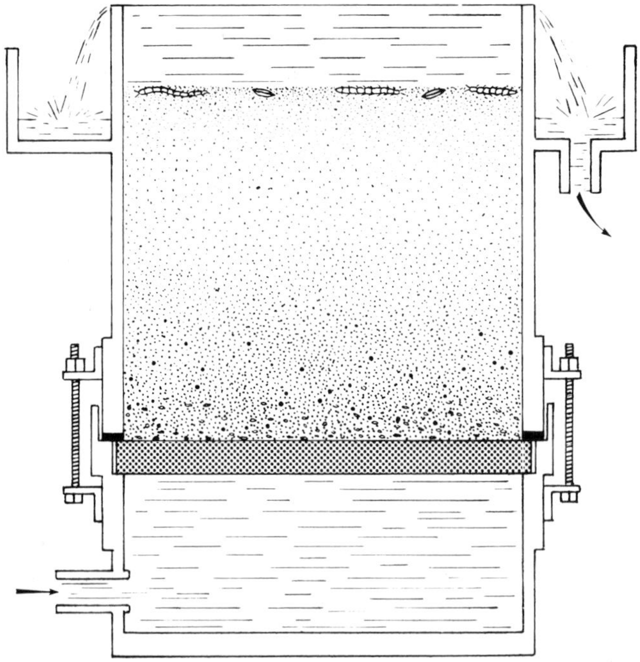

Fig. 6.29. Barnett's fluidized sand bath. Water entering at the bottom passes upward via a sintered ceramic sheet through the sand bed. A sieved sample tipped into the top is separated so that nearly all the organisms either float on the surface of the sand, from where they can be skimmed off, or pass with the overflowing water through a sieve of suitable mesh. (Drawn from information supplied by P.R.O. Barnett.)

Lewis, 1966), sugar solutions (Anderson, 1959; Kajak *et al.*, 1968; Fast, 1970; Lackey & May, 1971), $ZnCl_2$ (Sellmer, 1956; Mattheisen, 1960) and many others have been applied with varying degrees of success. Unfortunately organic detritus also floats, making these methods unsuitable for muds and silts without adjustment of the specific gravity of the flotation medium. Other disadvantages are that most of these techniques are messy, and, in the case of carbon tetrachloride, dangerous if inhalation of vapour occurs (Dillon, 1964). Thus the liquid must always be covered with water and the operation must be carried out with adequate ventilation. De Jonge & Bouwman (1977) used a colloidal silica polymer (Ludox-TM) to separate nematodes from sediment and detritus (see Chapter 7); this technique has also been used for the successful separation of macrofauna species.

Elutriation involves passing a continuous upward stream of water through the sample in a container with an overflow to a fine collecting screen. The flow rate is adjusted so that the water agitates the sample and carries up the small animals but not the sediment. Several apparati suitable for macrofauna (Lauff *et al.*, 1961; Pauly, 1973; Worswick & Barbour, 1974) utilize both water and air jets. Barnett's fluidized sand bath (Fig. 6.29) uses an upward current of water to separate animals from large quantities (up to 20 l) of sediment. Sand uniformly fluidized by the passage of water provides a dense flotation medium into which the sample is tipped. Organisms float at the sand/water interface and are collected with a special sieve while the lightest ones are retained on the fine screen in the overflow. The average separation time is approximately 10 minutes and the claimed efficiency, 98–100%.

Data recording

Where large amounts of information on species occurrence, with supporting environmental data, are collected, there is likely to be a need for use of a computerized system for information storage, sorting and retrieval. Following recent developments in microprocessors, much can now be achieved on microcomputers which would formerly have needed the use of a main-frame system. Although the data manipulation required may be of a relatively unsophisticated nature, the quantity of data to be handled is likely to be considerable, so that it is important to employ a system having adequate storage capability. It is true that with skilful programming much can be stored on a limited system but such a computer may have limited search facilities, and is likely to take a long time to process the data. Microcomputers are evolving at such a speed that any recommendation would rapidly become out of date, so current advice and specifications should be sought. It seems likely that microcomputers, with word-processing facilities, will be increasingly used, at least for making initial entries and correction of records, the data being transferred where necessary to a main-frame computer for long term storage and manipulation.

Computer storage typically requires a preliminary streamlining of data so that it can be recorded in standardized format. Early decisions must be made on such aspects as species names, scales of density, geographical co-ordinates and grids for mapping schemes, and on techniques for measurement and presentation of environmental parameters. By its very nature biological data does not lend itself to presentation in standardized form: nearly every record needs some qualification, and here a choice must be made between having sufficient fields to cover all possible circumstances, space for entry of additional data in longhand on the computer file, or instructions to refer back to original notes not entered into the computer. 'Comprehensive' recording schemes devised at centralized data banks tend to include all possible fields but these require so much data to be entered that individual workers have devised their own schemes to fit the requirements of specific scientific programmes. However, there is much to be said in favour of a degree of standardization which will allow ready exchange of information between different users.

Because of possible loss or damage to records on magnetic files a copy should always be made, and retention of original notes, entry sheets etc. is essential.

Interpretation will, in the first place, be based on print-out of records, sorted according to such attributes as species, geographical area, depth, or time period. Seldom are the data sufficiently complete for complex analyses, and care must be taken to ensure that records lacking complete supporting parameters (e.g. where they are taken from the literature), are not missed during a 'sort' programme. Print-outs in map form (whether based on point positions or grid squares) are useful for summarizing progress, but reference must frequently be made back to the original records to confirm for example the validity of occurrences in unexpected areas.

Standardization of records requires in the first place agreement on the scientific names for genus and species, if in doubt 'splitting' rather than 'lumping' possible subspecies. Reference can be made to names used in a standard text, or a list, with author references, must be specially prepared. Species abundance is likely to cause problems, particularly when records are to be added from the literature. The simplest situation is when 'presence' alone is recorded, without recourse to density estimates. It is, however, usually possible to make a rough estimate of density, and the abundance codes given in Heath & Scott (1977) (Table 6.5) may prove an acceptable basis. 'Absence' records are particularly difficult to assess, and some recorders may consider 'Absent or Very Rare' a more appropriate category when no specimens are found, but this will depend on the particular taxon and the method of sampling.

Table 6.5. Table of abundance codes. One category is selected from each of the two scales. (From Heath & Scott, 1977.)

Quantity	Unit
1. Present (no numerical data)	1. $/cm^2$
2. <1	2. $/m^2$
3. 1–10	3. $/cm^3$
4. 11–100	4. $/m^3$
5. 101–1000	5. per trawl haul
6. 1001–10 000	6. per dredge haul
7. >10 000	7. per 15 min search on foot or diving
8. Absent (none found during careful search of suitable habitat)	8. per 30 min search on foot or diving
9. Other	9. Other

Records may be conveniently summarized on a map in which occurrence within squares of a mapping grid is indicated (Clayton, 1971; Humphris, 1971). A grid should be based on latitude/longitude co-ordinates, although for near-shore and estuarine surveys a grid could be based on that of local land maps (e.g. Universal Transverse Mercator—see Day, 1973). In northern Europe it is convenient to use ICES statistical squares, based on 1° of longitude and $\frac{1}{2}$° of latitude, which gives a true square at 60° North (Griffith, 1974; Holme, 1974; Petersen, 1977). Where a systematic grade analysis is not to be carried out, sediments can be classified by reference to a series of standards of known particle size, for example the grain-size comparator described by Kirby (1973). Descriptions of habitats are more difficult to systematize, but for seashores, both rocky and sedimentary, a method based on field cards is given by Holme & Nichols (1980).

It will be appreciated that whatever computer-based system is adopted, the possibilities for modification rapidly diminish as entries are commenced. It is, therefore, particularly important to take up-to-date advice based on the computer which will be available to carry out the processing.

Examples of computer-based schemes include Fredj (1972, 1973), Kohlenstein (1972), Swartz (1972), Brogden *et al.* (1974), Earll (1980), Knight & Mitchell (1980) and Domanski (1981).

Acknowledgements

We are indebted to the following for permission to reproduce copyright material: DAFS Laboratory, Aberdeen (Figs 6.2, 6.14); Professor P. Brunel, Montreal (Fig. 6.3); Institute of Oceanographic Sciences, Wormley (Figs 6.5, 6.28); Columbia University Press (Fig. 6.6); NERC Research Vessel Base,

Barry (Fig. 6.16); Dr C. Emig, Marseille (Fig. 6.21); Dr M. Mulder, Texel, Netherlands Journal of Sea Research (Fig. 6.22); Ministry of Agriculture, Fisheries and Food, Lowestoft (Fig. 6.23); Dr L. Laubier, Centre Océanologique de Bretagne (Fig. 6.24); and Dr P.R.O. Barnett, SMBA, Oban (Fig. 6.29). We are indebted to M. J-C. Dauvin, Roscoff, for help with preparation of Fig. 6.17. This chapter includes material written for Chapters 8, 9 and 10 of the first edition of the Handbook by L. Birkett, A.D. McIntyre and E.I.S. Rees.

References

Admiralty, 1948. *Admiralty Manual of Hydrographic Surveying*, 2nd ed. Hydrographic Department, Admiralty, London, 572 pp.

Agassiz, A., 1888. Three cruises of the United States Coast and Geodetic Survey Steamer "Blake". Volume 1. *Bulletin of the Museum of Comparative Zoology at Harvard College, in Cambridge*, **14**, 314 pp.

Albert, Prince of Monaco, 1932. Sur l'emploi de nasses pour des recherches zoologiques en eau profonde. *Résultats des Campagnes Scientifiques du Prince de Monaco*, **84**, 176–178.

Aldred, R.G., Thurston, M.H., Rice, A.L. & Morley, D.R., 1976. An acoustically monitored opening and closing epibenthic sledge. *Deep-Sea Research*, **23**, 167–174.

Allen, D.M. & Hudson, J.H., 1970. A sled-mounted suction sampler for benthic organisms. *United States Fish and Wildlife Service Special Scientific Report—Fisheries*, **614**, 5 pp.

Andersin, A.B. & Sandler, H., 1981. Comparison of the sampling efficiency of two van Veen grabs. *Finnish Marine Research*, **248**, 137–142.

Anderson, R.O., 1959. A modified flotation technique for sorting bottom fauna samples. *Limnology and Oceanography*, **4**, 223–225.

Andreev, N.N., 1962. *Handbook of fishing gear and its rigging*. English translation from Russian. Israel Program for Scientific Translations, Jerusalem, 1966, 454 pp.

Aneer, A. & Nellbring, S., 1977. A drop-trap investigation of the abundance of fish in very shallow water in the Askö area, northern Baltic proper. pp. 21–30 in B.F. Keegan, P. ÓCéidigh & P.J.S. Boaden (eds) *Biology of Benthic Organisms. Eleventh European Symposium on Marine Biology. Galway, October 1976.* Pergamon Press, Oxford, 630 pp.

Ankar, S., 1977a. Digging profile and penetration of the van Veen grab in different sediment types. *Contributions from the Askö Laboratory, University of Stockholm, Sweden*, **16**, 22 pp.

Ankar, S., 1977b. The soft bottom ecosystem of the Northern Baltic proper with special reference to the macrofauna. *Contributions from the Askö Laboratory, University of Stockholm, Sweden*, **19**, 62 pp.

Ankar, S., Andersin, A-B, Lassig, J., Norling, L. & Sandler, H., 1979. Methods for studying benthic macrofauna. An intercalibration between two laboratories in the Baltic Sea. *Finnish Marine Research*, **246**, 147–160.

Ankar, S., Cederwall, H., Lagzdins, G. & Norling, L., 1978. Comparison between Soviet and Swedish methods of sampling and treating soft bottom macrofauna. Final report from the Soviet–Swedish expert meeting on intercalibration of biological methods and analyses. Askö, July 5–12, 1975. *Contributions from the Askö Laboratory, University of Stockholm, Sweden*, **23**, 38 pp.

Arkel, M.A. van. & Mulder, M., 1975. A device for quantitative sampling of benthic organisms in shallow water by means of a flushing technique. *Netherlands Journal of Sea Research*, **9**, 365–370.

Backus, R.H., 1966. The 'pinger' as an aid in deep trawling. *Journal du Conseil Permanent international pour l'Exploration de la Mer*, **30**, 270–277.

Baird, R.H., 1955. A preliminary report on a new type of commercial escallop dredge. *Journal du Conseil Permanent international pour l'Exploration de la Mer*, **20**, 290–294.

Baird, R.H., 1958. A preliminary account of a new half square metre bottom sampler. *International Council for the Exploration of the Sea, Shellfish Committee*, **C.M. 1958/70**, 4 pp.

Baird, R.H., 1959. Factors affecting the efficiency of dredges. pp. 222–4 in H. Kristjonnson (ed.) *Modern Fishing Gear of the World*. Fishing News (Books), London, 607 pp.

Bakanov, A.I., 1977. Comparative evaluation of the effectiveness of different dredges. *Gidrobiologicheskii Zhurnal, Kiev*, **13(2)**, 97–103.

Bandy, O.L., 1965. The pinger as a deep-water grab control. *Undersea Technology*, **6(3)**, 36.

Barnett, P.R.O., 1969. A stabilizing framework for the Knudsen bottom sampler. *Limnology and Oceanography*, **14**, 648–649.

Barnett, P.R.O. & Hardy, B.L.S., 1967. A diver-operated quantitative bottom sampler for sand macrofaunas. *Helgoländer wissenschaftliche Meeresuntersuchungen*, **15**, 390–398.

Beamish, R.J., 1972. Design of a trapnet for sampling shallow-water habitats. *Technical Report, Fisheries Research Board of Canada*, **305**, 14 pp.

Belyaev, G.M. & Sokolova, M.N., 1960. On methods of quantitative investigation of deep-water benthos. *Trudy̆ Instituta Okeanologii, Akademiya nauk SSSR*, **39**, 96–100 (in Russian).

Beukema, J.J., 1974. The efficiency of the van Veen grab compared with the Reineck box sampler. *Journal du Conseil Permanent international pour l'Exploration de la Mer*, **35**, 319–327.

Beyer, F., 1958. A new bottom living Trachymedusa from the Oslo Fjord. Description of the species, and a general discussion of the life conditions and fauna of the fjord deeps. *Nytt magasin for zoologi*, **6**, 121–143.

Bhaud, M. & Duchêne, J-C., 1977. Observations sur l'efficacité comparée de deux bennes. *Vie et Milieu, Ser. A*, **27**, 35–54.

Bieri, R. & Tokioka, T., 1968. Dragonet II, an opening–closing quantitative trawl for the study of microvertical distribution of zooplankton and the meio-epibenthos. *Publications of the Seto Marine Biological Laboratory*, **15**, 373–390.

Birkett, L., 1957. Flotation technique for sorting grab samples. *Journal du Conseil Permanent international pour l'Exploration de la Mer*, **22**, 289–292.

Birkett, L., 1958. A basis for comparing grabs. *Journal du Conseil Permanent international pour l'Exploration de la Mer*, **23**, 202–207.

Boalch, G.T., Holme, N.A., Jephson, N.A. & Sidwell, J.M.C., 1974. A resurvey of Colman's intertidal traverses at Wembury, South Devon. *Journal of the Marine Biological Association of the United Kingdom*, **54**, 551–553.

Boillot, G., 1964. Géologie de la Manche occidentale. Fonds rocheux, dépôts quaternaires, sédiments actuels. *Annales de l'Institut Océanographique*, **42**, 1–219.

Bossanyi, J., 1951. An apparatus for the collection of plankton in the immediate neighbourhood of the sea-bottom. *Journal of the Marine Biological Association of the United Kingdom*, **30**, 265–270.

Bouma, A.H., 1969. *Methods for the Study of Sedimentary Structures*. John Wiley & Sons, New York, 458 pp.

Bouma, A.H. & Marshall, N.F., 1964. A method for obtaining and analysing undisturbed oceanic bed samples. *Marine Geology*, **2**, 81–99.

Bourcier, M., Massé, H., Plante, R., Reys, J.P. & Tahvildari, B., 1975. Note préliminaire sur l'étude comparative des bennes Smith–McIntyre et Briba–Reys. *Rapports et procès-verbaux des Réunions. Commission Internationale pour l'Exploration sciéntifique de la Mer Méditerranée*, **23** (**2**), 155–156.

Briba, C. & Reys, J.P., 1966. Modifications d'une benne 'orange peel' pour les prélèvements quantitatifs du benthos de substrats meubles. *Recueil des Travaux de la Station Marine d'Endoume*, **41**, 57, 117–121.

Brogden, W.B., Cech, J.J. & Oppenheimer, C.H., 1974. A computerized system for the organized retrieval of life history information. *Chesapeake Science*, **15**, 250–254.

Brunel, P., Besner, M., Messier, D., Poirier, L., Granger, D. & Weinstein, M., 1978, Le traîneau suprabenthique Macer-GIROQ; appareil amélioré pour l'echantillonage quantitatif etagé de la petite faune nageuse au voisinage du fond. *Internationale Revue der Gesamten Hydrobiologie*, **63**, 815–829.

Burke, J.C., 1968. A sediment coring device of 21-cm diameter with sphincter core retainer. *Limnology and Oceanography*, **13**, 714–718. (*Collected Reprints, Woods Hole Oceanographic Institution*, **2149**.)

Carey, A.G. & Hancock, D.R., 1965. An anchor-box dredge for deep-sea sampling. *Deep-Sea Research*, **12**, 983–984.

Carey, A.G. & Heyamoto, H., 1972. Techniques and equipment for sampling benthic organisms. pp. 378–408 in A.T. Pruter & D.L. Alverson (eds) *The Columbia River Estuary and Adjacent Ocean Waters: Bioenvironmental Studies*. University of Washington Press, 882 pp.

Carey, A.G. & Paul, R.R., 1968. A modification of the Smith–McIntyre grab for simultaneous collection of sediment and bottom water. *Limnology and Oceanography*, **13**, 545–549.

Carney, R.S. & Carey, A.G. Jr, 1980. Effectiveness of metering wheels for measurement of area sampled by beam trawls. *Fishery Bulletin, National Marine Fisheries Service, NOAA, Seattle, USA*, **78**, 791–796.

Chapman, C.J., Mason, J. & Kinnear, J.A.M., 1977. Diving observations on the efficiency of dredges used in the Scottish fishery for the scallop, *Pecten maximus* (L.). *Scottish Fisheries Research Report*, **10**, 15 pp.

Christie, N.D., 1975. Relationship between sediment texture, species richness and volume of sediment sampled by a grab. *Marine Biology*, **30**, 89–96.

Christie, N.D. & Allen, J.C., 1972. A self-contained diver-operated quantitative sampler for investigating the macrofauna of soft substrates. *Transactions of the Royal Society of South Africa*, **40**, 299–307.

Clarke, A.H., 1972. The arctic dredge, a benthic biological sampler for mixed boulder and mud substrates. *Journal of the Fisheries Research Board of Canada*, **29**, 1503–05.

Clayton, G., 1971. Geographical reference systems. *The Geographical Journal*, **137**, 1–12.

Cleve, R. van, Ting, R.Y. & Kent, J.C., 1966. A new device for sampling marine demersal animals for ecological study. *Limnology and Oceanography*. **11**, 438–443.

Colman, J.S. & Segrove, F., 1955. The tidal plankton over Stoupe Beck Sands, Robin Hood's Bay (Yorkshire, North Riding). *Journal of Animal Ecology*, **24**, 445–462.

Dahl, E., Laubier, L., Sibuet, M. & Strömberg, J-O., 1976. Some quantitative results on benthic communities of the deep Norwegian Sea. *Astarte*, **9**, 61–79.

Dauvin, J-C., 1979. *Recherches Quantitatives sur le Peuplement des Sables Fins de la Pierre Noire, Baie de Morlaix, et sur la Perturbation par les Hydrocarbures de l'AMOCO-CADIZ'*. Thèse du Diplome de Docteur de 3e Cycle, Université Pierre et Marie Curie, Paris, 251 pp.

Davis, F.M., 1925. Quantitative studies on the fauna of the sea bottom. No. 2. Results of investigations in the southern North Sea, 1921–24. *Fishery Investigations. Ministry of Agriculture Fisheries and Food*, Series 2, **8** (4), 50 pp.

Davis, F.M., 1958. An account of the fishing gear of England and Wales (4th ed.). *Fisheries Investigations, London, Ser 2*, **8(4)**, 50 pp.

Day, G.A., 1973. The IGS Universal Transverse Mercator mapping scheme. *Marine Geophysics Unit*, **Report 16,** 8 pp + maps. Institute of Geological Sciences, Edinburgh.

Day, G.F., (in preparation). The Day Grab. A simple sea bed sampler. *Institute of Oceanographic Sciences*, **52** (Unpublished Manuscript).

Dayton, P.K. & Hessler, R.R., 1972. Role of biological disturbance in maintaining diversity in the deep sea. *Deep-Sea Research*, **19**, 199–208.

De Jonge, V.N. & Bouwman, L.A., 1977. A simple density separation technique for quantitative isolation of meiobenthos using the colloidal silica Ludox-TM. *Marine Biology*, **42**, 143–148.

Dickie, L.M., 1955. Fluctuations in abundance of the giant scallop *Placopecten magellanicus* (Gmelin) in the Digby area of the Bay of Fundy. *Journal of the Fisheries Research Board of Canada*, **12**, 797–857.

Dickinson, J.J. & Carey, A.G. Jr, 1975. A comparison of two benthic infaunal samplers. *Limnology and Oceanography*, **20**, 900–902.

Dillon, W.P., 1964. Flotation technique for separating faecal pellets and small marine organisms from sand. *Limnology and Oceanography*, **9**, 601–602.

Domanski, P., 1981. BIOS a database for marine biological data. *Journal of Plankton Research*, **3**, 475–490.

Doodson, A.T. & Warburg, H.D., 1941. *Admiralty Manual of Tides*. H.M.S.O., London, 270 pp.

Drake, C.M. & Elliott, J.M., 1982. A comparative study of three air-lift samplers used for sampling benthic macro-invertebrates in rivers. *Freshwater Biology*, **12**, 511–533.

Driscoll, A.L., 1964. Relationship of mesh opening to faunal counts in a quantitative benthic study of Hadley Harbor. *Biological Bulletin, Marine Biological Laboratory Woods Hole*, **127**, 368.

Dybern, B.I., Ackefors, H. & Elmgren, R., 1976. Recommendations on methods for marine biological studies in the Baltic Sea. *The Baltic Marine Biologists*, **1**, 98 pp.

Dyer, M.F., Fry, W.G., Fry, P.D. & Cranmer, G.J., 1982. A series of North Sea benthos surveys with trawl and headline camera. *Journal of the Marine Biological Association of the United Kingdom*, **62**, 297–313.

Dyer, M.F., Pope, J.G., Fry, P.D., Law, R.J. & Portmann, J.E., 1983. Changes in fish and benthos catches off the Danish coast in September 1981. *Journal of the Marine Biological Association of the United Kingdom*, **63**, 767–775.

Eagle, R.A., Norton, M.G., Nunny, R.S. & Rolfe, M.S., 1978. The field assessment of effects of dumping wastes at sea: 2. Methods. *Fisheries Research Technical Report, Lowestoft*, **47**, 24 pp.

Earll, R.C., 1980. The development and use of a computer-based system for handling habitat and species information from the sublittoral environment. pp. 285–302 in J.H. Price, D.E.G. Irvine & W.F. Farnham (eds) *The Shore Environment, Volume 1, Methods*, Systematics Association Special Volume 17(a). Academic Press, London, 321 + 41 pp.

Edwards, R. & Steele, J.H., 1968. The ecology of O-group plaice and common dabs at Loch Ewe. I. Population and food. *Journal of Experimental Marine Biology and Ecology*, **2**, 215–238.

Eleftheriou, A. & McIntyre, A.D., 1976. The intertidal fauna of sandy beaches—a survey of the Scottish coast. *Scottish Fisheries Research Report*, **6**, 61 pp.

Elliott, J.M. & Drake, C.M., 1981. A comparative study of seven grabs used for sampling benthic macroinvertebrates in rivers. *Freshwater Biology*, **11**, 99–120.

Elliott, J.M. & Tullett, P.A., 1978. A bibliography of samplers for benthic invertebrates. *Freshwater Biological Association, Occasional Publication*, **4**, 61 pp.

Emery, K.O., 1961. A simple method of measuring beach profiles. *Limnology and Oceanography*, **6**, 90–93.

Emery, K.O., Merrill, A.S. & Trumbull, V.A., 1965. Geology and biology of the sea floor as deduced from simultaneous photographs and samples. *Limnology and Oceanography*, **10**, 1–21. (*Collected Reprints, Woods Hole Oceanographic Institute*, **1965(1), No. 1508**.)

Emig, C.C., 1977. Un nouvel aspirateur sous-marins, à air comprimé. *Marine Biology*, **43**, 379–380.

Emig, C.C. & Lienhart, R., 1971. Principe de l'aspirateur sous-marin automatique pour sédiments meubles. *Vie et Milieu* (Suppl.), **22**, 573–578.

Ervin, J.L. & Haines, T.A., 1972. Using light to collect and separate zooplankton. *Progressive Fish Culturist*, **34**, 171–174.

Espinosa, L.R. & Clark, W.E., 1972. A polypropylene light trap for aquatic invertebrates. *California Fish and Game*, **58**, 149–152.

FAO, 1972. *Catalogue of fishing gear designs*. Fishing News (Books), London, 155 pp.

Farris, R.A. & Crezée, M., 1976. An improved Reineck box for sampling coarse sand. *Internationale Revue der Gesamten Hydrobiologie*, **61**, 703–705.

Fast, A.W., 1970. An evaluation of the efficiency of zoobenthos separation by sugar flotation. *Progressive Fish Culturist*, **32**, 212–216.

Finnish IBP-PM Group, 1969. Quantitative sampling equipment for the littoral benthos. *Internationale Revue der Gesamten Hydrobiologie*, **54**, 185–193.

Flannagan, J.F., 1970. Efficiencies of various grabs and corers in sampling freshwater benthos. *Journal of the Fisheries Research Board of Canada*, **27**, 1691–1700.

Forster, G.R., 1953. A new dredge for collecting burrowing animals. *Journal of the Marine Biological Association of the United Kingdom*, **32**, 193–198.

Fredj, G., 1972. Stockage et exploitation des données en écologie marine. A. Un fichier sur ordinateur des invertebrés macrobenthiques. *Memoires de l'Institut Océanographique, Monaco*, **5**, 60 pp.

Fredj, G., 1973. Stockage et exploitation des données en écologie marine. B.—Collecte des données en mer et archivage (Stations ponctuelles dans l'espace et dans le temps). *Memoires de l'Institut Océanographique, Monaco*, **5**, 60 pp.

Frolander, H.F. & Pratt, I., 1962. A bottom skimmer. *Limnology and Oceanography*, **7**, 104–106.

Gage, J.D., 1975. A comparison of the deep sea epibenthic sledge and anchor-box dredge samplers with the van Veen grab and hand coring by divers. *Deep-Sea Research*, **22**, 693–702.

Gallardo, V.A., 1965. Observations on the biting profiles of three $0{\cdot}1\,m^2$ bottom-samplers. *Ophelia*, **2**, 319–322.

Garner, J., 1962. *How to Make and Set Nets; or, the Technology of Netting*. Fishing News (Books), London, 95 pp.

Garner, J., 1973. *Modern Inshore Fishing Gear. Rigging and Mending*. Fishing News (Books), West Byfleet, Surrey, 67 pp.

Garner, J., 1977. *Modern Deep Sea Trawling Gear*. Fishing News (Books), Farnham, Surrey, 83 pp.

Gaunt, D.I. & Wilson, J.B., 1975. Acoustic monitoring of dredge behaviour on the sea floor. *Deep-Sea Research*, **22**, 91–97.

Gibson, R.N., 1967. The use of the anaesthetic quinaldine in fish ecology. *Journal of Animal Ecology*, **36**, 295–301.

Gilat, E., 1964. The macrobenthonic invertebrate communities on the Mediterranean continental shelf of Israel. *Bulletin de l'Institut Océanographique, Monaco*, **62** (1290), 46 pp.

Gilat, E., 1968. Methods of study in marine benthonic ecology. *Colloque Commité du Benthos, Commité international pour l'Exploration de la Mer Mediterranée, Marseille, November 1963*, 7–13.

Gonor, J.J. & Kemp, P.F., 1978. Procedures for quantitative ecological assessments in intertidal environments. *U.S. Environmental Protection Agency, Office of Research and Development, Ecological Research Series*, **EPA-600/3-78-087.** Corvallis, Oregon, 103 pp.

Grange, K.R. & Anderson, P.W., 1976. A soft-sediment sampler for the collection of biological specimens. *Records New Zealand Oceanographic Institute*, **3**, 9–13.

Griffith, D. de G., 1974. A description of the ICES statistical area (North), statistical sub-areas, divisions and sub-divisions. *International Council for the Exploration of the Sea Statistics Committee*, **C.M. 1974/D:9**, 10 pp.

Grussendorf, M.J., 1981. A flushing-coring device for collecting deep-burrowing infaunal bivalves in intertidal sand. *Fishery Bulletin, National Marine Fisheries Service, NOAA, Seattle, USA*, **79**, 383–385.

Gulland, J.A., 1966. Manual of sampling and statistical methods for fisheries biology. Part 1. Sampling Methods. *FAO Manual of Fisheries Science*, **3**, Foreword, etc., 6 pp.; Fasc. 1, 13 pp.; Fasc. 2, 16 pp.; Fasc. 3, 35 pp.; Fasc. 4, 20 pp.; Fasc. 5, 3 pp.

Gunter, G., 1957. Dredges and trawls. *Memoirs of the Geological Society of America*, **67(1)**, 73–78.

Haahtela, I., 1969. Methods for sampling scavenging benthic Crustacea especially the isopod *Mesidotea entomon* (L.) in the Baltic. *Annales Zoologici Fennici*, **15**, 182–185.

Hamilton, A.C., 1969. A method of separating invertebrates from sediments using longwave ultraviolet light and fluorescent dyes. *Journal of the Fisheries Research Board of Canada*, **26**, 1667–1672.

Hardy, A.C., 1959. *The Open Sea: its Natural History. Part II. Fish and Fisheries.* New Naturalist Series, 37. Collins, London, 322 pp.

Hartman, O., 1955. Quantitative survey of the benthos of San Pedro Basin, Southern California. Part I. Preliminary results. *Allan Hancock Pacific Expeditions*, **19(1)**, 185 pp.

Heath, J. & Scott, D., 1977. *Instructions for Recorders*, 2nd ed. Biological Records Centre, Monks Wood Experimental Station, Huntingdon, 28 pp.

Hessler, R.R., Ingram, C.L., Yayanos, A.A. & Burnett, B.R., 1978. Scavenging amphipods from the floor of the Philippine Trench. *Deep-Sea Research*, **25**, 1029–1047.

Hessler, R.R. & Jumars, P.A., 1974. Abyssal community analysis from replicate box cores in the central North Pacific. *Deep-Sea Research*, **21**, 185–209.

Hessler, R.R. & Sanders, H.L., 1967. Faunal diversity in the deep-sea. *Deep-Sea Research*, **14**, 65–78.

Higgins, R.C., 1972. Comparative efficiencies of the Smith–McIntyre and Baird grabs in collecting *Echinocardium cordatum* (Pennant) (Echinoidea: Spatangoida) from a muddy substrate. *Records New Zealand Oceanographic Institute*, **1**, 135–140.

Holme, N.A., 1949. A new bottom-sampler. *Journal of the Marine Biological Association of the United Kingdom*, **28**, 323–332.

Holme, N.A., 1953. The biomass of the bottom fauna in the English Channel off Plymouth. *Journal of the Marine Biological Association of the United Kingdom*, **32**, 1–49.

Holme, N.A., 1959. A hopper for use when sieving bottom samples at sea. *Journal of the Marine Biological Association of the United Kingdom*, **38**, 525–529.

Holme, N.A., 1961. The bottom fauna of the English Channel. *Journal of the Marine Biological Association of the United Kingdom*, **41**, 397–461.

Holme, N.A., 1964. Methods of sampling the benthos. *Advances in Marine Biology*, **2**, 171–260.

Holme, N.A., 1974. Recording schemes for benthic macrofauna. *International Council for the Exploration of the Sea*, **CM 1974/K:17**, 5 pp.

Holme, N.A. & Barrett, R.L., 1977. A sledge with television and photographic cameras for quantitative investigation of the epifauna on the continental shelf. *Journal of the Marine Biological Association of the United Kingdom*, **57**, 391–403.

Holme, N.A. & Nichols, D., 1980. Habitat survey cards for the shores of the British Isles. *Occasional Publications of the Field Studies Council*, **2**, 16 pp.

Hopkins, T.L., 1964. A survey of marine bottom samplers. *Progress in Oceanography*, **2**, 215–256.

Huberdeau, L. & Brunel, P., 1982. Efficacité et sélectivité faunistique comparée de quatre appareils de prélèvements endo-, épi- et suprabenthiques sur deux types de fonds. *Marine Biology*, **69**, 331–343.

Hudson, P.L., 1970. Quantitative sampling with three benthic dredges. *Transactions of the American Fisheries Society*, **99**, 603–607.

Humphris, N.W., 1971. Marine Planning. Its incorporation into town and country planning. *Surveyor, Local Government Technology*, February 19, 1971, 37–46 + Appendix.

Hunter, W. & Simpson, A.E., 1976. A benthic grab designed for easy operation and durability. *Journal of the Marine Biological Association of the United Kingdom*, **56**, 951–957.

Ingham, A.E. (ed.), 1975. *Sea Surveying*. John Wiley & Sons, London, text, 306 pp., illustrations, 233 pp.

Isaacs, J.D. & Schick, G.B., 1960. Deep-sea free instrument vehicle. *Deep-Sea Research*, **7**, 61–67.

Isaacs, J.D. & Schwartzlose, A., 1975. Active animals of the deep-sea floor. *Scientific American*, **233(4)**, 85–91.

Ivanov, A.I., 1965. Underwater observations of the functioning of sampling equipment for benthos collections (Petersen, Okean dredges). *Okeanologiia*, **5(5)**, 917–23 (in Russian); *Oceanology*, **5**, 119–24 (English translation).

Jensen, K., 1981. Comparison of two bottom samplers for benthos monitoring. *Environmental Technology Letters*, **2**, 81–84.

Johansen, A.C., 1927. Preliminary experiments with Knudsen's bottom sampler for hard bottom. *Meddelelser fra Kommissionen for Havundersøgelser, Ser. Fisk*, **8(4)**, 6 pp.

Jonasson, A. & Olausson, E., 1966. New devices for sediment sampling. *Marine Geology*, **4**, 365–372.

Jónasson, P.M., 1955. The efficiency of sieving techniques for sampling freshwater bottom fauna. *Oikos*, **6**, 183–207.

Jones, R. (ed.), 1979. Materials and methods used in marking experiments in fishery research. *FAO Fisheries Technical Paper*, **190**, 134 pp.

Jones, W.E., 1980. Field teaching methods in shore ecology. Chapter 2, pp. 19–44 in J.H. Price, D.E.G. Irvine & W.F. Farnham (eds) *The Shore Environment. Volume 1: Methods*, Systematics Association Special Volume 17(a). Academic Press, London, 321 + 41 pp.

Jumars, P.A., 1975. Methods for measurement of community structure in deep-sea macrobenthos. *Marine Biology*, **30**, 245–252.

Kain, J.M., 1958. Observations on the littoral algae of the Isle of Wight. *Journal of the Marine Biological Association of the United Kingdom*, **37**, 769–780.

Kajak, Z., 1971. Benthos of standing water. pp. 25–65 in W.T.Edmondson & G.G. Winberg (eds) *A Manual on Methods for the Assessment of Secondary Productivity in Fresh Waters*. International Biological Programme Handbook No. 17. Blackwell Scientific Publications, Oxford, 358 pp.

Kajak, Z., Dusoge, K. & Prejs, A., 1968. Application of the flotation technique to assessment of absolute numbers of benthos. *Ekologia Polska, Ser. A*, **29**, 607–619.

Kanneworff, P., 1979. Density of shrimp (*Pandalus borealis*) in Greenland waters observed by means of photography. *Rapports et Procès-verbaux des Réunions, Conseil International pour l'Exploration de la Mer*, **175**, 134–138.

Kanneworff, E. & Nicolaisen, W., 1973. 'The Haps'. A frame-supported bottom corer. *Ophelia*, **10**, 119–128.

Kaplan, E.H., Welker, J.R. & Krause, M.G., 1974. A shallow-water system for sampling macrobenthic infauna. *Limnology and Oceanography*, **19**, 346–350.

Kawamura, G. & Bagarinao, T., 1980. Fishing methods and gears in Panay Island, Philippines. *Memoirs of Faculty of Fisheries Kagoshima University*, **29**, 81–121.

Kirby, R., 1973. The U.C.S. grain-size comparator disc. *Marine Geology*, **14**, M11–14.

Kissam, P., 1956. *Surveying*, 2nd ed. McGraw-Hill Book Company, New York, 482 pp.

Knight, S.J.T. & Mitchell, R., 1980. The survey and nature conservation assessment of littoral areas. pp. 303–321 in J.H. Price, D.E.G. Irvine & W.F. Farnham (eds) *The Shore Environment, Volume 1: Methods*, Systematics Association Special Volume 17(a). Academic Press, London, 321 + 41 pp.

Knudsen, M., 1927. A bottom sampler for hard bottom. *Meddelelser fra Komissionen for Havundersøgelser, ser Fisk*, **8(3)**, 4 pp.

Kohlenstein, L.C., 1972. Systems for storage, retrieval and analysis of data. *Chesapeake Science*, **13**, S157–168.

Kristjonsson, H. (ed.), 1959. *Modern Fishing Gear of the World*. Fishing News (Books), London, 607 pp.

Kullenberg, B., 1951. On the shape and the length of the cable during a deep-sea trawling. *Reports of the Swedish Deep-Sea Expedition, 1947–1948*, **2(1)**, 29–44.

Kushlan, J.A., 1981. Sampling characteristics of enclosure fish traps. *Transactions of the American Fisheries Society*, **110**, 557–562.

Kutty, M.K. & Desai, B.N., 1968. A comparison of the efficiency of the bottom samplers used in benthic studies off Cochin. *Marine Biology*, **1**, 168–171.

Lackey, R.T. & May, B.E., 1971. Use of sugar flotation and dye to sort benthic samples. *Transactions of the American Fisheries Society*, **100**, 794–797.

Lagler, K.F., 1978. Capture, sampling and examination of fishes. Chapter 2, pp. 6–47 in T. Bagenal (ed.) *Methods for Assessment of Fish Production in Fresh Waters*. International Biological Programme Handbook No. 3 (3rd ed). Blackwell Scientific Publications, Oxford, 365 pp.

Lamotte, M. & Bourlière, F., 1971. *Problèmes d'Ecologie: l'echantillonage des peuplements animaux des milieux aquatiques*. Comité francais du Programme Biologique International. Masson et Cie, Paris, 294 pp.

Larsen, P.F., 1974. A remotely operated shallow water benthic suction sampler. *Chesapeake Science*, **15**, 176–178.

Lassig, J., 1965. An improvement to the van Veen bottom grab. *Journal du Conseil Permanent International pour l'Exploration de la Mer*, **29**, 352–353.

Laubier, L., Martinais, J. & Reyss, D., 1971. Operations de dragages en mer profonde. Optimisation du traict et determination des trajectoires grace aux techniques ultrasonores. *Rapport Scientifiques et Techniques, Centre National Pour l'Exploitation des Oceans*, **3**, 23 pp. (French and English texts.)

Lauff, G.M., Cummins, K.W., Eriksen, C.H. & Parker, M., 1961. A method for sorting bottom fauna samples by elutriation. *Limnology and Oceanography*, **6**, 462–466.

Laughton, A.S., 1967. Dredging. *International Dictionary of Geophysics*, pp. 261–262. Pergamon Press, Oxford.

Lewis, F.G. III & Stoner, A.W., 1981. An examination of methods for sampling macrobenthos in seagrass meadows. *Bulletin of Marine Science*, **31**, 116–124.

Lewis, J.R., 1964. *The Ecology of Rocky Shores*. English Universities Press, London, 323 pp.

Lie, U. & Pamatmat, M.M., 1965. Digging characteristics and sampling efficiency of the 0·1 m^2 van Veen grab. *Limnology and Oceanography*, **10**, 379–384.

Lincoln, R.J. & Sheals, J.G., 1979. *Invertebrate animals. Collection and Preservation*. British Museum (Natural History), 150 pp.

Lisitsin, A.I. & Udintsev, G.B., 1955. A new type of grab. *Trudÿ Vsesoyuznogo gidrobiologicheskogo obshchestva*, **6**, 217–222 (in Russian).

Little, F.J. & Mullins, B., 1964. Diving plate modification of Blake (beam) trawl for deep-sea sampling. *Limnology and Oceanography*, **9**, 148–150.

Longhurst, A.R., 1964. A review of the present situation in benthic synecology. *Bulletin de l'Institut Océanographique de Monaco*. **63,** No. 1317, 54 pp.

Macer, T.C., 1967. A new bottom-plankton sampler. *Journal du Conseil Permanent International pour l'Exploration de la Mer*, **31**, 158–163.

MacGinitie, G.E., 1935. Ecological aspects of a California marine estuary. *American Midland Naturalist*, **16**, 629–765.

MacGinitie, G.E., 1939. Littoral marine communities. *American Midland Naturalist*, **21**, 28–55.

Mason, J., Chapman, C.J. & Kinnear, J.A.M., 1979. Population abundance and dredge efficiency studies on the scallop, *Pecten maximus* (L.). *Rapports et procès verbaux des réunions. Conseil International pour l'Exploration de la Mer*, **175**, 91–96.

Massé, H., 1967. Emploi d'une suceuse hydraulique transformée pour les prélèvements dans les substrats meubles infralittoraux. *Helgoländer wissenschaftliche Meeresuntersuchungen*, **15**, 500–505.

Massé, H., Plante, R. & Reys, J.P., 1977. Etude comparative de l'efficacité de deux bennes et d'une suceuse en fonction de la nature du fond. pp. 433–41 in B.F. Keegan, P. ÓCéidigh & P.J.S. Boaden (ed) *Biology of Benthic Organisms. Eleventh European Symposium on Marine Biology. Galway, October 1976*. Pergamon Press, Oxford, 630 pp.

Mattheisen, G.C., 1960. Intertidal zonation in populations of *Mya arenaria*. *Limnology and Oceanography*, **5**, 381–388.

McAllister, R.F., 1957. Photography of submerged vertical structures. *Transactions of the American Geophysical Union*, **38**, 314–319. (*Contributions of the Scripps Institute of Oceanography*, 1957, **No. 924.**)

McIntyre, A.D., 1956. The use of trawl, grab and camera in estimating marine benthos. *Journal of the Marine Biological Association of the United Kingdom*, **35**, 419–429.

McIntyre, A.D., 1961. Quantitative differences in the fauna of boreal mud associations. *Journal of the Marine Biological Association of the United Kingdom*, **41**, 599–616.

McIntyre, A.D. & Eleftheriou, A., 1968. The bottom fauna of a flatfish nursery ground. *Journal of the Marine Biological Association of the United Kingdom*, **48**, 113–142.

McManus, D.A., 1965. A large-diameter coring device. *Deep-Sea Research*, **12**, 227–232.

McNeeley, R.L. & Pereyra, W.T., 1961. A simple screening device for the separation of benthic samples at sea. *Journal du Conseil Permanent International pour l'Exploration de la Mer*, **26**, 259–262.

Menzies, R.J., 1962. The isopods of abyssal depths in the Atlantic Ocean, pp. 79–206 in J.L. Barnard, R.J. Menzies & M.C. Bacescu (eds) *Abyssal Crustacea*, Vema Research Series, 1. Columbia University Press, New York, 223 pp.

Menzies, R.J., 1964. Improved techniques for benthic trawling at depths greater than 2000 metres. *Antarctic Research Series*, **1**, 93–109.

Menzies, R.J., 1972. Current deep benthic sampling techniques from surface vessels. pp. 164–9 in R.W. Brauer (ed.) *Barobiology and the Experimental Biology of the Deep Sea*. University of North Carolina, 428 pp.

Menzies, R.J., George, R.Y. & Rowe, G.T., 1973. *Abyssal Environment and Ecology of the World Oceans.* John Wiley & Sons, New York, 488 pp.

Menzies, R.J. & Rowe, G.T., 1968. The LUBS, a large undisturbed bottom sampler. *Limnology and Oceanography*, **13**, 708–714. (*Contribution from Florida State University*, 224.)

Menzies, R.J., Smith, L. & Emery, K.O., 1963. A combined underwater camera and bottom grab: a new tool for investigation of deep-sea benthos. *Internationale Revue der Gesamten Hydrobiologie*, **48**, 529–545.

Morgans, J.F.C., 1965. A simple method for determining levels along seashore transects. *Tuakara*, **13**, pt. 3.

Motoh, H., 1980. Fishing gear for prawn and shrimp used in the Philippines today. *Technical Report Aquaculture Department, Southeast Asian Fisheries Development Centre*, **5**, 43 pp.

Mulder, M. & Arkel, M.A., Van, 1980. An improved system for quantitative sampling of benthos in shallow water using the flushing technique. *Netherlands Journal of Sea Research*, **14**, 119–122.

Myers, A.C., 1979. Summer and winter burrows of a mantis shrimp, *Squilla empusa*, in Narragansett Bay, Rhode Island (U.S.A.). *Estuarine and Coastal Marine Science*, **8**, 87–98.

Myers, E.H., 1942. Rate at which Foraminifera are contributed to marine sediments. *Journal of Sedimentary Petrology*, **12**, 92–95. (*Collected Reprints, Woods Hole Oceanographic Institution*, **314**.)

Nalepa, T.F. & Robertson, A., 1981. Screen mesh size affects estimates of macro- and meio-benthos abundance and biomass in the Great Lakes. *Canadian Journal of Fisheries and Aquatic Sciences*, **38**, 1027–1034.

Nalwalk, A.J., Hersey, J.B., Reitzel, J.S. & Edgerton, H.E., 1962. Improved techniques of deep-sea rock dredging. *Deep-Sea Research*, **8**, 301–302. (*Collected Reprints, Woods Hole Oceanographic Institution*, **1216**.)

Nelson-Smith, A., 1979. Monitoring the effect of oil pollution on rocky sea shores. pp. 25–53 in D. Nichols (ed.) *Monitoring the Marine Environment*, Institute of Biology Symposia No. 24. Institute of Biology, London, 205 pp.

Nikolskii, G.V., 1969. *Theory of Fish Population Dynamics as the Biological Background for Rational Exploitation and Management of Fishery Resources.* Oliver and Boyd, Edinburgh, 323 pp. (Translated from Russian.)

Nybelin, O., 1951. Introduction and station list. *Reports of the Swedish Deep-Sea Expedition 1947–1948*, **2**(**1**), 3–28.

Ockelmann, K.W., 1964. An improved detritus-sledge for collecting meiobenthos. *Ophelia*, **1**, 217–222.

Owen, D.M., Sanders, H.L. & Hessler, R.R., 1967. Bottom photography as a tool for estimating benthic populations. Chapter 21, pp. 229–34, in J.B. Hersey (ed.) *Deep-Sea Photography*. The Johns Hopkins Oceanographic Studies, **3**, 310 pp.

Pamatmat, M.M., 1968. Ecology and metabolism of a benthic community on an intertidal sandflat. *Internationale Revue der Gesamten Hydrobiologie*, **53**, 211–298. (*Department of Oceanography, University of Washington, Contribution* **427**.)

Paterson, C.G. & Fernando, C.H., 1971. A comparison of a simple corer and an Ekman grab for sampling shallow water benthos. *Journal of the Fisheries Board of Canada*, **28**, 365–368.

Paul, A.Z., 1973. Trapping and recovery of living deep-sea amphipods from the Arctic Ocean floor. *Deep-Sea Research*, **20**, 289–290.

Pauly, D., 1973. Über ein Gerät zur Vorsortierung von Benthosproben. *Berichte der Deutschen Wissenschaftlichen Kommission für Meeresforschung*, **22**, 458–460.

Pearcy, W.G., 1972. Distribution and diel changes in the behavior of pink shrimp, *Pandalus jordani*, off Oregon. *Proceedings of the National Shellfisheries Association*, **62**, 15–20.

Pedrick, R.A., 1974. Nonmetallic elutriation and sieving device for benthic macrofauna. *Limnology and Oceanography*, **19**, 535–538.

Pemberton, G.S., Risk, M.J. & Buckley, D.E., 1976. Supershrimp: deep bioturbation in the Strait of Canso, Nova Scotia. *Science, New York*, **192**, 790–791.

Peters, R.D., Timmins, N.T., Calvert, S.E. & Morris, R.J., 1980. The IOS box corer: its design, development, operation and sampling. *Institute of Oceanographic Sciences*, **106**, 7 pp. + figs. (Unpublished manuscript.)

Petersen, C.G.J. & Boysen Jensen, P., 1911. Valuation of the sea. I. Animal life of the sea bottom, its food and quantity. *Report from the Danish Biological Station*, **20**, 81 pp.

Petersen, G.H., 1977. The density, biomass and origin of the bivalves of the central North Sea. *Meddelelser fra Danmarks fiskeri-og Havundersøgelser*, **N.S. 7**, 221–273.

Phillips, B.F. & Scolaro, A.B., 1980. An electrofishing apparatus for sampling sublittoral benthic marine habitats. *Journal of Experimental Marine Biology and Ecology*, **47**, 69–75.

Pickett, G., 1973. The impact of mechanical harvesting on the Thames estuary cockle fishery. *Laboratory Leaflet, MAFF*, **29**, 21 pp.

Powers, C.F. & Robertson, A., 1967. Design and evaluation of an all-purpose benthos sampler. *Great Lakes Research Division, Special Report*, **30**, 126–131.

Price, J.H., Irvine, D.E.G. & Farnham, W.F. (eds), 1980. *The Shore Environment. Volume 1. Methods*, Systematics Association Special Volume 17(a). Academic Press, London, 321 + 41 pp.

Pugh, J.C., 1975. *Surveying for Field Scientists*. Methuen, London, 230 pp.

Pullen, E.J., Mock, C.R. & Ringo, R.D., 1968. A net for sampling the intertidal zone of an estuary. *Limnology and Oceanography*, **13**, 200–202.

Reineck, H.E., 1958. Kastengreifer und Lotröhre 'Schnepfe', Gerate zur Entnahme ungestörter, orientierter Meeresgrundproben. *Senckenbergiana lethaea*, **39**, 42–48; 54–56.

Reineck, H.E., 1963. Der Kastengreifer. *Natur und Museum*, **93**, 102–108.

Reineck, H.E., 1967. Ein Kolbenlot mit Plastik-Rohren. *Senckenbergiana lethaea*, **48**, 285–289.

Reish, D.J., 1959a. A discussion of the importance of screen size in washing quantitative marine bottom samples. *Ecology*, **40**, 307–309.

Reish, D.J., 1959b. Modification of the Hayward orange-peel bucket for bottom-sampling. *Ecology*, **40**, 502–503.

Reys, J.-P., 1964. Les prélèvements quantitatifs du benthos de substrats meubles. *Terre et la Vie*, **1**, 94–105. (*Recueil des Travaux de la Station Marine d' Endoume*, **46(18)**.)

Reys, J.-P. & Salvat, B., 1971. L'enchantillonage de la macrofaune des sediments meubles marins. pp. 185–242 in M. Lamotte & F. Bourlière (eds) *Problèmes d'Ecologie: l'echantillonage des peuplements animaux des milieux aquatiques*, Comité francais du Programme Biologique International. Masson et Cie, Paris, 294 pp.

Rice, A.L., Aldred, R.G., Darlington, E. & Wild, R.A., 1982. A quantitative estimation of the deep-sea megabenthos; a new approach to an old problem. *Oceanologica Acta*, **5**, 63–72.

Richards, S.W. & Riley, G.A., 1967. The benthic epifauna of Long Island Sound. *Bulletin of the Bingham Oceanographic Collection*, **19**, 89–135.

Ricker, W.E., 1975. Computation and interpretation of biological statistics of fish population. *Bulletin of the Fisheries Research Board of Canada*, **191**, 382 pp.

Riedl, R., 1961. Études des fonds vaseux de l'Adriatique. Méthodes et résultats. *Recueil des Travaux de la Station Marine d'Endoume*, **23**, 161–169.

Riley, J.D. & Corlett, J., 1965. The numbers of O-group plaice in Port Erin Bay. *Report of the Marine Biological Station, Port Erin*, **78**, 51–56.

Rowe, G.T. & Clifford, C.H., 1973. Modification of the Birge–Ekman box corer for use with SCUBA or deep submergence research vessels. *Limnology and Oceanography*, **18**, 172–175.

Rowe, G.T. & Menzies, R.J., 1967. Use of sonic techniques and tension recordings as improvements in abyssal trawling. *Deep-Sea Research*, **14**, 271–274.

Russell, B.C., Talbot, F.H., Anderson, G.R.V. & Goldman, B., 1978. Collection and sampling of reef fishes. pp. 329–345 in D.R. Stoddart & R.E. Johannes (eds) Coral reefs: research methods. *UNESCO Monographics on Oceanographic Methodology*, **5**, 581 pp.

Sanders, H.L., 1956. Oceanography of Long Island Sound, 1952–1954. X. The biology of marine bottom communities. *Bulletins of the Bingham Oceanographic Collection*, **15**, 345–414.

Sanders, H.L., Hessler, R.R. & Hampson, G.R., 1965. An introduction to the study of deep-sea benthic faunal assemblages along the Gay Head–Bermuda transect. *Deep-Sea Research*, **12**, 845–867. (*Collected Reprints, Woods Hole Oceanographic Institution*, **1666**.)

Sellmer, G.P., 1956. A method for the separation of small bivalve molluscs from sediments. *Ecology*, **37**, 206.

Shepard, F.P., 1965. Submarine canyons explored by Cousteau's diving saucer. pp. 303–309 in W.F. Whittard & R. Bradshaw (eds) *Submarine Geology and Geophysics*, Proceedings of the 17th Symposium of the Colston Research Society, University of Bristol, April 5th–9th 1965. Butterworths, London, 464 pp.

Shepard, F.P., Curry, J.R., Inman, D.L., Murray, E.A., Winterer, E.L., & Dill, R.F., 1964. Submarine geology by diving saucer. *Science, New York*, **145**, 1042–1046.

Shulenberger, E. & Hessler, R.R., 1974. Scavenging abyssal benthic amphipods trapped under oligotrophic central North Pacific Gyre waters. *Marine Biology*, **28**, 185–187.

Sjölund, T. & Purasjoki, K.J., 1979. A new mechanism for the van Veen grab. *Finnish Marine Research*, **246**, 143–146.

Sly, P.G., 1969. Bottom sediment sampling. *Proceedings of the 12th Conference on Great Lakes Research*, 883–898. (*Collected Reprints, Canada Center for Inland Waters*, **2–23**.)

Smith, K.L. & Howard, J.D., 1972. Comparison of a grab sampler and large volume corer. *Limnology and Oceanography*, **17**, 142–145.

Smith, W. & McIntyre, A.D., 1954. A spring-loaded bottom sampler. *Journal of the Marine Biological Association of the United Kingdom*, **33**, 257–264.

Southward, A.J., 1965. *Life on the Sea-Shore*. Heinemann, London, 153 pp.

Steedman, H.F. (ed.), 1976. Zooplankton fixation and preservation. *Monographs on oceanographic methodology, UNESCO, Paris*, **4**, 350 pp.

Stephen, W.J., 1977. A one-man profiling method for beach studies. *Journal of Sedimentary Petrology*, **47**, 860–863.

Steven, G.A., 1952. *Nets. How to Make, Mend and Preserve Them*. Routledge and Kegan Paul, London, 128 pp.

Stoddart, D.R. & Johannes, R.E. (eds), 1978. Coral reefs: research methods. *Monographs on oceanographic methodology, UNESCO, Paris*, **5**, 581 pp.

Strange, E.S., 1977. An introduction to commercial fishing gear and methods used in Scotland. *DAFS, Scottish Fisheries Information Pamphlet*, **1**, 34 pp.

Swartz, R.C., 1972. A preliminary design of an information storage system for biological collection data. *Chesapeake Science*, **13**, S191–197.

Thamdrup, H.M., 1938. Der van Veen-Bodengreifer. Vergleichsversuche über die Leistungsfähigkeit des van Veen- und des Petersen-Bodengreifers. *Journal du Conseil Permanent International pour l'Exploration de la Mer*, **13**, 206–212.

Thayer, G.W., Williams, R.B., Price, T.J. & Colby, D.R., 1975. A large corer for quantitatively sampling benthos in shallow water. *Limnology and Oceanography*, **20**, 474–480.

Thomas, M.L.H., 1960. A modified anchor dredge for collecting burrowing animals. *Journal of the Fisheries Research Board of Canada*, **17**, 591–594.

Thomas, M.L.H. & Jelley, E., 1972. Benthos trapped leaving the bottom in Bideford River, Prince Edward Island. *Journal of the Fisheries Research Board of Canada*, **29**, 1234–1237.

Thorson, G.E., 1957. Sampling the benthos. *Memoirs of the Geological Society of America*, **67(1)**, 61–73.

Thurston, M.H., 1979. Scavenging abyssal amphipods from the north-east Atlantic Ocean. *Marine Biology*, **51**, 55–68.

True, M.A., Reys, J.-P. & Delauze, H., 1968. Progress in sampling the benthos: the benthic suction sampler. *Deep-Sea Research*, **15**, 239–242.

Tyler, P. & Shackley, S.E., 1978. Comparative efficiency of the Day and Smith–McIntyre grabs. *Estuarine and Coastal Marine Science*, **6**, 439–445.

Ursin, E., 1954. Efficiency of marine bottom samplers of the van Veen and Petersen types. *Meddeleleser Fra Danmarks Fiskeri-og Havundersøgelser*, **N.S. 1(7)**, 8 pp.

Ursin, E., 1956. Efficiency of marine bottom samplers with special reference to the Knudsen sampler. *Meddeleleser Fra Danmarks Fiskeri-og Havundersøgelser*, **N.S. 1(14)**, 6 pp.

Uzmann, J.R., Cooper, R.A., Theroux, R.B. & Wigley, R.L., 1977. Synoptic comparison of three sampling techniques for estimating abundance and distribution of selected megafauna: submersible vs camera sled vs otter trawl. *Marine Fisheries Review*, **39(12)**, 11–19.

van Veen, J., 1933. Onderzoek naar het zandtransport von rivieren. *De Ingenieur*, **48**, 151–159.

von Brandt, A., 1972. *Fish catching methods of the world*. Fishing News (Books), West Byfleet, Surrey, 240 pp.

von Brandt, A., 1978. Bibliography for fishermen's training. *FAO Fisheries Technical Paper*, **184**, 176 pp.

Watkin, E.E., 1941. Observations on the night tidal migrant Crustacea of Kames Bay. *Journal of the Marine Biological Association of the United Kingdom*, **26**, 81–96.

Whitehouse, J.W. & Lewis, B.G., 1966. The separation of benthos from stream samples by flotation with carbon tetrachloride. *Limnology and Oceanography*, **11**, 124–126.

Wickstead, J., 1953. A new apparatus for the collection of bottom plankton. *Journal of the Marine Biological Association of the United Kingdom*, **32**, 347–355.

Wigley, R.L., 1967. Comparative efficiencies of van Veen and Smith–McIntyre grab samplers as revealed by motion pictures. *Ecology*, **48**, 168–169.

Wigley, R.L. & Emery, K.O., 1967. Benthic animals, particularly *Hyalinoecia* (Annelida) and *Ophiomusium* (Echinodermata), in sea-bottom photographs from the continental slope. Chapter 22, pp. 235–349 in J.B. Hersey (ed.) *Deep-Sea Photography*. The Johns Hopkins Oceanographic Studies, **3**, 310 pp.

Wigley, R.L. & Theroux, R.B., 1970. Sea-bottom photographs and macrobenthos collections from the continental shelf off Massachusetts. *United States Fish and Wildlife Service, Special Scientific Report—Fisheries*, **613**, 12 pp.

Wildish, D.J., 1978. Sublittoral macro-infaunal grab sampling reproducibility and cost. *Technical Report, Fisheries and Marine Service, Canada*, **770**, 4 pp.

Williams, D.D. & Williams, N.E., 1974. A counterstaining technique for use in sorting benthic samples. *Limnology and Oceanography*, **19**, 152–154.

Wolff, T., 1961. Animal life from a single abyssal trawling. *Galathea Reports*, **5**, 129–162.

Word, J.Q., 1976. Biological comparison of grab sampling devices. *Southern California Coastal Water Research Project. Annual report for the year ending 30 June 1976*, pp. 189–194.

Worswick, J.M. & Barbour, M.T., 1974. An elutriation apparatus for macroinvertebrates. *Limnology and Oceanography*, **19**, 538–540.

Zinn, D.J., 1969. An inclinometer for measuring beach slopes. *Marine Biology*, **2**, 132–134.

Zismann, L., 1969. A light-trap for sampling aquatic organisms. *Israel Journal of Zoology*, **18**, 343–348.

Chapter 7: Meiofauna Techniques

A. D. MCINTYRE AND R. M. WARWICK

Sampling

Any instrument or method suitable for sampling macrofauna will, in principle, also be suitable for the smaller organisms. Chapter 6 in this Handbook is therefore relevant, and sampling of small metazoans is discussed in a review by McIntyre (1969), and in Hulings & Gray (1971). The main difference between sampling for macrofauna and meiofauna is that, because of the much higher numerical density of the latter, smaller samples are usually adequate. These can be obtained by sub-sampling from a larger volume but this may introduce errors or inaccuracies, so there are advantages in collecting a sample which can be examined entire.

If specific animals are required in large numbers, or if only particular groups are to be studied, it may be useful to deal with quite large volumes of deposit. Thus Wells & Clark (1965), working on intertidal sand, dug out 3000 ml of sand at each station to sample copepods, and Higgins (1964) used a sledge-dredge on subtidal mud for collecting kinorhynchs. For qualitative sampling on intertidal sediments, a general impression of the meiofauna can be obtained by simply digging a trench and allowing water to accumulate in it. Actively swimming animals are then scooped from the water with a fine plankton net, and the more sedentary or adhesive forms are sampled by collecting sand grains from the trench. A refined version of this procedure, the 'methode des sondages' is often used on tideless beaches to concentrate animals from the level of the water table (Delamare-Deboutteville, 1960). Alternatively, sand can be added to filtered sea water (preferably containing anaesthetic) in a bucket, stirred, and poured through a fine sieve. Such techniques are useful in the field and deal qualitatively with quite large areas or volumes, but for quantitative work small samples are usually required, and these are most conveniently obtained in core samples.

Core sampling

Intertidal work

In intertidal areas, or in very shallow water where the apparatus can be operated by hand, a tube or pipe of any available rigid material may be adapted for use, the diameter chosen depending on the volume and depth of sample required. The tube should have a smooth internal surface to facilitate core removal. In many sandy intertidal habitats a diameter of 2–4 cm has been found to give samples which can be sorted entire, and cores of about 1 cm diameter are suitable for estuarine muds where densities are usually high. Tubes made of transparent material such as Perspex are useful because the intact sample can be seen throughout its entire length and the distribution of the soil components as well as any changes in the packing of the core can be observed. In collecting the sample, the tube is pushed or tapped gently into the sediment, the top is plugged, and the tube withdrawn, the bottom then being plugged if the sample is to be transported. This simple method usually produces a sample of 30–50 cm length, but if the sediment core tends to slide or break on withdrawal of the tube it may be necessary to excavate a trench round the tube and plug the lower end before raising it. In some deposits, forcing the tube into the bottom results in shortening of the core and distortion of the sample, especially if a narrow tube is used. If this occurs, a slight application of suction to the tube can help it to slide into the sediment and in intertidal situations this can sometimes be done by mouth. It may be convenient to attain the depth to be sampled by a series of shorter cores, a trench being excavated to the lower end of each successive sample, exposing a new horizontal level at which the next core is inserted. In this way deep layers may be sampled step-wise, as described by Bush (1966). To obtain optimum results it may be appropriate to use both vertical and horizontal coring in a single study (Hummon, 1975). In some deposits, where stones or shells are present, or when very deep cores are required in single lengths, the tube may need to be hammered into the bottom, and a metal corer is then probably required. McIntyre (1968) used a metal tube halved longitudinally, with the two parts held together by a collar so that the entire core could eventually be exposed for examination and processing. For very deep work Renaud-Debyser (1957) devised a completely demountable corer, square in cross-section. Each of the four sides consisted of a metal sheet, 100 cm long, 5 cm wide and 1 mm thick, and each was slotted so as to slide within the other sides. Since the sides can be inserted one by one into the sand, little pressure is required, and shortening of the sample does not occur. Cores as long as 1 m can be obtained, and the removal of the sides gives easy access to the sample.

In removing the sample from a standard core tube, it is best to allow the core to slide down into the collecting vessel, since this causes less disturbance than pushing it out with a piston. Fenchel (1967) describes a method by which the top stopper of the core tube is replaced by a bored cork fitted with a short glass tube attached to a length of rubber tubing closed by a clamp. Opening and closing the clamp allows the core to slide down inside the tube step-wise. Alternatively, more control may be obtained by compressing and releasing the top rubber bung with the fingers. If the deposit is of dry sand and the column tends to break rather than slide evenly, a small quantity of filtered sea water carefully added to the top helps to keep the column intact. In samples from muddy ground, the lower part of the core may form a plug of clay, and the sample must then be pushed out from below with a piston. This method is less satisfactory, since it compresses the core, and in most samples causes water and particles to mix from one layer to another. These effects can be minimized by selecting a tube length as close as possible to the sediment length required. Short cores are usually adequate on muddy grounds, since most of the animals live in the upper layers, and there is little life below 6–8 cm (Rees, 1940; McIntyre, 1968). On sand, however, with interstitial species extending to great depth, cores of 20–30 cm may be required to collect the bulk of the fauna, and animals are known to exist in considerable numbers below 50 cm (Renaud-Debyser, 1963), and even below 1 m in exposed beaches (McLachlan *et al.*, 1979), where the species are likely to be different from those in the surface layers.

Sublittoral work

In subtidal areas, core tubes again provide the most suitable means of general meiofauna sampling. Subsamples may be taken with core tubes from a grab, or gravity corers may be used directly. However, both these methods have the serious disadvantage that the downwash which occurs as the sampler descends displaces the surface layers of the sediment where the meiofauna is usually concentrated (McIntyre, 1971). This may lead to a gross under-estimate of meiofauna numbers. In shallow water, where SCUBA divers can work, some of the intertidal techniques can be highly satisfactory. In deeper water, or when divers are not available, samplers which can be lowered from a boat are required.

If there is no alternative to using gravity corers, the open-barrel gravity corer is the most suitable type, but these, often designed for geological work, are usually large and heavy. A useful, light instrument, such as that described by Moore & Neill (1930), is simply a tube constructed to fall vertically and penetrate by its own weight, and can be adapted to take one or more samples of various diameters (illustrated by Holme, 1964, Fig. 27). On soft sediments, especially where a layer of clay is present to plug the bottom of the tube, good samples up to 40 cm long can be obtained. On very soft bottoms the corer may fill completely or even bury itself in the deposit, spoiling the sample. In such areas the instrument should be lowered to a few metres above the bottom and then the warp should be paid out quickly so that the corer falls free under its own weight into the sediment. The optimum distance for this free fall, which may be determined by trial and error, will depend on the instrument as well as on local conditions. Burns (1966) showed experimentally that for corers of about 25–50 kg weight and approximately 4 cm inside diameter, the optimum free fall distance was 2–3 m.

Coring on sand is more difficult with a light instrument, and a retainer, such as that described by Mills (1961) or Fenchel (1967) is usually required to prevent loss of the sample on hauling.

A peculiarity of gravity open-barrel corers is that the length of the core retrieved may be considerably less than the penetration depth of the tube into the sediment. The shortening ratio (core length: penetration distance) may be $\leq 0{\cdot}5$ for long cores, and this ratio depends on the weight and dimensions of the corer as well as on the type of sediment. The work of Emery & Dietz (1941) suggests that the shortening is due to wall friction and that some time after initial penetration the corer begins to act as a solid rod, pushing the sediments in front of it downwards or sideways so that they eventually enter the core tube either incompletely or in layers of reduced thickness. Shortening may have important implications with regard to animal counts. Provided that the core is sufficiently long to completely sample the strata where most of the animals are concentrated, which will almost certainly be the case on soft grounds, this will not affect estimates of total fauna. Shortening may, however, affect studies of the vertical distribution of the fauna, since segments cut from successive levels down the core may not correspond to the same levels on the sea bed. The problem of core shortening may be more serious in subtidal work, if only because it could pass unnoticed.

Although various types of gravity corer have been shown to give underestimates of meiofaunal abundance (McIntyre, 1971; Elmgren, 1973), there is some evidence that, in certain situations, remote cores with a large tube (7–8 cm internal diameter) may give true estimates of abundance, whereas smaller cores are less efficient (Holopainen & Sarvala, 1975). When using this type of corer it may, therefore, sometimes be best to opt for the largest acceptable core size, and to subsample the core for the meiofauna. This has some statistical advantages when estimating meiofaunal numbers (see p. 233.

Corers have been designed which push the core-tube very gently into the sediment and these are potentially as efficient as cores taken by divers. The simplest amd most extensively tested is the Craib or Millport corer (Craib, 1965), which is essentially a frame-mounted core tube, 5–7 cm in diameter, which is forced into the bottom by weights (Fig. 7.1). The rate of penetration is controlled by a hydraulic damper and this, combined with a slow initial approach, ensures minimum disturbance of the light superficial layer of sediment. Samples of 15 cm length are produced, and because of the spherical closing device which retains the core, good samples have been reported from sediments ranging from hard sand to soft mud. The apparatus weighs 44 kg and can be used from a small boat. Its main disadvantage is that, since it must stand upright for a short period while the tube is penetrating, there may be difficulties in bad weather, on uneven bottoms, or in deep water.

The principle of the Craib corer has been further developed by the Scottish Marine Biological Association as a much larger multiple corer (Fig. 7.2) to take undisturbed samples of deep sea sediments (P.R.O. Barnett, J. Watson & A.G.H. Connelly, personal communication). This consists of an outer framework constructed from galvanized steel scaffold tubing which supports a weighted assembly of plastic core tubes on a water-filled dashpot. When the corer is lowered to the sea bed the framework (A) rests on the bottom. As the wire to the ship slackens, the dashpot (C) slowly lowers the coring assembly (B) so that the core tubes (D) penetrate the sea bed. When the wire to the ship is heaved, a special mechanism (not shown) first closes the valves (E) on the top of each core tube, and releases bottom valves (not shown). Continued heaving pulls the core tubes out of the bottom, the mud cores being retained by the seal of each top valve (E). As the tubes clear the bottom, the bottom valves drop into place and help retain the cores during recovery to the surface. A reversing thermometer also inverts at this point, giving bottom temperature. Core tubes can be varied in number from 2 to 12. The corer is proving to be reliable and undisturbed core samples have now been taken from many stations in the

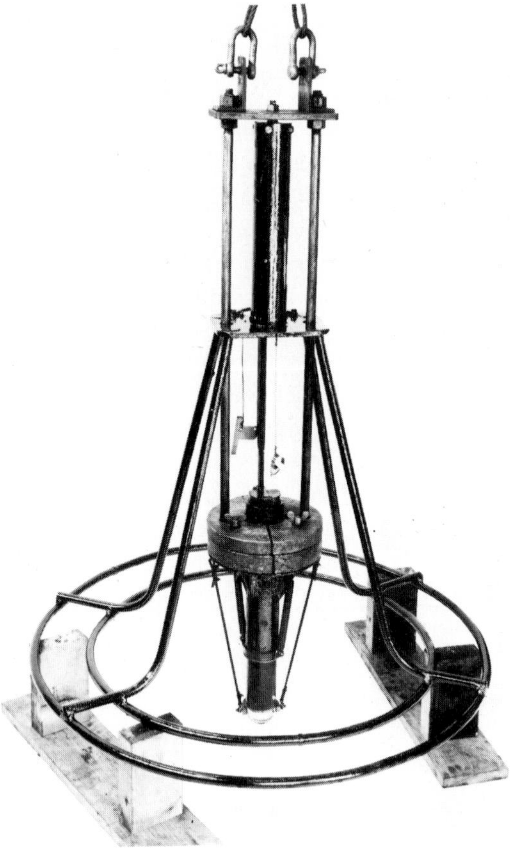

Fig. 7.1. Craib (1965) corer, with ball closing device in position. (Photograph J.S. Craib).

North Atlantic, down to depths of 5000 m. It has also taken cores successfully in sands on the continental shelf.

In the absence of a suitable corer, it may be acceptable to collect cores from a grab sample. The grab must be such that it can be opened from the top, allowing vertical cores to be taken by hand. Cores should be taken only when it has been possible to keep the grab upright with little disturbance to the surface sediment of the sample. The Bacescu sampler (Bacescu, 1957), which is a grab with a built-in core tube, was designed for such work. This procedure is useful when comparison is being made of macro- and meiofauna from the same area. Subsamples taken from box-type samplers (pp. 167–9) are better than those taken from conventional grabs, e.g. the van Veen, because the vertical structure of the sample is less disturbed (Elmgren, 1973). Thiel

Fig. 7.2. SMBA multiple corer. For explanation see p. 221. (Photograph R. Summers.)

(1966) describes a device (the 'meiostecher') for subsampling from a large sampler.

Epibenthic sampling

The epibenthic meiofauna, and species living in the surface centimetre of flocculent bottom material, can be collected non-quantitatively in large numbers for taxonomic and other purposes with epibenthic sledges. This type of apparatus can also be used in a semi-quantitative way to investigate the seasonal appearance of the so-called 'temporary meiofauna'—the juvenile stages of the macrofauna, which are often patchily distributed and usually restricted to the superficial deposits. These sledges have wide runners (Fig. 7.3), designed to disturb and skim off the sediment surface, which is then collected in a fine net (Mortensen, 1925; Ockelmann, 1964). For collecting kinorhynchs, Higgins (1964) fitted a rake-like apparatus at the leading edge behind which was a plane-blade which, like a carpenter's plane, removed the upper layer of substratum. A sledge of this kind is at best only semi-quantitative, although a 'super-gadgeted' version described by Bieri & Tokioka (1968) and fitted with a flow meter and odometer seems to represent a

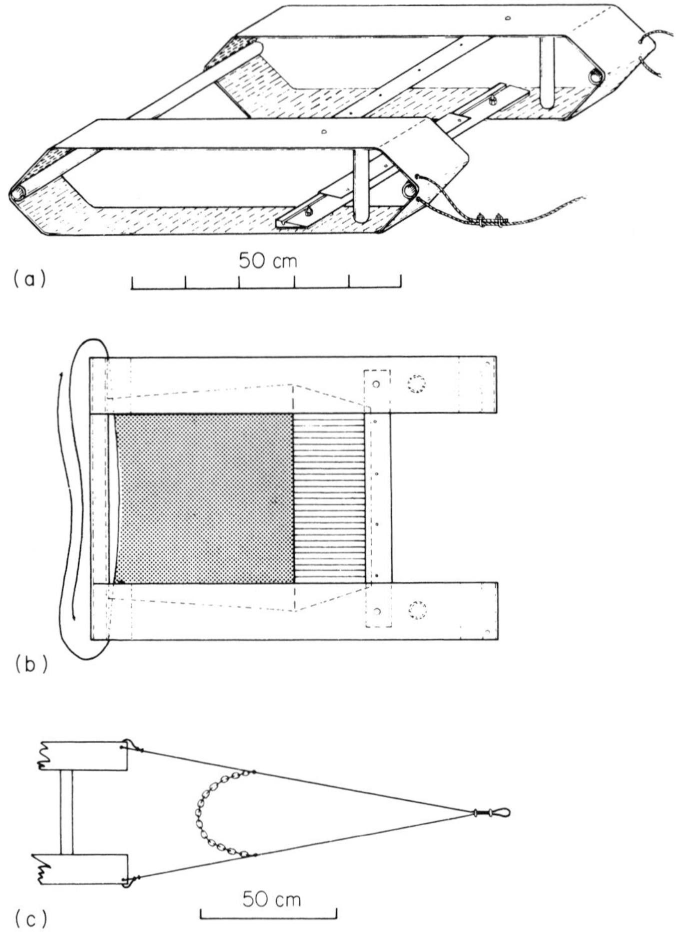

Fig. 7.3. Ocklemann (1964) detritus-sledge. (a) Parallel-perspective view (no net); (b) view from above; (c) view from above, showing wire bridles and tickler chain. (Redrawn from Ocklemann, 1964.)

considerable advance. If fully quantitative data are required, a known surface area and depth must be sampled. While this can be done by sub-sampling from a standard grab haul, some devices have been designed with quantitative collection of small macrofauna particularly in mind. Muus (1964) described an instrument which cuts out the top 2–3 cm of deposit below a surface area of 150 cm^2, and collects it in a nylon bag. In operation, this 'mouse-trap' releases and closes after penetrating the bottom and it is claimed that no downwash is caused which might disturb the surface deposit. Its total weight is 25–30 kg, and it can be used from a small boat. Sarvala & Ranta (1977) found that the Muus sampler was as efficient as diver-collected cores for surface organisms, but Corey & Craib (1966) considered that it was not suitable for such animals as amphipods, isopods and cumaceans which became free-swimming when disturbed. To overcome this difficulty they describe a more complicated device which samples a surface area of 0·05 m^2 to a depth of 3 cm. It releases and closes automatically when the lowering rope slackens and the instrument is designed so as to prevent any disturbance of the sea bed during initial penetration.

Treatment and sorting of samples

In order to examine and count meiofauna, the animals must be extracted from the sediment. Extraction methods vary according to the type of sediment and depend on whether extraction is to be qualitative—to obtain representative specimens—or quantitative, i.e. to extract every organism possible for detailed counts. Extraction can be done on either fresh or preserved samples. If the vertical distribution of the fauna is to be studied it is essential that the sample should be divided into appropriate sections immediately on collection, since changes within the sample (e.g. in packing, water content, temperature) can produce rapid alterations in the vertical distribution of the fauna.

Sediment samples are usually brought back to the laboratory for extraction. Preservation and extraction techniques depend on the degree to which taxa are to be identified. Certain groups of 'hard' meiofauna such as nematodes, copepods, ostracods and kinorhynchs remain identifiable after rough preservation in the sediment using 4 % formaldehyde. 'Soft' meiofauna (gastrotrichs, turbellarians, etc.) are, however, difficult to recognize after this treatment, and for these, live extraction and examination are essential.

For qualitative extraction the sample can be allowed to stand in sea water in the laboratory and the deposit examined at intervals when organisms which come to the surface (often aggregated away from the light) can be pipetted off. Stirring the sample or bubbling air through it, brings certain types of animals (such as most kinorhynchs and some small crustaceans which have a hydrophobic cuticle) to the surface film of the water, from where they can be scooped off using a wire loop or blotting paper. A bicycle pump attached to an aeration block by flexible tubing can be used in the field to bubble samples. For deep-living fauna the sample may be shaken and decanted, preferably using an anaesthetic as described below.

Techniques for quantitative extraction fall into two broad categories: first, those like decantation, elutriation and flotation which rely on the differential rates of sedimentation of organisms and sediment particles and which are suitable for both living and preserved material, and second, techniques which employ an environmental gradient to drive living animals out of the sediment. Recommended extraction methods for different situations are given in Table 7.1, and are detailed below.

Table 7.1. Summary of recommended methods for extracting meiofauna from sediments.

	Coarse sediments	Fine sediments
Live material	1. Simple decantation with anaesthetic 2. Elutriation 3. Sea-water ice	1. Sieve out fine fraction, examine sieve residue 2. Centrifugation in silica sol–sorbitol mixture
Preserved material	Pretreatment: stain with Rose Bengal 1. Simple decantation 2. Elutriation	1. Sieve out fine fraction, examine sieve residue 2. Flotation or centrifugation in colloidal silica (Ludox)

Extraction from coarse sediments

Live material

Where the sediment particles are heavier than the animals living in the interstices, decantation techniques may be used, and these may be simple or elaborate. It is important always to use filtered water to prevent contamination of the sample with organisms which might be present in, for example, the deck water supply of a research vessel.

For simple decantation, a good procedure is as follows:

1. Wash the sample into a 1 litre capacity stoppered measuring cylinder and make up to 800 ml with sea water, giving a sedimentation height of about 30 cm.
2. Invert the cylinder several times to suspend the sediment and then leave until the sand particles sediment out (about 60 seconds).
3. Decant the supernatant through a sieve with a pore size of $\leq 63\ \mu m$;
4. Repeat several times (three is usually sufficient);
5. Wash the material off the sieve into a petri-dish for counting.

Best results are obtained if the whole sample is first treated with an anaesthetic, since many interstitial animals tend to attach to sand grains when motion occurs. A solution of magnesium chloride isotonic with sea water is widely used—75·25 g $MgCl_2.6H_2O$ dissolved in 1 litre distilled water is approximately isotonic with sea water of 35 per mille salinity. (Since this salt is very hygroscopic, a solution is made up more accurately by heating the salt to constant weight at 500–600 °C, 35·24 g of the anhydrous salt being used to make 1 litre of solution). This is sufficient to relax the fauna without adverse effects after several minutes' treatment, but the addition of 10 % alcohol is also satisfactory.

Elutriation is a more sophisticated, and perhaps more efficient, version of

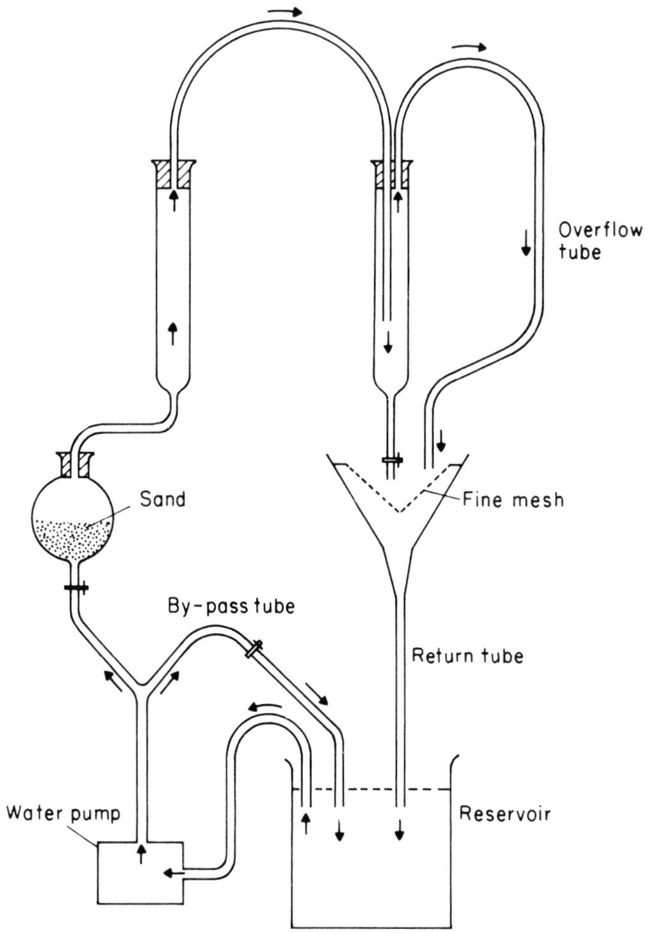

Fig. 7.4. Boisseau-type apparatus for elutriation of sand; closed-system arrangement. For discussion see text.

the decanting procedure. It has been used for macrofauna separation (p. 199), and Boisseau (1957) introduced it for meiofauna work. A possible set-up, illustrated in Fig. 7.4, involves the use of a small pump with a by-pass to control the pressure. Alternatively, pressure can be provided by raising the reservoir high enough to provide a head, or by direct attachment to a suitable tap, and the water need not be recirculated. Filtered water should be used in the reservoir, or a filter must be introduced into the direct system to prevent contamination of the sample. The time required for elutriation depends on the size and nature of the sample: samples of 20 ml can usually be completed in a few minutes. The fauna is collected on a fine sieve and the flow should be arranged to meet the sieve surface at an oblique angle to minimize the loss of animals of roughly the same cross-section diameter as the sieve mesh. Any such loss may be further reduced by the addition of a second sieve below the first. Elutriation carries over most of the light fauna, but the residue of sediment should be examined for heavier organisms (molluscs, ostracods) which may have been left behind and which can be extracted by hand. Tiemann & Betz (1979) used a long, narrow, elutriation vessel, 40 cm in height

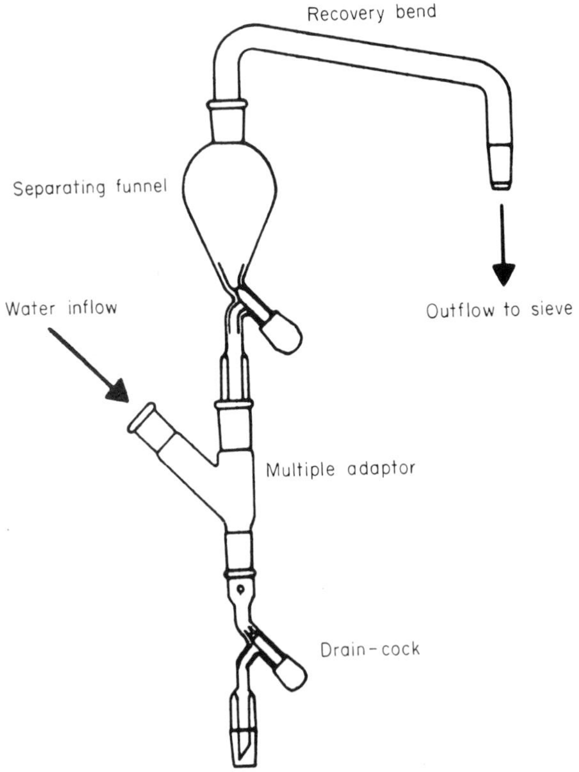

Fig. 7.5. Elutriation apparatus using Quickfit glassware. (From Hockin, 1981.)

and 8 cm at its widest part. They emphasize the value of this shape in reducing turbulence and achieving better animal/sediment separation.

A simpler version of this system has been described by Hockin (1981). It is constructed from readily available 'Quickfit' laboratory glassware (Fig. 7.5). An apparatus for elutriation in warm water, described by Uhlig *et al.* (1973), has improved efficiency for nematodes, which stretch out under these conditions, and for ostracods which open their shells, resulting in higher water resistance.

Behavioural methods for extraction of living animals were originally developed for terrestrial work, and make use of the activity of the organisms in response to changes imposed by the operator on their physical conditions. A technique was developed by Uhlig (1966, 1968) specifically for marine work. The sediment is placed in a tube on a nylon gauze base which just dips into filtered sea water in a collection dish (Fig. 7.6). Crushed sea-water ice is added to a layer of cotton wool on top of the sediment, and as the ice melts, organisms move down into the collecting dish. Uhlig considered that movement of the animals is due more to the salinity and the streaming action of the water than to the temperature gradient. The method has certain limitations and can be applied only to sandy sediments, which have a capillary structure.

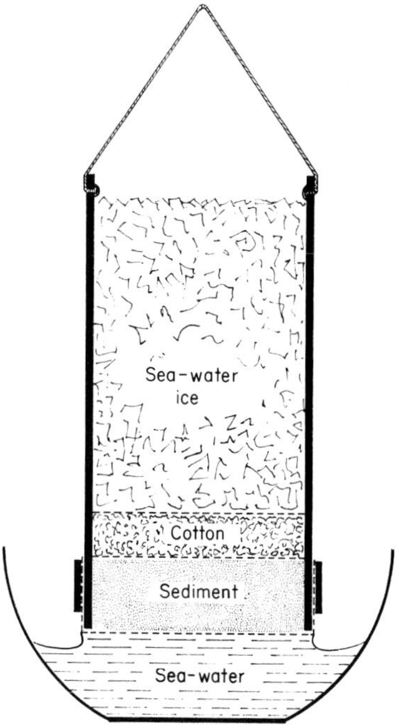

Fig. 7.6. Apparatus for separating interstitial fauna from sand by sea-water ice method. (Redrawn from Uhlig, 1968.)

Most of the smaller forms leave the deposit and concentrate in the collecting dish but nematodes and some of the larger animals are not extracted quantitatively, and the sediment should later be elutriated or decanted to obtain complete extraction. The technique is, however, well suited to 'soft' meiofauna (and also to ciliates and flagellates).

Preserved material

While live extraction is the ideal, the number of samples involved, distance of the laboratory from the sampling site or the time constraints imposed, may make this impossible, and in such circumstances the entire sample (or the sections into which it has been divided) should be preserved in 4–5% formaldehyde for storage or transport.

Most of the decantation and elutriation techniques available for live samples can also be applied to preserved material, but tap water can, of course, replace sea water for these procedures. Indeed, extraction of preserved material may be more efficient because preserved animals tend to sediment out of suspension more slowly (many live animals curl into a tight ball) and because the thread-like species are less likely to pass through the meshes of a sieve, which they can do by active burrowing when alive.

The final process of sorting preserved material tends to be less certain because the smallest animals are difficult to detect when not in motion. It is therefore useful to stain the entire sample before extraction. Rose Bengal has frequently been used for this purpose, and a solution of 1 g in 1 litre of water, formalin or alcohol imparts sufficient colour after several hours' immersion. Thiel (1966) advocates the addition of 5% phenol in aqueous solution to eliminate differences in staining efficiency caused by varying pH. Hamilton (1969) describes a method, developed in freshwater work, by which the samples are stained in rhodamine B and examined under ultraviolet light. Most invertebrates fluoresce a brilliant orange, and in the presence of mud sediment and debris, the time required to sort small organisms can often be reduced to about one-third.

Extraction from fine sediments

Live material

For quantitative extraction of living material from fine muds and silts, it is usually convenient to divide the sample into two size fractions using a fine sieve. A mesh of 62 or 50 μm is appropriate since one of these is usually accepted as defining the upper limit of the silt-clay fraction of the sediment, but even finer meshes (30 or 40 μm) are often used to ensure that most of the fauna is retained in the sieve residue.

The residue is hand-sorted under a stereoscopic microscope, attention always being paid to the surface film of the overlying water. This is of particular importance in the case of some groups such as Kinorhyncha, which tend to be caught by surface tension. Most of the silt-clay fraction of the deposit, along with the smallest meiofauna—mainly juvenile stages—passes through the sieve. Normal hand sorting from such fine material is difficult, and an appropriate subsampling technique can be employed (see p. 233). The volume thus subsampled should be related to the size of the sorting dish so that the settled subsample covers the bottom of the dish with a layer only a few grains thick. When viewed by light from below, living animals can easily be seen either directly as they move or by the track they leave. Counts from the subsample are adjusted to give an estimate for the total volume of water, and this is added to the sieve residue count to obtain a value for the original sample. The section of the sample awaiting examination must be kept under suitable conditions of temperature and restricted light so that the animals remain in good condition. The whole procedure is time-consuming, and restricts the number of samples which can be dealt with, but it does permit an accurate count of the fauna in fine muddy deposits.

With many organic muds large quantities of light debris present in the sieve residue severely hamper the sorting of meiofauna. Until recently, density separation techniques used to separate preserved animals from such debris (p. 232) have not been employed for living meiofauna because the flotation media are usually toxic, or cause distortion of the organisms. However, Schwinghamer (1981) has described a method for the extraction of living organisms from such sediments using centrifugation in a buffered mixture of sorbitol and the silica sol 'Percoll' (Pharmacia Fine Chemicals AB, Uppsala, Sweden), which is non-toxic.

The 'Percoll' sorbitol mixture is prepared by first dissolving 91·1 g sorbitol, 2·64 g Tris-HCl, and 4·03 g Tris base in 0·5 litre 'Percoll'. Then 1·43 g $MgCl_2 \cdot 6H_2O$ is dissolved in 100 ml filtered sea water, and this is added to the 'Percoll' mixture. The volume is made up to 1000 ml with 'Percoll'. The mixture should be filtered and stored in a refrigerator, as it is a good growth medium for bacteria and diatoms. The density of the mixture should be 1·15 and this should be checked regularly. 5 ml of untreated sediment or sediment plus an equal volume of 7% $MgCl_2$ is added to 20 ml of Percoll-sorbitol in a 50 ml centrifuge tube (round bottom is preferable), mixed gently by inversion or vortex mixer and centrifuged at 200–300 r.p.m. for 15–30 min. The speed and time of centrifugation, which should be sufficient to form a compact plug and sediment-free supernatant, but allow for easy resuspension of the plug for a second extraction, must be determined for different sediment types. The residue of the first extraction usually needs to be resuspended in 20 ml of Percoll-sorbitol by inversion or vortex mixer and recentrifuged in order to obtain an adequate representation of all the taxa present. The supernatant can be collected by decantation through a suitable mesh screen. Two such extractions have been found sufficient for removal of all the organisms extractable in this medium. The method does not damage soft meiofauna, ciliates or flagellates.

Preserved material

A problem with extracting meiofauna from fine sediments is the presence of cohesive lumps of clay and faecal pellets, which are difficult to sieve out and which may conceal animals. With preserved samples, pretreatment to improve the sieving efficiency is possible. Thiel *et al.* (1975) used ultrasonic treatment, by means of either an ultrasonic bath or a probe, which effectively broke up sediment aggregations and improved sieving efficiency without undue damage to the meiofauna. Barnett (1980) found that freezing in a domestic freezer for 24 h, followed by thawing, broke down resistant sediments into small particles which could then be further fragmented with the water softening agent 'Calgon'. A solution made by adding approximately 100 g of Calgon to 500 ml of the sample in formalin was effective, the mixture being stored for 24 h with intermittent shaking.

After sieve separation, the fine fraction is usually discarded, since hand sorting of small animals from a silt-clay mixture, even when stained, is extremely time-consuming. Although these techniques for preserved samples are generally adequate, occasional live samples should be examined to help with interpretation of the preserved material, and to allow identification of the more delicate forms.

If there is a large volume of sieve residue, animals can be extracted by density separation techniques (see also pp. 198–9). Various media have been used—sugar solutions (Anderson, 1959; Heip *et al.*, 1974; Higgins, 1977), magnesium sulphate (Lydell, 1936), sodium chloride (Lyman, 1943), zinc chloride (Sellmer, 1956) and carbon tetrachloride (Dillon, 1964). These media have now been largely replaced by the use of the colloidal silica polymer 'Ludox' (Du Pont, 1973).

De Jonge & Bouwman (1977) describe a technique using Ludox-TM (specific gravity 1·39 g), which is toxic to all living organisms. The sediment sample is suspended in a beaker of 25 % (V/V) Ludox-TM made up with water. The surface of the Ludox is then gently flooded with distilled water to prevent desiccation, which would transform the Ludox into gel form. (This precaution is not necessary for short duration separations). After 16 h the upper layer of Ludox containing the meiofauna is sucked off and washed through a 35 μm sieve, the organisms being rinsed with distilled water to remove aggregates which may adhere to them.

A more rapid method involving centrifugation has been developed at the Laboratoria voor Morfologie en Systematiek in Ghent, Belgium, using 'Ludox-TM', and is now deployed as a highly successful routine in several European Laboratories. The procedure is as follows:

1. Remove any large particles, pieces of shell etc, and the fine silt fraction from the sediment sample by decantation and sieve separation.
2. Wash the light fraction with tap water from the meiofauna sieve into centrifuge containers (50 ml tubes are suitable for small samples, 250 ml pots for larger samples).
3. Add a quantity of kaolin powder to each container (a heaped teaspoon is sufficient for the 250 ml pots). Shake vigorously, balance and spin for 7 minutes at 6000 r.p.m. in a high speed centrifuge.
4. Pour off the water; the kaolin settles last and forms a plug over the sediment, preventing it from being poured out.
5. Add Ludox-TM made up to a specific gravity of 1·15, until the pots are nearly full, and then resuspend the sediment by vigorous agitation. Stirring with a spatula may be necessary to break up the kaolin plug.
6. Rebalance the pots and spin again for 7 minutes at 6000 r.p.m.
7. Pour off the supernatant through the meiofauna sieve and wash thoroughly with tap water before washing into the sorting dish.
8. Repeat the extraction twice more.

This separates all but a few animals, virtually free of detritus and sediment. Although this method is only suitable for preserved material, it may be possible to detoxify the Ludox by dialysis (de Jonge and Bouwman, 1977), and use it with living specimens.

It is a wise precaution to wear rubber gloves when handling Ludox. Glassware can be cleaned by washing in dilute NaOH. The disadvantage of Ludox is that, at the present time, it is only possible to purchase it in large quantities (300 kg drums).

Subsampling

Because of the problem of small-scale patchiness in the distribution of meiofauna, there is something to be said for collecting large samples then subsampling from a homogeneous suspension of the sample if too many animals are present for practical counting, although the dangers of subsampling, already referred to, must be recognized. For fine sediments, subsampling can be undertaken before extraction of the fauna; alternatively the sieve residue or completely extracted samples of the meiofauna can be subsampled.

Elmgren (1973) built a special sample divider (Fig. 7.7), consisting of a

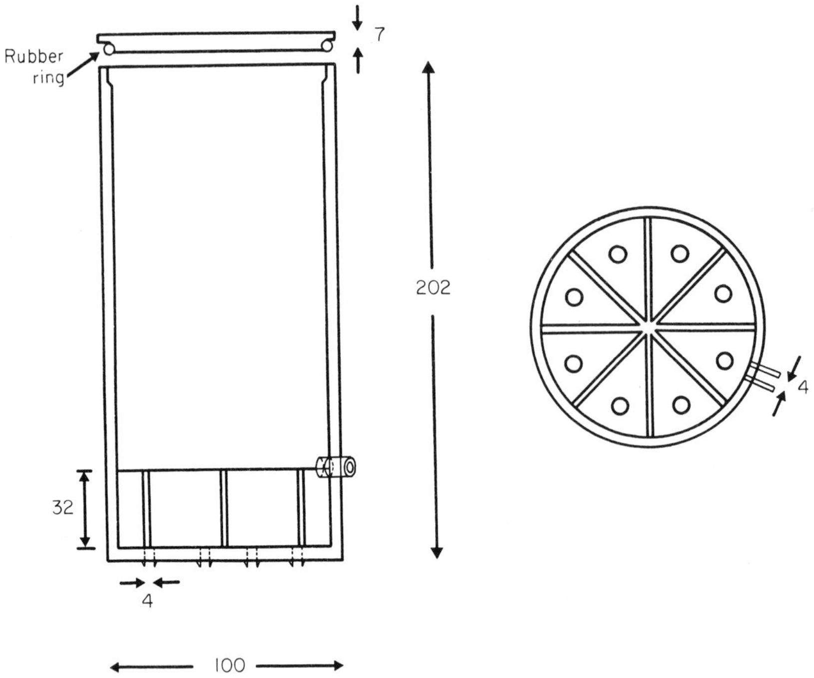

Fig. 7.7. Plexiglass (Perspex) sample divider: section (left) and plan of base (right) (measurements are in mm). (From Elmgren, 1973.)

plexiglass cylinder with its bottom divided into eight equal chambers. A preserved sample is poured into the sample divider, a little detergent added (to prevent copepods and ostracods adhering to the water surface), the volume made up to 1 litre and a tightly fitting lid applied. The sample divider is then inverted and vigorously shaken for a short while. After the sample has been given about one hour for settlement to the bottom, during which time a few twists of the sample divider make material sedimenting on the dividing walls fall down into the chambers, the water is slowly drained off through a tap, until it reaches the level of the dividing walls. The rubber stoppers in the bottoms of the chambers are removed and the subsamples drained into eight small containers. Sediment remaining is washed out by a gentle jet of water.

Rather more simple techniques have also proved adequate. The sample can be washed into a tall-form beaker and made up to a known volume with tap water. It is then vigorously agitated into an even suspension and a subsample is rapidly withdrawn. This should be done with a sampler of known volume such as a Stempel or Hensen pipette (available from Hydro-Bios, Kiel, West Germany). It is convenient to adjust the volume in the beaker so that a one-tenth or one-twentieth subsample is withdrawn each time.

Alternatively, the sample may be washed into a measuring cylinder, made up to a known volume with tap water, agitated by vigorous bubbling induced by blowing through a graduated pipette (with a fairly wide mouth) and then quickly sucking up a measured volume of suspension.

Examination and counting

Animals are sorted and counted under a stereoscopic microscope in petri-dishes, which can be marked with parallel lines or squares etched on to the bottom to aid scanning. For identification to species level under the high power microscope, the animals must be picked out for mounting on slides using a fine pipette, fine tungsten needle, sharpened quill or similar instrument. Counting of higher taxa (e.g. nematodes, copepods) *in situ* is far less time-consuming and in this case the use of a moveable stereo-microscope on an extended arm is advantageous. The petri-dish can be fixed in position (with a piece of graph paper beneath it) and the animals are not disturbed during the scanning of the dish bottom and the surface film (Uhlig *et al.*, 1973).

The soft meiofauna can usually only be identified with certainty when alive, but it may be helpful to reduce or eliminate mobility by the use of a compression chamber such as the rotary microcompressor described by Spoon (1978) or the special chamber (available from Hydro-Bios) described by Uhlig & Heimberg (1981), which can also be used with inverted microscopes (Fig. 7.8). With such an instrument it is possible gradually to reduce the gap between the glass base-plate and cover slip (or between two cover slips), and so squeeze the animal until it can no longer move.

For the hard meiofauna, it is possible to make permanent slides which can be examined at leisure and retained as a reference collection. Details of the special requirements of all meiofaunal groups are given in Hulings & Gray (1971); some notes on recommended methods for the most important hard taxa are provided below.

Nematoda

These are the dominant meiofauna in most habitats. Ideally, living specimens should be observed before making permanent mounts as the mounting procedure often dissolves pigments such as the eyespots which are found in certain species. If animals are extracted alive from the sediment they should be transferred to a cavity block in a small quantity of sea water, heat-killed and fixed by pouring on a hot (near-boiling) mixture of 4 % formaldehyde and 1 % glacial acetic acid. This results in most specimens being uncoiled which facilitates examination. The worms are left in this solution for 24 h and then transferred to glycerine by the slow evaporation method of Seinhorst (1959).

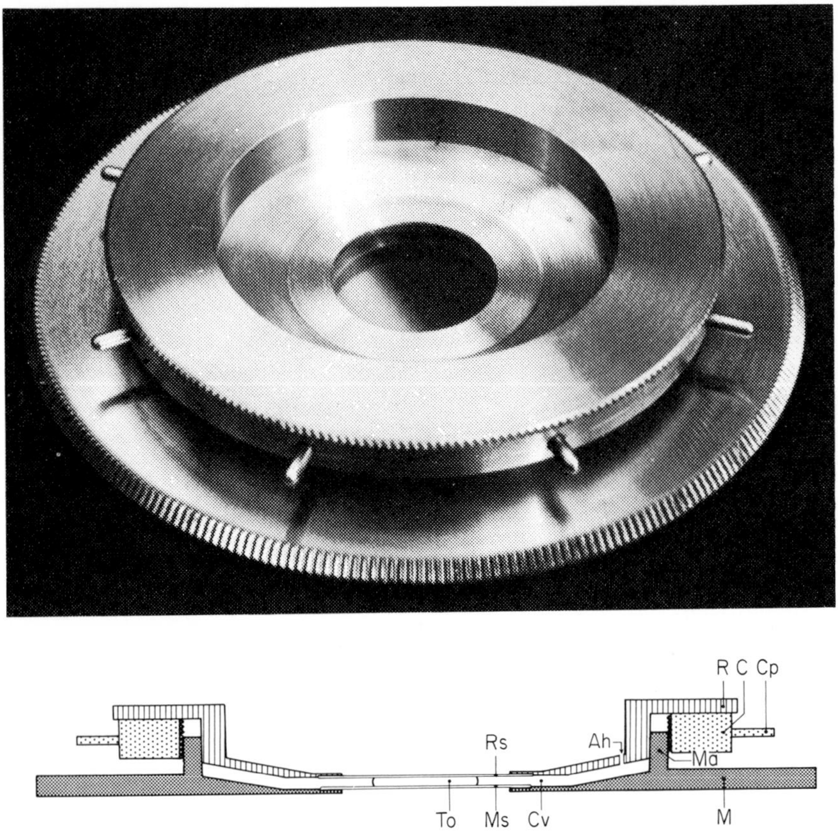

Fig. 7.8. Uhlig & Heimberg compression chamber for examining living meiofauna: assembled for use (above) and cross section (below). Ah, air-hole; C, compression-ring; Cp, pin for turning compression ring; Cv, chamber volume; M, baseplate; Ma, head-piece of the baseplate; Ms, lower coverslip; Rs, upper coverslip; R, rotor; To, compressed medium with organisms. (From Uhlig & Heimberg, 1981.)

The fixative is carefully pipetted off, and a mixture of 9 parts 50 % ethanol or distilled water and 1 part glycerine pipetted into the watchglass. The watchglass is then placed in a desiccator to evaporate off the ethanol and water and leave the nematodes in pure glycerine. This process takes several days, but if more rapid transfer is required the watchglass may be placed in an oven at 50 °C, when evaporation is completed in a matter of hours. The nematodes are then transferred to small drops of anhydrous glycerine on glass slides. For temporary mounts a cover-slip may be propped up along one edge with a small piece of lens tissue to prevent flattening of the worms.

Permanent mounts can be prepared by supporting the cover-slip with fine glass beads (ballotini) or rods of appropriate diameter, and sealing the edges with 'Glyceel', 'Clearseal' plus 'Bioseal' or another appropriate sealant for fluid mounts. Identification often depends on making measurements, which can be achieved from camera-lucida drawings, the scale of which is calibrated with a stage micrometer. The lengths of curved structures can be determined by running an opisthometer (map measurer) along the drawing, and then running it back to zero along a scale drawn from the stage micrometer.

Copepoda

Harpacticoid copepods are usually the second most abundant group of meiobenthos. Hamond (1969) and Coull (1977) give useful procedures. For whole mounts the same technique as outlined for nematodes is acceptable, or mounts can be made in a matrix which sets hard, such as Hoyer's medium. (Dissolve 8 g gum arabic in 10 ml distilled water; add 75 g chloral hydrate, 5 ml glycerine and 3 ml glacial acetic acid; strain through clear muslin or glass wool), or Reyne's medium (dissolve 10 g chloral hydrate in 10 ml distilled water; add 2·5 ml glycerine and mix, add 6 g gum arabic and stir very cautiously—avoid bubbles; allow to stand for one week. No filtering is necessary). Reyne's medium hardens in 1–2 days and the slide need not be ringed; Hoyer's takes 1–2 weeks and should be ringed for permanent mounts.

The form and setation of the limbs are important features in copepod taxonomy. 5–10 animals randomly orientated on the slide will often allow examination of all of the relevant body parts, but ideally the animal should be dissected and each pair of limbs examined separately. The animal is placed in some mounting medium on a slide and each somite, with its associated appendages, is cut off from anterior to posterior, one by one, using fine tungsten needles. These needles should be about 0·2 mm diameter, and sharpened either by dipping into molten sodium nitrite or by 6 V electrolysis in a weak solution of potassium hydroxide, and mounted in a suitable handle. After each somite is removed it is mounted on a slide, the urosome being mounted whole. Hamond (1969) arranges the limbs in sequence in a small streak of mountant, and covers the whole with a single 13 mm square cover-slip, whereas Coull (1977) mounts all the limbs on the same slide but with each under a separate micro-cover-slip. The streak of mountant containing the limbs is allowed to harden somewhat before the cover-slip, bearing additional mountant, is applied.

Ostracoda

Sealed glycerine mounts, as used for nematodes, are satisfactory, and dissected appendages can also be mounted in the same way. The use of

Reyne's or Hoyer's mounting medium, as for copepods, is also likely to be appropriate. Because micropalaentologists are interested in this group, much of the taxonomy is based on valve shape and ornamentation. Dry shells are usually glued to micropalaeontological slides with a little water-soluble gum tragacanth, to which a few drops of formalin or phenol have been added to inhibit fungal growth.

Kinorhyncha

Again, sealed glycerine mounts are satisfactory. Alternatively, the animals can be mounted in Hoyer's medium after evaporation into glycerine (Higgins, 1977).

Tardigrada

Mount in anhydrous glycerine as for nematodes.

Determination of biomass

The dry weight of some larger meiofauna species can be determined directly on an electrobalance sensitive to $\pm 0{\cdot}1\ \mu g$. For small species, it is more usual to make estimates of weight from volume determinations. For taxa such as Nematoda which can be assumed to have a regular cross-section (in this case circular), the volume can be determined from scale drawings made under the camera-lucida.

The animal shape can be approximated to geometrical figures (Andrássy, 1956), or by using Simpsons' first rule taking equally spaced co-ordinates along the body length (Warwick & Price, 1979). A rough estimate of volume (V) can be obtained by measuring the body length (l) and maximum width (W) and applying a formula for an 'average' nematode such as $V = lW^2/16 \times 10^5$, where V is in nl and W and l in μm (Andrássy, 1956) or $V = 530\ lW^2$ where V is in nl and W and l in mm (Warwick & Price, 1979). Volume can be approximately translated to dry weight by assuming a specific gravity of 1·13 (Wieser, 1960), estimates of dry weight for nematodes varying between 20 % and 25 % of wet weight (Myers, 1967; Wieser, 1960).

For irregularly shaped animals such as copepods, it may be necessary to make models based on camera-lucida drawings or photographs from different aspects to determine the relationship between body dimension(s) and volume. The volume can be determined by immersing the models in a measuring cylinder of water. The body volume of soft meiofauna can be determined by measuring the area of specimens under the camera-lucida when squashed to uniform thickness in a microcompression chamber (see (p. 235).The distances between the glasses of the chamber can be determined by measuring the focal distance from top to bottom of the specimen using the calibration on the fine focus of a good microscope.

Holter (1945) described a method of direct body volume determination. The animal is drawn up into a capillary tube of known bore in a column of dye solution of known density. The length of the dye column is measured and the dye emptied into a known volume of water. The dye concentration is determined colorimetrically in a microcuvette and compared with a standard curve made from solutions of dye columns alone.

Specific gravity and wet:dry weight ratios have not been determined for most meiofaunal groups, but values used for nematodes should give a reasonable approximation. Specific gravity can be determined by flotation in different mixtures of kerosene and bromobenzene (Wieser, 1960) or in a gradient column of bromobenzene and xylene (Low & Richards, 1952).

Energy flow measurements

The general principles of energy flow measurement are described in Chapter 9, but the practical details in that chapter are often only applicable to macrobenthic species. The following notes indicate some of the techniques available for meiobenthos. Space does not permit detailed methodological descriptions but reference is made to appropriate review articles and research papers.

Production

By and large, meiofauna species have continuous reproduction, although this is sometimes curtailed during certain seasons. Development times are short and generations overlap so that it is impossible to follow the history of recognizable cohorts in a time-series of samples. In most cases, therefore, it will be necessary to estimate the size-specific growth rate and apply this to measured standing stocks in the field (method 3A in Chapter 9). This usually involves the establishment of cultures in the laboratory, and every attempt should be made to ensure that the culture conditions resemble those in the field as closely as possible with respect to temperature, available food etc. A comprehensive review of the methods available for culturing meiofaunal taxa is given by Kinne (1977).

Respiration

Unless large numbers of animals are available, for example from laboratory cultures, conventional manometric methods for determining oxygen consumption are not sufficiently sensitive. Small differential volumetric respirometers, such as that described by Dixon (1979) have, however, been successfully used for cultured nematodes (Warwick, 1981) and meiofaunal polychaetes (V.A. Tennant, personal communication). More sensitive techniques must be employed for single animals or low numbers of animals. Lasserre (1974) gives a useful review of microrespirometers available for use in the meiofauna. Cartesian divers of the standard or stoppered type have been most widely used. The non-Cartesian gradient diver is in many ways less exacting to use and is more sensitive than the Cartesian diver, but each diver can be used only once and the technique has rarely been employed for meiofauna. The use of oxygen electrodes has also been limited, although Atkinson & Smith (1973) devised an oxygen electrode microrespirometer which was used successfully for relatively large nematodes (Atkinson, 1973). Finally, Lasker *et al.* (1970) used the highly sensitive reference diver of Scholander *et al.* (1952) to measure the respiration of a small harpacticoid copepod: this method is not suitable for active animals which will interfere with the reference bubble, which is attached to a hydrophobic weight and floats in the same chamber as the animal.

Ingestion, assimilation and faeces production

The lack of reliable techniques for determining rates of ingestion, assimilation and egestion has hampered the completion of energy budgets for the meiofauna. Assimilation has often been estimated by summing respiration and growth, but this ignores any production of dissolved organics or mucus, the latter being potentially important in some meiofaunal groups.

Techniques employing bacterial or algal foods labelled with ^{14}C or ^{32}P are well developed (Sorokin, 1968), but there are problems in the interpretation of results (Conover & Francis, 1973). Radio-tracers have been used in food preference experiments; they have been less commonly applied to quantify rates of ingestion, assimilation or faeces production. Coulter counters are potentially useful for measuring ingestion rates if the species investigated can be induced to feed on particulate food in suspension, but for meiobenthos this is not often the case. On the other hand, it is quite a simple matter to quantify ingestion in carnivorous species which engulf large food items, e.g. *Protohydra* (Muus, 1966). Clearly, much more research is required before special guidelines can be given on these components of the energy budget.

Acknowledgements

We are indebted to D.J. Murison, DAFS Laboratory, Aberdeen, for helpful criticism of the manuscript for this chapter.

References

Anderson, R.O., 1959, A modified flotation technique for sorting bottom fauna samples. *Limnology and Oceanography*, **4**, 223–225.

Andrássy, I., 1956. Die Rauminhalts—und Gewichtsbestimmung der Fadenwürmer (Namatoden). *Acta Zoologica Academiae Scientiarum Hungaricae*, **2**, 1–15.

Atkinson, H.J., 1973. The respiratory physiology of the marine nematodes *Enoplus brevis* (Bastian) and *E. communis* (Bastian). I. The influence of oxygen tension and body size. *Journal of Experimental Biology*, **59**, 255–266.

Atkinson, H.J. & Smith, L., 1973. An oxygen electrode microrespirometer. *Journal of Experimental Biology*, **59**, 247–253.

Bacescu, M., 1957. Apucatorul-sonda pentru studiul cantitativ al organismelor de fundun aparat mixt pentru colectarea simulanta a macro-si microbentosulia. (La bennesonde pour l'étude quantitative des organismes benthiques—un appareil mixte pour prelever à la fois le macro-et le micro-benthos.) *Buletinul Institutului de Cercetări Piscicole*, **16(2)**, 69–82.

Barnett, B.E., 1980. A physico-chemical method for the extraction of marine and estuarine benthos from clays and resistant muds. *Journal of the Marine Biological Association of the United Kingdom*, **60**, 255.

Bieri, R. & Tokioka, T., 1968. Dragonet 11, an opening–closing quantitative trawl for the study of micro-vertical distribution of zooplankton and the meio-epibenthos. *Publications of the Seto Marine Biological Laboratory*, **15**, 373–390.

Boisseau, J-P., 1957. Technique pour l'étude quantitative de la faune interstitielle des sables. *Comptes Rendus du Congrés des Sociétés Savantes de Paris et des Départments*, 1957, 117–119.

Burns, R.E., 1966. Free-fall behaviour of small, light-weight gravity corers. *Marine Geology*, **4**, 1–9.

Bush, L.F., 1966. Distribution of sand fauna on beaches at Miami, Florida. *Bulletin of Marine Science*, **16**, 58–75.

Conover, R.J. & Francis, V., 1973. The use of radioactive isotopes to measure the transfer of materials in aquatic food chains. *Marine Biology*, **18**, 272–283.

Corey, S. & Craib, J.S., 1966. A new quantitative bottom sampler for microfauna. *Journal du Conseil Permanent International pour l'Exploration de la Mer*, **30**, 346–353.

Coull, B.C., 1977. Marine flora and fauna of the Northeastern United States. Copepoda: Harpacticoida. *NOAA Technical Report NHFS Circular*, **399**, 1–48.

Craib, J.S., 1965. A sampler for taking short undisturbed cores. *Journal du Conseil Permanent International pour l'Exploration de la Mer*, **30**, 34–39.

Delamare-Deboutteville, C., 1960. *Biologie des Eaux Souterraines Littorales et Continentales*. Hermann, Paris, 740 pp.

De Jonge, V.N. & Bouwman, L.A., 1977. A simple density separation technique for quantitative isolation of meiobenthos using colloidal silica Ludox-TM. *Marine Biology*, **42**, 143–148.

Dillon, W.P., 1964. Flotation technique for separating fecal pellets and small marine organisms from sand. *Limnology and Oceanography*, **9**, 601–602.

Dixon, D.R., 1979. A differential volumetric micro-respirometer for use with small aquatic organisms. *Journal of Experimental Biology*, **82**, 379–384.

Du Pont, 1973. *Ludox Colloidal Silica*. Product Information Bulletin, E.I. Du

Pont de Nemours & Co. (Inc.), Industrial Chemicals Department, Wilmington, Delaware, U.S.A., 20 pp.

Elmgren, R., 1973. Methods of sampling sublittoral soft bottom meiofauna. *Oikos*, **Suppl. 15**, 112–120.

Emery, K.O. & Dietz, R.S., 1941. Gravity coring instruments and mechanics of sediment coring. *Bulletin of the Geological Society of America*, **52**, 1685–1714. (*Contribution, Scripps Institution of Oceanography*, 1941, **148**.)

Fenchel, T., 1967. The ecology of marine microbenthos. I. The quantitative importance of ciliates as compared with metazoans in various types of sediments. *Ophelia*, **4**, 121–137.

Hamilton, A.L., 1969. A method of separating invertebrates from sediments using long wave ultra-violet light and fluorescent dyes. *Journal of the Fisheries Research Board of Canada*, **26**, 1667–1672.

Hamond, R., 1969. Methods of studying the copepods. *Microscopy*, **31**, 137–149.

Heip, C., Smol, N. & Hautekiet, W., 1974. A rapid method for extracting meiobenthic nematodes and copepods from mud and detritus. *Marine Biology*, **28**, 79–81.

Higgins, R.P., 1964. Three new kinorhynchs from the North Carolina coast. *Bulletin of Marine Science of the Gulf and Caribbean*, **14**, 479–493.

Higgins, R.P., 1977. Two new species of *Echinoderes* (Kinorhyncha) from South Carolina. *Transactions of the American Microscopical Society*, **96**, 340–354.

Hockin, D.C., 1981. A simple elutriator for extracting meiofauna from sediment matrices. *Marine Ecology Progress Series*, **4**, 241–242.

Holme, N.A., 1964. Methods of sampling the benthos. *Advances in Marine Biology*, **2**, 171–260.

Holopainen, I.J. & Sarvala, J., 1975. Efficiencies of two corers in sampling soft-bottom invertebrates. *Annales Zoologici Fennici*, **12**, 280–284.

Holter, H., 1945. A colorimetric method for measuring the volume of large amoebae. *Comptes-rendus des Travaux du Laboratoire de Carlsberg*, Série Chimie, **25**, 156–167.

Hulings, N.C. & Gray, J.S., 1971. A manual for the study of meiofauna. *Smithsonian Contributions to Zoology*, **78**, 84 pp.

Hummon, W.D., 1975. Habitat suitability and the ideal free distribution of Gastrotricha in a cyclic environment. pp. 495–525 in H. Barnes (ed.) *Ninth European Marine Biology Symposium*. Aberdeen University Press, Aberdeen, 760 pp.

Kinne, O., 1977. *Marine Ecology, Volume III. Cultivation*, part 2, pp. 579–1293. John Wiley & Sons, Chichester.

Lasker, R., Wells, J.B.J. & McIntyre, A.D., 1970. Growth, reproduction, respiration and carbon utilization of the sand-dwelling harpacticoid copepod, *Asellopis intermedia*. *Journal of the Marine Biological Association of the United Kingdom*, **50**, 147–160.

Lasserre, P., 1974. Metabolic activities of benthic microfauna and meiofauna. Recent advances and review of suitable methods of analysis. pp. 95–142 in I.N. McCave (ed.) *The Benthic Boundary Layer*. Plenum Press, New York, 323 pp.

Low, B.W. & Richards, F.M., 1952. The use of gradient tube for the determination of crystal densities. *Journal of the American Chemical Society*, **74**, 1660–1666.

Lydell, W.R.S., 1936. A new apparatus for separating insects and other arthropods from soil. *Annals of Applied Biology*, **23**, 862–879.

Lyman, F.E., 1943. A pre-impoundment bottom fauna study of Watts Bar Reservoir area (Tennessee). *Transactions of the American Fisheries Society*, **72**, 52–62.

McIntyre, A.D., 1968. The meiofauna and microfauna of some tropical beaches. *Journal of Zoology*, **156**, 377–392.

McIntyre, A.D., 1969. Ecology of marine meiobenthos. *Biological Reviews of the Cambridge Philosophical Society*, **44**, 245–290.

McIntyre, A.D., 1971. Deficiency of gravity corers for sampling meiobenthos and sediments. *Nature, London*, **231**, 260.

McLachlan, A., Dye, A.H. & Van Der Ryst, P., 1979. Vertical gradients in the fauna and oxidation of two exposed sandy beaches. *South African Journal of Zoology*, **14**, 43–47.

Mills, A.A., 1961. An external core retainer. *Deep-Sea Research*, **7**, 4.

Moore, H.B. & Neill, R.G., 1930. An instrument for sampling marine muds. *Journal of the Marine Biological Association of the United Kingdom*, **16**, 589–594.

Mortensen, R.H., 1925. An apparatus for catching the micro-fauna of the sea bottom. *Videnskabelige Meddelelser fra Dansk naturhistorik Forening i Kjøbenhavn*, **80**, 445–451.

Muus, B., 1964. A new quantitative sampler for the meiobenthos. *Ophelia*, **1**, 209–216.

Muus, K., 1966. Notes on the biology of *Protohydra leuckarti* Greef (Hydroidea, Protohydridae). *Ophelia*, **3**, 141–150.

Myers, R.F., 1967. Osmoregulation in *Panagrellus redivivus* and *Aphelenchus avenae*. *Nematologica*, **12**, 579–586.

Ockelmann, K.W., 1964. An improved detritus-sledge for collecting meiobenthos. *Ophelia*, **1**, 217–222.

Rees, C.B., 1940. A preliminary study of the ecology of a mud-flat. *Journal of the Marine Biological Association of the United Kingdom*, **24**, 185–199.

Renaud-Debyser, J., 1957. Description d'un carrotier adapte aux prelèvements des sables de plate. *Revue de l'Institut Français du Pétrole*, **12**, 501–502.

Renaud-Debyser, J., 1963. Recherches Ecologiques sur la fauna interstitielle des sables (Bassin d'Arcachon, Ile de Bimini, Bahamas). *Vie et Milieu*, **Suppl. 15**, 157 pp.

Sarvala, J. & Ranta, E., 1977. Performance of the Muus sampler in surveys of the brackish-water macrofauna. *Annales Zoologica Fennici*, **14**, 191–197.

Scholander, P.F., Claff, C.L. & Sveinsson, S.L., 1952. Respiratory studies of single cells. I. Methods. *Biological Bulletin, Marine Biological Laboratory, Woods Hole*, **102**, 157–177.

Schwinghamer, P., 1981. Extraction of living meiofauna from marine sediments by centrifugation in a silica sol-sorbital mixture. *Canadian Journal of Fisheries and Aquatic Sciences*, **38**, 476–478.

Seinhorst, J.W., 1959. A rapid method for the transfer of nematodes from fixative to anhydrous glycerine. *Nematologica*, **4**, 67–69.

Sellmer, G.P., 1956. A method for the separation of small bivalve molluscs from sediments. *Ecology*, **37**, 206.

Spoon, D.M., 1978. A new rotary microcompressor. *Transactions of the American Microscopical Society*, **97**, 412–416.

Sorokin, J.I., 1968. The use of ^{14}C in the study of nutrition of aquatic animals. *Internationale Vereinigung für Theoretische und Angewandte Limnologie*, **16**, 41 pp.

Thiel, H. Von, 1966. Quantitative Untersuchungen über die Meiofauna des Tiefseebodens. *Veröffentlichungen des Instituts für Meeresforschung in Bremerhaven*, Sonderband II, 131–147.

Thiel, H. Von, Thistle, D. & Wilson, G.D., 1975. Ultrasonic treatment of sediment samples for more efficient sorting of meiofauna. *Limnology and Oceanography*, **20**, 472–473.

Tiemann, H. & Betz, K-M., 1979. Elutriation: theoretical considerations and methodological improvements. *Marine Ecology*, **1**, 277–281.

Uhlig, G., 1966. Untersuchungen zur Extraktion der vagilen Mikrofauna aus marinen Sedimenten. *Zoologischer Anzeiger*, **Suppl. 29**, 151–157.

Uhlig, G., 1968. Quantitative methods in the study of interstitial fauna. *Transactions of the American Microscopical Society*, **87**, 226–232.

Uhlig, G., Thiel, H. & Gray, J.S., 1973. The quantitative separation of meiofauna. A comparison of methods. *Helgoländer wissenschaftliche Meeresuntersuchungen*, **25**, 173–195.

Uhlig, G. & Heimberg, S.H.H., 1981. A new versatile compression chamber for examination of living microorganisms. *Helgoländer Meeresuntersuchungen*, **34**, 251–256.

Warwick, R.M., 1981. The influence of temperature and salinity on energy partitioning in the marine nematode *Diplolaimelloides bruciei*. *Oecologia* (*Berlin*), **51**, 318–325.

Warwick, R.M. & Price, R., 1979. Ecological and metabolic studies on free-living nematodes from an estuarine mud-flat. *Estuarine and Coastal Marine Science*, **9**, 257–271.

Wells, J.B.J. & Clark, M.E., 1965. The interstitial crustacea of two beaches in Portugal. *Revista de Biologia*, **5**, 87–108.

Wieser, W., 1960. Benthic studies in Buzzards Bay. II. The meiofauna. *Limnology and Oceanography*, **5**, 121–137. (*Collected Reprints, Woods Hole Oceanographic Institution*, 1054.)

Chapter 8. Phytobenthos Sampling and Estimation of Primary Production

F. E. ROUND AND M. HICKMAN

Primary production

Primary production may be defined as the formation of organic molecules from relatively simple inorganic sources utilizing either light energy (performed by some bacteria, algae, lichens and angiosperms in the marine environment) or chemical energy (certain bacteria only).

The marine phytobenthos includes all of the associations of organisms involved in the above processes and living at or closely associated with the many solid/liquid interfaces. The bulk of phytobenthos occurs in the intertidal and shallow subtidal regions, where both photosynthetic and chemosynthetic associations are involved. Below the limit of penetration of photosynthetically usable light only the chemosynthetic bacteria are involved. The primary producers live on or in the silt, sand or rock, attached to the macroscopic algae, angiosperms and animals associated with these interfaces, or even on the under surface of permanent sea ice.

The greater part of the primary synthesis is probably achieved by various algal groups, although bacteria can occasionally be significant, as on some sandy sediments. Angiosperms are relatively scarce except in tropical regions, where they may form dense beds in shallow bays. Lichens are mostly confined to the upper intertidal zone. Relatively few studies of marine primary productivity by the phytobenthos have been made; hence we shall also refer to techniques designed for freshwater studies wherever applicable, since the same principles are involved. Considerable testing and development work is required on all aspects and the following should be treated merely as a working outline around which studies can be planned. In order to enable comparison of primary productivity in different regions it is essential that the habitats and plant associations are adequately defined, and if two or more primary producing associations live in close association these should either be separated or studied as an aggregate. It is essential that methods should be standardized or the data presented in such a way that valid comparisons can be made. Unfortunately some studies have neglected to define or sample discrete habitats, although the situation is now improving.

Definition of habitats

Habitats themselves can be defined from a physicochemical standpoint without reference to the organisms. Habitats occupied by the phytobenthos start on the uppermost parts of the shore, where sufficient sea spray is deposited to form substrata suitable for colonization by halophytic plants. Obviously, in some localities, this supratidal zone merges into the realm of the terrestrial ecologist and an upper limit must be set in each individual project. The intertidal zone tends to form the next easily definable gross habitat and can be relatively accurately defined from tide tables and direct surveying of the shore (see pp. 140–4). The subtidal zone may be divided into that occupied by the photosynthetic plants (down to the limit of penetration of photosynthetically usable light), and the effectively dark regions occupied only by chemosynthetic plants. The boundary between these will vary with season and turbidity of the water column. For precision and ease of comparison the substratum should be clearly defined, and as much information as possible should be gathered on angle of slope, aspect, length of tidal exposure, etc. For a discussion of estuarine situations, see Round (1979a).

Inorganic substrata

Hard—Rock surfaces, intertidal and subtidal

Geological formation is rarely quoted, hence it is difficult to assess the importance of this factor, but topography, angle and degree of bedding, friability, etc. are important in the description of this habitat. Rock faces and rock pools are quite separate habitats and subject to different environmental stresses. Hard artificial substrata, e.g. concrete, may be of importance at some sites. Luther (1976) and Harlin & Lindbergh (1977) have clearly shown that the type of rock surface, particularly the granule size, affects initial colonization. Hardness affects production, since the turnover of plants is greater on soft rock. On some shores loose stones, which may be moved to varying degrees by wave action, constitute a further habitat for primary producers. These are undoubtedly difficult to study but they are often productive of both microscopic (e.g. diatoms and Cyanophyta) and macroscopic associations (e.g. large Chlorophyta, Phaeophyta and Rhodophyta). (Aleem, 1950a; Castenholz, 1963).

Along many tropical shores the debris from coral reefs is cemented by algal growth and this forms a primary producing community to which macroscopic producers attach. Forming beach rock is also a rich carbon fixing community along tropical shores.

Soft or mobile—sand, silt and various intermediates, intertidal and subtidal

Again some information on geological origin, composition, grain size and organic content is desirable where possible, and it is important to distinguish between sand and silt. Beds of biogenic origin, such as those derived from shell fragments or calcareous remains of algae, are also colonized by primary producers, and these are very common in some shallow tropical seas. Calcareous algae detached from their substratum are known to continue photosynthesis and this must be taken into account. Microscopic associations are primarily of diatoms and dinoflagellates but some Cyanophyta and Chrysophyceae may also occur. (Brockmann, 1935; Aleem, 1951; von Stosch, 1956; Round, 1960; Bodeanu, 1964). Macroscopic siphonaceous Chlorophyta are especially common on sediments in the tropics.

Ice

This forms a relatively stable habitat for algal primary producers, mainly diatoms (Bunt, 1963; Apollonio, 1965). Burkholder & Mandelli (1965) estimated a massive contribution of carbon from this source around Antarctica. Care is needed to distinguish between plankton frozen into ice and the under-ice flora which attaches to the bottom of sea ice and can form a layer 2–3 cm thick (Horner & Alexander, 1972).

Organic substrata

Algae

Some microscopic and most macroscopic algae frequently support a rich flora. The host plants are either attached to the hard inorganic substrata or grow in silt and sand, but a few are free-living (e.g. some populations of *Sargassum*). They are abundant in intertidal and subtidal habitats.

Angiosperms

Although only a small number of angiosperms grow in the subtidal zone they can be very abundant and form a stable habitat for algae, as do plants of salt marshes and mangrove swamps (e.g. *Enteromorpha*, *Caloglossa*, *Bostrychia*). There is now a certain amount of data on primary production of host and of epiphytes (Penhale, 1977, Thayer & Adams, 1975).

Epibiotic on fauna

Primary producing algae live attached externally. Many sedentary animals (e.g. limpets) and a few motile species living associated with the above inorganic and organic substrata support dense microscopic and macroscopic algal floras.

Endobiotic in fauna

This habitat is occupied by a number of algae and is of great importance in some regions. The host may be associated with sand/silt shores, e.g. *Convoluta*, or with hard inorganic substrata, e.g. corals.

There is considerable carbon fixation by the symbionts of sea anemones, etc. and more recently saccoglossan molluscs have been shown to 'farm' chloroplasts ingested into the gut.

In some instances the above gross habitats occupy considerable uninterrupted areas but they usually tend to be intermixed, adding complexity to the study. Furthermore, there are smaller niches within these habitats and some will be considered in the next section.

All descriptions should state exactly where the host substratum is located, e.g. on rock face, in rock pool, permanently submerged, etc.

Definition and recognition of primary production associations

The term association is here used in the sense defined by Hutchinson (1967): 'an assemblage of species that recurs under comparable ecological conditions in different places'. Two vital concepts are involved, firstly the recognition of these assemblages and secondly definition of 'comparable ecological conditions'. To some extent the recognition of the assemblages is simplified if the habitats are clearly defined as above and even more precisely if the individual niches are recognized. The assemblage can, however, be defined without recourse to descriptions of the habitat. For example, there are associations attached to inorganic sand both in the intertidal and subtidal zone, and these are recognizable from habitats throughout the world. There is another totally different series of associations which live amongst the sand but which are unattached to the sand grains and capable of movement through the deposit. The definition and recognition of these two groups of associations is important in considering their primary productivity which can be estimated separately for the two associations at any one site, or combined. If the latter approach is employed it is important to recognise that the data from two totally unlike associations are being combined. The species composition of, for example, associations attached to sand grains will vary somewhat from place to place according to local conditions, but the life form of the species, and probably many of the species themselves, will be common.

The number of associations recognised will increase as detailed work expands and, strictly speaking, they are definable only from the dominants, sub-dominants and other species present. They can, however, be classified according to niches which they occupy in the gross habitats defined above and this is extremely helpful. See also Round (1981) for further detail.

On hard inorganic substrata

These are the *epilithic* and *endolithic* associations (*epilithon* or *endolithon*).

A series of epilithic associations which may contain pure stands of microscopic species of encrusting Cyanophyta, Chrysophyta, Chlorophyta, Phaeophyta, Rhodophyta, diatoms or mixtures of these groups coats the surface of rocks. Some may be confined to sloping rock surfaces, others to rock pools. Data for these in the intertidal are scarce and even scarcer for the subtidal.

Macrophytic growths of Chlorophyta, Phaeophyta, Rhodophyta and lichens occur attached to rock surfaces. These associations are fairly well described in intertidal habitats and are now being increasingly documented for the subtidal by divers.

Data on carbon fixation of these associations has been obtained by Mann (1973) for north eastern Atlantic sites, Johnston (1969) for the Canary Islands, Wanders (1976) for the West Indies, Littler (1973) for Hawaii, and Littler & Murray (1974) for California (for further details see Round, 1981).

Endolithic associations occur particularly on and penetrating into soft (often calcareous) rock, but even the pores and crevices in hard rock and corals are colonized. These associations, which comprise certain Cyanophyta, diatoms and Rhodophyta, have received little attention.

On soft or mobile inorganic substrata

Associations attached to sand particles (*epipsammic*) consist mainly of attached diatoms and bacteria with occasional Cyanophyta and Chlorophyta. The species involved are often minute and taxonomically difficult. Similar associations probably grow on calcareous shell and algal fragments but, as far as we are aware, have hardly been investigated (but see papers of Scoffin, 1970; Neumann & Land, 1969; Neumann *et al.*, 1970).

Associations moving on and in the surface sediments (*epipelic*) consist primarily of motile diatoms, but occasionally also contain filamentous or coccoid Cyanophyta, dinoflagellate and euglenoid flagellates. They grow on sand and silt, both in the intertidal and subtidal. Wherever silt accumulates on other substrata, e.g. on stones, rocks or plants, the epipelic association is liable to exist as a contaminant or may completely replace the true association of the substrata. The productivity of this association on salt marshes has been measured in North America by Valiela & Teal (1974), Gallagher & Daiber (1973) and Carpenter *et al.* (1978). In Holland, Admiraal (1977a, b) has made a detailed study of this association. The species often undergo rhythmic vertical movements (Pomeroy, 1959; Bodeanu, 1964; Round & Palmer, 1966; Round, 1966; Palmer & Round 1967; Eaton & Simpson 1979).

Associations living within the sediment (*endopelic*) are relatively unstudied—they include diatoms occurring in mucilage tubes and motile or semi-motile species living beneath the surface on sandy shores and not moving vertically (Williams, 1963; Round, 1979b).

Associations of larger algae or angiosperms rooted in sediments (*rhizobenthic*) consist of species of Chlorophyta, Phaeophyta, Rhodophyta or angiosperms. These are more common in tropical zones.

On other living organisms

The *epiphytic* associations which are found attached to plants are composed of species of Chlorophyta, Phaeophyta, Rhodophyta, Cyanophyta and Bacillariophyta. In some habitats, species from other associations may become entangled amongst these—this flora has been termed metaphyton in freshwater studies and the term is equally applicable to the marine environment. Thalloid algae are often coated with diatom epiphytes and measurement of the primary productivity of e.g. *Ceramium* spp. growing on *Chorda*, *Laminaria* etc. could often include a considerable contribution from the diatoms.

Similar species of algae occur in the *epizoic* associations which are found attached to shells and the surfaces of animals. These have been little investigated.

Endophytic species which grow within other algae or angiosperms consist of Chlorophyta, Phaeophyta and Rhodophyta. Primary production is likely to be estimated together with that of the host, but densely colonized hosts may not assimilate carbon at the same rate as uncolonised hosts. As far as we know no studies have been undertaken on these combinations of species.

Endozoic associations, actually growing within animals, have been studied because of an intrinsic interest in the kinds of symbiotic organisms and also because in the economy of coral reefs the endozoic algae contribute considerably to primary production.

Bacteria are likely to be involved in many of the above associations—a conspicuous example being the occurrence of purple sulphur bacteria on intertidal sands.

Field sampling techniques

The pre-requisite of field sampling is that the algal associations and the niches they occupy be recognized (see previous section) in order that well-defined associations can be collected wherever possible and mixed associations can be recognized when field separation is impossible, e.g. epiphytes on larger algae. Preliminary field surveys may be necessary to determine the extent and heterogeneity of the associations, and to fix sampling stations.

Biomass studies will rarely be possible on plants left *in situ* in the habitat.

Macrophytes

In most instances these will have to be removed to the laboratory. Removal of whole plants attached to rock is difficult but if removal of individual pieces of rock is possible and does not drastically affect the sampling area, this is desirable since losses are more likely to occur if sampling is undertaken in the field. Holdfasts must be included in the samples. In the case of plants embedded in silt the underground 'rhizomes' should be removed at the same time, as should the root systems of angiosperms such as *Zostera*. Plants can be sampled completely for weight, volume, chemical component analysis, or individual plants can be marked and growth measured directly. Rate of production can be related to rate of colonization of cleaned areas but this is a somewhat artificial measure. Sampling of epiphytes usually requires the removal of the host plant to the laboratory. Some are loosely attached and samples should therefore be taken with the minimum of disturbance, preferably by enclosing the plant in a sampling device while still in position under the water. Even in apparently rough water there is often an easily lost diatom flora, especially of long chain forming species, e.g. *Rhabdonema*. Transport to the laboratory may free other species from the substrata and these need to be added to the biomass still attached to the plant.

Sediments

Sampling of sand and silt in the intertidal is relatively simple since cores can be removed and transported to the laboratory. In some instances the motile epipelic flora can be removed in the field by the use of fabric or lens tissue as described on p. 253. The proportion of the total flora removed by this technique requires determination for each sediment type. Cores can also be obtained from underwater sediments, but care should be taken to avoid water movement as the sampling apparatus approaches the surface of the sediment since some of the surface forms may be washed away (see p. 185). Sampling devices for soft and hard bottoms are described by Kangas (1972)—see also Chapters 6 and 7.

Hard surfaces

Microscopic, epilithic species can be scraped off rocks, but rough surfaces are difficult to deal with and underwater surfaces do not appear to have been sampled. Much development work is needed before satisfactory sampling techniques can be recommended. Films of plastic material, e.g. collodion, have been used (Margalef, 1949) but the algae so collected are liable to be damaged and therefore difficult to identify.

Contamination of all benthic associations by phytoplankton is likely, although this may only be serious when there are massive phytoplankton blooms. Under normal circumstances the contaminating cells from the water bathing the benthic associations will be negligible since the density of cells at the surface/liquid interfaces is many times greater than the density of phytoplankton in the same volume of water. However, during decline of heavy phytoplankton populations, dying and dead cells can accumulate rapidly and may bias cell counts unless recognized and discounted. They will also contribute pigments, though these may be degraded forms rather than the active forms (see p. 258). Owing to the special environmental conditions in the benthos, phytoplankton remains rarely stay active for more than a few days.

Areal sampling

Since, by definition, the benthic associations occur at interfacial niches it is obviously desirable to express results on an area basis which for productivity measurements should be related to sea surface area (p. 275). Sampling of known areas of sediments is relatively simple in the intertidal but difficult in the subtidal. Dredges are too crude for almost all algal populations and coring devices sampling known areas are essential (see relevant sections in Chapters 5, 6 and 7). Macrophytic growths need to be sampled from marked quadrats, and the angle of slope, degree of shading, degree of wave action etc. should be recorded. On ice surfaces the algal associations penetrate the ice and volumes must also be sampled. Sampling of both rocky and soft bottoms have yielded data for both plants and animals on a weight per area basis (Hällfors *et al.*, 1975; Lappalainen *et al.*, 1977). Similar biomass estimations were used by Bellamy *et al.* (1973).

Artificial substrata

Numerous experiments, frequently related to fouling and pollution, have employed artificial substrata (glass, plastic, concrete, etc.) to study rates of colonization under different ecological conditions. The algal flora tends, however, to be a somewhat specialized one, although similar associations can sometimes be recognized on adjacent natural substrata. Undoubtedly much valuable information can be gained from this approach although it cannot be directly translated into productivity of the adjacent surfaces which, over a period of time, may support rather different associations growing at differing rates.

Separation of live populations from substratum and estimation of biomass from cell counts

Epipelic algae

Diatoms can be isolated from damp intertidal sediments by placing a piece of fabric on the surface, and allowing time for the phototactic algae to move up into it. Eaton & Moss (1966) used lens tissue for this purpose, and they also review earlier work. Another method frequently used in the laboratory is to make a suspension of the algae and sediment and then count the organisms in an aliquot of the suspension in a counting chamber. This direct method has been used by Aleem (1950b), and by Hopkins (1963) for core samples. Plante (1966) used a technique which involved separation of diatoms and fine detrital material gravimetrically by ascensional water flows, followed by differential centrifugation in liquids of differing densities. Several important studies on this association are given in Admiraal (1980).

In another technique for epipelic algae, a sample is mixed in the laboratory to disperse the algae. It is then poured into dishes, and coverglasses are placed on the damp sediment surface. The algae move up on to the underside of the coverglass, which can be removed from the sediment surface and placed on a slide for counting and identification of the algae (Round, 1953).

Eaton & Moss (1966) developed a quantitative technique of defined accuracy for sampling and estimating epipelic populations under a wide range of field conditions. Methods for sampling shallow and deep water were developed and three possible methods of estimating numbers of algae in field samples were studied: (1) direct counting; (2) use of coverglasses, and (3) lens tissues.

Samples can be taken from sediments in shallow intertidal pools etc. using a glass tube—diameter 0·5 cm—stoppered by a finger at the upper end and with the lower end placed on the surface of the mud. By releasing the finger, and simultaneously drawing the lower end of the tube across the mud surface, the tube fills with a mixture of surface and overlying material. Results with this method are qualitatively reproducible but the area sampled is not known (Round, 1953). Cores have been used in some cases. Aleem (1950a), Hopkins (1963) and Round & Palmer (1966) used short open-ended cylinders and took cores by hand from intertidal areas. For shallow waters, Perspex (acrylic plastic) cylinders, 9 cm diameter and 35 cm long (or long enough to reach above the water surface), can be pushed into the sediment to delineate a known area. The sediment from within the cylinder can be removed by suction into a polyethylene bottle attached to a glass tube which is moved over the sediment surface. It was found that the suction prevented preferential uptake of either small or large sediment particles. Samples could be easily obtained from substrata of widely differing textures. In deeper waters, mechanical corers have to be used before the surface sediment can be removed by suction. The main requirement is that the surface of the core should be disturbed as little as possible.

Samples from underwater sites often contain far more water than sediment, and to separate this the sample is allowed to settle in the laboratory in the dark. This ensures that the flagellate groups, which tend to be phototactically attracted to the water surface, also sediment out. After 5–7 h the supernatant water is removed, using a pipette or pump. Losses here were estimated to be $<1\%$ of the total number of diatoms present. The sample is then vigorously shaken to thoroughly mix the sediment and algae, and the mixture is poured into a petri-dish or other suitable vessel of known area to give a depth of 0·5–1·0 cm, the remainder of the sample being dried to a constant weight at 105 °C. Squares of lens-cleaning or other suitable tissues (experimentation may be needed to find the ideal tissue and size for each particular sediment) are placed on the surface and the sample left for at least 12 h.

The time of final harvesting is important because the epipelic populations show strong diurnal vertical migration rhythms (Round & Palmer, 1966; Palmer & Round, 1967; Eaton & Simpson, 1979; Round, 1978, 1979a). It has also been found that where the algal populations migrate vertically under laboratory conditions in phase with tidal cycles occurring simultaneously in the field, the timing of the migratory cycle must be determined by preliminary sampling and, hence, harvesting time adjusted accordingly. Samples in the laboratory should not be subjected to artificial illumination at night; instead they should experience a normal light/dark cycle (Round & Eaton, 1966). As mentioned above, coverglasses can be placed on to the surface of sediment which has been poured into a suitable container of known surface area, and removed with forceps the following day. Counting can be achieved by placing the coverglass onto a slide; however, since the majority of the epipelic algae are motile, it is not satisfactory to count them alive. Placing the coverglass on to a drop of 40 % glycerol prevents movement and overcomes this difficulty without rendering the algae unrecognizable. The algae can then be identified and counted. Numbers increase toward the edge of the coverglass, because the algae tend to migrate whilst it is on the mud surface. This would suggest that some deoxygenation occurs under the middle of the coverglass. Also, any sediment sticking to the underside of the coverglass will obscure the algae, thus making counting difficult, and also inaccurate. When counting by this method one should use standard traverses to ensure reproducibility, although Eaton & Moss (1966) found that even this did not totally eliminate error.

Williams (1963) used silk fabric to trap epipelic algae from marine mudflats, then separated them for counting by washing the fabric with water. This is satisfactory for such organisms as pennate diatoms, which readily wash off, but flagellates etc. may not, and while suitable for field use, the technique described below is more suitable for laboratory studies.

Eaton & Moss (1966) investigated the possibility of using cellulose tissues instead of coverglasses or fabric to trap the algae. They studied eight commercially available types of cellulose. It was found that a double layer of Grade 105 lens tissue (J. Barcham Green Ltd, Hayle Mill, Maidstone,

England) trapped twice as many algae as the others from all types of sediment. This type of tissue is recommended for all such work. Results showed that up to 87·5 % of algae were removed by the tissues. Samples containing euglenoid flagellates can also be harvested readily, as can samples containing mixtures of diatoms, volvocalean and euglenoid flagellates and blue-green filamentous algae. This tissue method was found to be much better than the coverglass method in efficiency of harvesting, estimation of species proportions and adaptability to different associations of algae. It is also readily adapted to production studies using both oxygen and ^{14}C (see pp. 262–8).

Eaton & Moss (1966) developed a technique for counting the algae. Tissues of 2 × 2 cm were placed on the sediment surface and removed the following day; these were then placed in 3 ml 40 % glycerol in Lugol's iodine solution which kills, stains and preserves the algae. When stored in the dark the algal pigments last longer. The addition of 5 % formaldehyde is desirable for long preservation, as the iodine will gradually lose its preserving power. Very delicate and small flagellates were distorted by glycerol and sometimes by iodine; thus other methods need to be developed for preservation and enumeration of the small and delicate algae. Tissues are broken up using a pair of dissecting needles. A known volume (0·020–0·025 ml) is transferred to a slide and counted under the microscope.

Squares of lens tissue (5 × 5 mm) mounted directly in a drop of 40 % glycerol have also been used. The lens tissue is first pulled apart with dissecting needles and coverglasses are then placed on the sample and the algae counted. This technique is useful where sites are relatively homogeneous. The use of larger lens tissue reduces the errors caused by heterogeneous populations.

If diatoms are present it is desirable to retain material for subsequent identification. Duplicate tissues can be removed from the sediment and placed in 10 ml conical tubes together with 5 ml dilute chromic acid. Tubes are then immersed in a water bath at 100 °C for 1 h. Lens tissue and other organic material is destroyed by the acid. This leaves only clean diatom valves and other mineral particulate matter. The residue is then washed by successive centrifugations and resuspensions until neutral to litmus paper. Afterwards it can be made up to 10 ml and 1 ml portions put on to 0·75 in (1·9 cm) circular coverglasses, allowed to dry, and the coverglasses mounted in Mikrops 163. Diatoms can then be identified and their numbers estimated, allowing for the fact that nearly every diatom is split into two separate valves during cleaning. The permanent mount method is subject to gross errors in estimation of quantity and quality of epipelic diatoms, and is, therefore, only useful for identification purposes (Eaton & Moss, 1966). Tissue traps and temporary mounts are recommended for further quantitative estimates.

As mentioned earlier, a method of estimation involving a suspension of the algae and associated sediment particles and counting of the organisms in an aliquot of the suspension in a counting chamber has been used. However, there are many problems and unless the sample is very rich in algae it is difficult to obtain a reliable count.

Epipsammic algae

In shallow water, the method of Eaton & Moss (1966) can be used to sample the surface sediment. The apparatus enables samples to be taken from a known area, delimited by a plastic cylinder pushed into the sediment, followed by removal of the sediment into a polythene bottle by suction. In deep waters mechanical corers can be used.

The non-epipsammic flora should be removed by repeatedly swirling the sediment with filtered water. The non-attached epipelic, planktonic algae and detrital material is swept into suspension and can be decanted off, whilst the epipsammic flora remains attached to the sand grains, which, being heavy compared with the epipelic algae and detrital material, readily sediments out. For cell counts of the epipsammic algae, a standard volume of this cleaned sand is taken, and placed in twice its volume of filtered water. This is then subjected to sonication in a Burndept BE 297 Ultrasonic cleaner for 10 minutes to remove the algae from the sand grains. It is advisable to cool the sonication bath with ice to prevent heating of the sample and subsequent cell breakage. Ten minutes has been found to be the optimum time for sonication, working with an epipsammic flora from a lake—90·5 % of the population is removed in this time with no breakage of cells. Sonication time beyond 10 minutes does remove more algae, but after 12–15 minutes cells tend to be broken up. Experimentation is needed to determine the optimal system for removing the algae from each sand type because some algae may adhere so strongly that a weak acid treatment may be necessary. The aqueous suspensions of the algae are next made up to a standard volume. Aliquots of known volume can then be placed on a slide and direct microscope counts can be made or a standard volume can be placed on a coverglass, dried and mounted in Microps, Hyrax, Naphthrax etc. Sediment used is dried to a constant weight at 105 °C so that all estimates of cell numbers can be related to the area sampled. (Round, 1965; Moss & Round, 1967.) Counting errors calculated by Moss & Round were small—the coefficient of variation ±7·9 % for a series of 10 replicate samples. As with epipelic algae, dead epipsammic cells can be distinguished from live cells. The dead cells either have no contents or the contents are considerably contracted. Only live cells should be counted for use in correlations with chlorophyll estimations, ^{14}C uptake, etc. For identification purposes the above method can still be used or the sand grains can be boiled in hydrogen peroxide or sulphuric acid to free the diatoms from the sand grains.

Epiphytic algae

The micro-epiphytic algae growing on the large algae can be removed by sonication as in the removal of the epipsammic flora. However, with this material a much higher percentage of the algae is left on the host plant. Sonication for periods longer than 10 minutes again tends to cause breakage of the cells. The algae growing on hosts such as *Laminaria*, *Fucus*, etc. could be scraped off using a sharp scalpel or razor, making sure that host epidermal cells are not removed in the process. If the population scraped off is very dense and clumped, these clumps can be broken down by sonication for 10 minutes. In both methods the algae are placed in known volumes of filtered sea water.

If difficulties arise in measuring the surface area of the host plants, the cell counts can be expressed on a cell/g dry weight of host basis or ash dry wt basis, even though this produces its own problems in that dry weights of different plants are not directly comparable. The simpler the shape of the host plant the easier it is to estimate the surface area.

Cell counts can be made by placing a known volume (0·020–0·030 ml) on a slide under a coverglass and counting all cells in a standard traverse across the coverglass. Alternatively, subsamples may be pipetted into a sedimentation tube and counted, after the addition of iodine, using an inverted microscope (Lund *et al.*, 1958). It is advisable to dilute the algae to a suitable degree since dense populations cannot be counted accurately. This method is hampered by any detrital material which also sediments.

The exact position and any inter-relationship between attached microflora can also be studied by scanning electron microscopy but the areas observed tend to be minute and many samples may be needed in order to obtain a detailed picture.

Estimation of biomass from chlorophyll extracts

Two techniques have been used, spectrophotometry and fluorimetry, but the former has been more popular.

The most commonly employed method for the estimation of pigments is that of absorption spectrophotometry, used by Richards with Thompson (1952). It is referred to as the 'trichromatic method', as it involves measurement of absorbency at three different wavelengths for the calculation of chlorophylls *a*, *b* and *c*. Creitz & Richards (1955) and Parsons & Strickland (1963) proposed a few minor changes in the equations used to calculate the amounts of pigments on the basis of re-determined specific absorption coefficients. However, in all these cases these equations depend for their validity on the absence of pheo-pigments in the extracts. According to Strickland & Parsons (1968), pheophytins are generally absent from the open ocean phytoplankton; however, large amounts of these compounds will probably be found when dealing with benthic communities. In these latter cases, therefore, the method described by Moss (1967a, 1967b) or that described by Lorenzen (1967) should be employed. These methods provide an estimate of the amount of chlorophyll *a* and pheophytin *a* in the extract.

The accuracy of the trichromatic equations is decreased by changes in the absorption spectra that occur on conversion of chlorophylls to pheophytins. These changes at 645 and 630 nm are fairly great and these are the principal wavelengths used in the calculation of chlorophylls *b* and *c*. Therefore, in the presence of pheophytins of chlorophylls *b* and *c*, the estimations of these chlorophylls by the trichromatic method will be unreliable. However, at 665 nm (wavelength for chlorophyll *a*) there is no appreciable change in the absorption in the presence of pheophytins *b* and *c*. Secondly, this wavelength is affected only slightly by the presence of different proportions of chlorophyll *b* and *c*. This has been recognized by Odum *et al.* (1958) and Talling & Driver (1963) who in their simplified equations used absorption only at 665 mμ to estimate chlorophyll *a* and pheophytin *a*, where the extract is assumed to contain only these two components. Moss (1967b) gives modified equations for a two component system in which no chemical reaction is occurring at the given wavelength; from these, the amounts of chlorophyll *a* and pheophytin *a* can be calculated.

As Moss (1967b) says, meaningful estimations of chlorophylls *b* and *c* in the presence of their pheophytins will be achieved by either extending the trichromatic method to a hexachromatic one, in which pheophytins, as well as chlorophylls are estimated, or by using separation methods such as that devised by Parsons (1963) for chlorophyll *c*.

It should be pointed out that the amount of organic material associated with a given quantity of plant pigment is very variable. Therefore, when relating cell numbers to chlorophyll *a* estimations, cell volume should also be taken into account.

Until methods are developed which will enable reliable results for chlorophyll *b* and *c* and their pheophytins to be obtained, the method of Moss (1967a and b) for chlorophyll *a* and pheophytin *a* and the method of Parsons (1963) for chlorophyll *c* are recommended. If chlorophyll *c* is to be determined, Millipore filters should be used (Strickland & Parsons, 1968).

A summary of the extraction procedure, etc. is given in Strickland & Parsons (1968) and the restrictions they list also apply to benthic algae.

Thin layer chromatographic separation of pigments has now been well developed for phytoplankton (Jeffrey, 1981 and earlier references referred to) and the technique should be tested and applied to benthic populations.

Routine methods for pigment estimations

Epipelic algae (adapted from the method of Eaton & Moss, 1966)

The algae have to be separated from the sediment for accurate estimations of chlorophyll *a* because large amounts of detrital pigment degradation products occur in many sediments compared with the amount in living epipelic algal populations. The tissue trapping technique can be used to trap the algal population prior to pigment extraction and estimation. Cropping can be achieved by placing large double layers of lens tissue, on the sediment surface in petri-dishes prepared as described on p. 255. Tissues are removed the following day and placed on small pieces of Whatman GF/C glass filter paper and left to air dry in the dark for approximately 3 h. They are then placed in ground-glass stoppered bottles, together with a small amount of analytical quantity anhydrous magnesium carbonate, plus a suitable volume of 90% acetone (aqueous, of analytical purity). Because of the risk of small amounts of water being left on paper and tissues, the volume of acetone added should be as large as practical to minimize this error. The addition of magnesium carbonate prevents degradation of the pigments during extraction. The samples are extracted at 3–4 °C in complete darkness for approximately 20 h. If large numbers of algae with thick cell walls or mucilage sheaths are present, grinding or homogenizing may be necessary.

The acetone plus extracted pigments is then centrifuged at 3000–4000 rev/min to sediment out any magnesium carbonate and other interfering particles. The absorbances at the selected wavelengths are then measured as quickly as possible on the spectrophotometer (Moss, 1967a, b: Lorenzen, 1967; Strickland & Parsons, 1968).

To each centrifuge tube 10 drops of 10% hydrochloric acid is then added, the tubes are shaken and left to stand in the dark for 10 min, before adding a small amount of magnesium carbonate and re-centrifuging. This process converts all the pigments to their degradation products. Each sample is then read again on the spectrophotometer at 430 nm and 410 nm (Moss, 1967a and b). (Using Lorenzen's method the samples are read again at 665 nm after acidification.)

The results for epipelic algae should be expressed as mg chlorophyll/m^2. The amount of chlorophyll *a* and pheophytin *a* can be calculated using the equations of Moss (1967a) and then converted to m^2 (see sampling section and section on separation of algae from substratum).

Epipsammic algae

The washed sand grains are extracted with 90% acetone as previously described and chlorophyll *a* and pheophytin *a* measured by the absorption spectrophotometric method of Moss (1967a and b). The coefficient of variation of replicate samples was ±4% for freshwater material (Moss & Round, 1967).

Epiphytic algae

For removal of algae see p. 258. Samples of algae which have been removed by either sonication or scraping are made up to a suitable volume before filtering through Whatman GF/C glass filter paper. A suitable aliquot of 90% acetone is then added. In most cases it is advisable to grind the filters in a homogenizer to ensure complete extraction because many epiphytic algae have thick mucilaginous sheaths. Then follow the procedure described for epipelic algae.

Cryophytic algae

Again, algal standing crop can be estimated using chlorophyll extraction by collecting ice and allowing it to melt, filtering, extracting, etc. (Apollonio, 1965; Bunt *et al.*, 1966).

Macro-algae and angiosperms

These can also be estimated by the chlorophyll technique. However, more macro-techniques would have to be employed, for example, grinding in 90% acetone and sand to extract the pigments. Only sections of the larger plants can be used, and the mucilaginous and rubbery nature of some of these algae may present problems.

Primary productivity measurement

The most direct, and in many instances perhaps the most satisfactory method of determining primary production is indirectly through measurement of increases or decreases in the standing crop determined over a defined time period. Here productivity is measured in forms of biomass changes and can be expressed in several ways, e.g.:

1 Increase in the area of substratum covered.
2 Increase in volume of organisms.
3. Increase in wet weight (i.e. fresh weight).
4. Increase in dry weight.
5. Increase in ash-free dry weight (i.e. loss on ignition).

Such methods have been applied successfully, in one form or another, to the study of the productivity of macroscopic marine plants, which undergo a regular growth cycle throughout the seasons. However, these methods are less suitable for measuring the productivity of the micro-algae of the epipelon, epipsammon, epiphyton and epilithion. Here pigment determinations (chlorophyll *a*) are more useful for standing crop (biomass) measurements. An excellent account of methods for the study of primary productivity of phytoplankton is available (Vollenweider, 1969), and we shall not repeat information in that publication except where essential for an understanding of the benthic situation.

Oxygen evolution

Direct primary productivity measurements are determined from metabolite exchange involving either oxygen evolution or ^{14}C uptake, (i.e., photosynthesis of the plants). The oxygen evolution technique originally described by Winkler (1888), and used later by Gaarder & Gran (1927) involves measuring the rate of change of dissolved oxygen in samples enclosed in incubation vessels. The length of the incubation period is determined by estimates of the intensity of the photosynthetic activity. Sufficient time must pass to allow for measurable changes in concentrations of dissolved oxygen to occur: 2–4 h is usually adequate, but longer is required if populations are small. Dissolved oxygen measurements may be made at the beginning of the incubation period (IB); then if both light and dark incubation vessels are used during the incubation the initial concentration of oxygen (IB) would decrease in the dark due to respiration (DB) while due to photosynthesis it would increase in the light (LB). Therefore, respiration and photosynthesis can be determined, and net and gross primary productivity calculated:

$$\mathrm{IB} - \mathrm{DB} = \text{respiratory activity}$$
$$\mathrm{LB} - \mathrm{IB} = \text{net photosynthetic activity (net productivity)}$$
$$(\mathrm{LB} - \mathrm{IB}) + (\mathrm{IB} - \mathrm{DB}) = \text{gross photosynthetic activity (gross productivity)}$$

Gross productivity refers to the gross synthesis of organic matter resulting from photosynthesis, and net productivity refers to the net formation of organic matter after losses due to respiration, extracellular release of soluble organic compounds, and other losses resulting from simultaneous metabolism.

This method provides an approximation to natural productivity because the incubation vessels can be placed back *in situ* and, therefore, receive the same illumination and temperature as the surrounding plants.

The Winkler method involves the oxidation of manganous hydroxide by the oxygen dissolved in the water, resulting in the formation of a tetravalent compound. When acidified, free iodine is liberated from the oxidation of potassium iodide, and is chemically equivalent to the amount of dissolved oxygen present in the sample. This is determined by titration with sodium thiosulphate.

The reagents involved are manganous sulphate, Winkler's reagent (a mixture of potassium hydroxide and potassium iodide) and sulphuric acid. Equal amounts of $MnSO_4$ and Winkler's reagent are added to the sample and the $MnSO_4$ reacts with the KOH-KI mixture producing a white flocculent precipitate of manganous hydroxide:

$$MnSO_4 + 2KOH \longrightarrow Mn(OH)_2 + K_2SO_4$$

If a white precipitate is formed no dissolved oxygen is present; a brown precipitate indicates that oxygen is present and reacting with the manganous hydroxide, forming manganic basic oxide:

$$2Mn(OH)_2 + O_2 \longrightarrow 2MnO(OH)_2$$

When H_2SO_4 is added this precipitate is dissolved forming manganic sulphate:

$$MnO(OH)_2 + 2H_2SO_4 \longrightarrow Mn(SO_4)_2 + 3H_2O$$

There follows an immediate reaction between manganic sulphate and KI which liberates iodine:

$$Mn(SO_4)_2 + 2KI \longrightarrow MnSO_4 + K_2SO_4 + I_2$$

The number of moles of iodine liberated is equivalent to the number of moles of oxygen present and the amount of iodine is determined by titration with sodium thiosulphate.

$$2Na_2S_2O_3 + I_2 \longrightarrow Na_2S_4O_6 + 2NaI$$

Using this method an oxygen content of 0·05 ml l^{-1} and assimilation of 0·02 mg l^{-1} can be detected with reasonable accuracy.

Problems with this method can arise through chemical interference from nitrite, ferrous ions and organic matter (Alsterburg, 1926; Theriault & McNamee, 1932; Pomeroy and Kirschman, 1945; Kliffmüller, 1959), and modifications may have to be made. Alsterberg (1925) used azide in the alkaline iodide solution to destroy nitrite in concentrations of $>0{\cdot}1$ mg l^{-1}. An acid-permanganate modification, introduced by Rideal & Stewart (1901), is suitable for dealing with nitrite and ferrous ions (>1 mg l^{-1} Fe^{2+}). Ferric ion (>10 mg l^{-1} Fe^{3+}), either present or produced by the permanganate treatment, can be removed by addition of potassium fluoride or phosphoric acid. If the interference stems from iodine-consuming compounds the iodine difference method of Ohle (1953) can be used, and that originating from organic matter can be minimized by avoiding delays during analysis (Alsterberg, 1926; Bruhns, 1916; Theriault & McNamee, 1932) and using a high iodide concentration (Pomeroy & Kirschman, 1945).

The steps in the technique are simple. To the incubation vessel add, gently and just below the water surface, first the solution of $MnSO_4$ and next the Winkler's reagent, being careful not to mix the pipettes. (It is advisable to use automatic pipettes.) If using 250–300 ml bottles, add 2 ml of each, then replace the stopper without introducing air bubbles, and mix well. Allow the precipitate to settle before adding 2 ml concentrated H_2SO_4, again with an automatic pipette, inserting the tip just below the water surface. Carefully replace the stopper and shake until all the precipitate has dissolved before pipetting out 200 ml of the sample into a 250 ml flat bottomed conical flask and titrating this against 0·025 N standardized sodium thiosulphate, mixing until a pale yellow colour forms. At this stage add 1 ml of starch indicator; mix and a blue colour forms. Then continue the titration to the colourless end point. The blue colour should return after 15–25 seconds. If it does not the end point has been exceeded. Now record the volume of titrant used, and for a 200 ml of the original sample, 1 ml of 0·025 N sodium thiosulphate is equivalent to 1 mg/litre oxygen dissolved in the sample.

Dissolved oxygen concentrations can also be determined using oxygen electrodes (Mancy & Jaffe, 1966). In all such experiments, whether using a titrametric or electrochemical method, the dissolved oxygen concentration is influenced by the respiration not only of the plants but also of the fauna, and the bacterial flora which may develop in water enclosed in bottles.

Reagents and materials for oxygen method

1. $MnSO_4$ (400 gm $MnSO_4.2H_2O$ m 1 litre distilled water).
2. Alkaline-iodide-azide reagent (dissolve 500 gm NaOH and 135 gm NaI in 1 litre distilled water; 10 gm NaN_3 dissolved in 40 ml distilled water and add to 960 ml of the alkali-iodide mixture).
3. Standardized sodium thiosulphate (0·025 N thiosulphate made by dissolving 6·205 gm $Na_2S_2O_3.5H_2O$ in 1 litre freshly boiled and cooled distilled water). 1 ml 0·025 N thiosulphate ≡ 0·2 mg O_2.

Standardization

(a) Dissolve 2 gm KI in 100–150 ml distilled water
(b) Add 10 ml of 1:9 H_2SO_4
(c) Add exactly 20 ml 0·02N KH $(IO_3)_2$

(Diluted from a 0·10 N stock solution of 3·249 gm KH $(IO_3)_2$ l^{-1})

(d) Dilute to 200 ml with distilled water
(e) Titrate with thiosulphate as described above
(f) Normality of thiosulphate can now be calculated

4. Starch indicator—(suspend 2·0 gm potato starch with 30 ml 20 % KOH in 400 ml distilled water, and stir until almost clear; stand for 1 h and neutralize with HCl; to preserve add 1 ml glacial acetic acid). A method to measure oxygen production using an oxygen electrode is described by Admiraal (1977).

Carbon-14 uptake

The more recently developed, much used and refined method of determining primary productivity involves measuring the uptake of radioactive ^{14}C by the plants (see Steeman Nielsen, 1952, 1953). Here the radioactive tracer is added in the form of $Na_2{}^{14}CO_3$, and if the total carbon dioxide content of the water is known and the ^{14}C content of the plants is measured after the incubation period, the total amount of carbon assimilated can be calculated by the proportional relationship:

$$\frac{^{14}\text{C available}}{^{14}\text{C assimilated}} = \frac{^{12}\text{C available}}{^{12}\text{C assimilated}}$$

According to Goldman (1963) 'the rate of carbon fixation at the level of primary producers currently provides the best assessment of the interactions of the host with the physical, chemical and biological factors determining the actual fertility of any environment'. Photosynthetic organisms are more important than chemosynthetic ones; the latter usually play a minor part in productivity (Kuznetsov, 1956; Steeman Nielsen, 1960).

As with the oxygen techniques, both light and dark incubation vessels are used which give a fractional correction roughly independent of the number of cells present, and which generally amount, with phytoplankton, to about 1 % of the rate of photosynthesis measured under optimum conditions. However, when studying the productivity of benthic communities the total uptake of ^{14}C in the dark is greater. Grøntved (1960) found that the ^{14}C uptake in the dark averaged approximately 10·5 % of the total ^{14}C uptake in the light in 2 h experiments using a mixture of sand, epipsammic algae and unfiltered sea water. In the suspended fraction of the sample the dark fixation was greater, averaging 13·6 % of the total fixation in light, whereas in the sand fraction it averaged 8·5 %. Grøntved (1960) also found that the dark fixation varied greatly in the various experiments, and the average figures from different fjords varied rather widely. This is not altogether surprising since the sample, besides containing autotrophic organisms, contains heterotrophic flagellates, fungi, and bacteria. In view of these features, the dark fixation of ^{14}C does not seem particularly large.

Routine methods

Epipsammon. Grøntved (1960, 1962), in his studies of the productivity of the epipsammic flora of a Danish fjord, devised an experimental technique whereby ^{14}C was added to a subsample of benthic algae in bottles which were then incubated. Since unfiltered sea water was used the fraction of the total assayed ^{14}C fixed by the epipsammic algae had to be calculated after allowing for fixation by the other algae present. Self-absorption of the radiation by the sediment particles was determined by several methods of radio-assay of filters before and after removal of large inorganic particles, and results were expressed as a potential productivity since the incubations were not performed *in situ*.

Steele & Baird (1968) and Baird & Wetzel (1968) have also developed techniques for studying the production of the epipsammon. As in the method of Moss & Round (1967), the detrital material and epipelic algae were removed by suspending the sand and attached epipsammic algae in filtered water, followed by decanting. This was performed once or twice. Sand samples are collected with a hand-held corer in shallow water or a remotely controlled corer in deeper water.

The incubation method involves adding about 15 g washed sand, 100 ml filtered sea water and 5 μCi. Na_2 $^{14}CO_3$ to 120 ml screw cap glass bottles. The jars are sealed, and gently mixed by inversion several times before leaving them in an inverted position on a suitable incubation tray, making sure that the sand forms an even layer upon the inside of the screw cap. Bottles for dark ^{14}C uptake are prepared in a similar manner. The incubation tray is placed *in situ*, and left for 4–6 h to incubate, after which most of the supernatant water is decanted off, and the rest is filtered into a Stefi filter funnel fitted with a glass fibre filter paper. The sand is then rinsed with three small quantities of filtered sea water, and excess moisture is removed under vacuum. A 1 % HCl wash may now be required if carbonate formation is suspected during the incubation. Two methods of determining the amount of incorporated ^{14}C can be used. In the first, an excess of sand is carefully placed in planchettes (aluminium, 23 mm diameter, 2 mm depth), and dried in a carbon dioxide-free desiccator before it is mixed and levelled to fill the planchette just to the rim. The incorporated activity is determined using a thin end-window Geiger-Müller counter. To correct for self-absorption, samples may be totally combusted, and a factor determined for self-absorption from the count of $^{14}CO_2$. Alternatively, a labelled culture of an alga may be mixed into a sand sample. The labelled culture is prepared by adding about 50–100 μCi Na_2 $^{14}CO_3$ to a 100 ml algal culture, followed by incubation for 12–24 h. The cells are then centrifuged and washed several times with fresh culture media, and finally resuspended in 100 ml fresh culture media. Subsamples (e.g. 1, 2, 4,

7 ml) of the suspension are filtered on to membrane filters which are counted following normal procedures. 5 ml subsamples are stirred into 20 g dry sand, and these samples are prepared as described for the epipsammic algae. The corrected and normalized counts for the filtered samples should be proportional to the volume filtered, indicating no self-absorption. The counts ml^{-1} min^{-1} (or zero thickness activity) can now be determined:

$$F = \frac{\text{(zero thickness (c.p.m.) added)(g dry sand)}^{-1}}{\text{observed (c.p.m.) in sand}}$$

then the uptake in mgC m^{-2} h^{-1} will be:

$$\frac{(\text{LB.c.p.m.} - \text{DB.c.p.m.})(1{\cdot}06)(\text{F})(\Sigma\text{C})(W)}{(t)(\text{added c.p.m.})(A)}$$

where 1·06 is an isotopic correction factor, ΣC = total inorganic carbon content as mgC, t = incubation period (time), A = area of jar lid in m^2, W = weight of dry sand added to jar.

The second technique for determining the incorporated ^{14}C involves scintillation counting. Several 1 gm subsamples of damp sand are weighed (by subtraction) into scintillation vials, 20 ml fluor is added to each and they are shaken (Fluor: PPO = 2,5-diphenyloxazole – 5.5 gm; POPOP = *p*-bis-[2-(5-phenyloxazolyl)] benzene – 50 mg to 1 litre with 1:2 v/v 2-ethoxy-ethanol, toluene). After standing for 24 h in the dark, any colour in the mixture is bleached out using 100 litre freshly prepared chlorine water before counting in a scintillation counter. The counts are corrected for efficiency of counting and uptake calculated in mgC m^{-2} h^{-1}.

Further work involving the measurement of epipsammic algal primary productivity has been done by Hickman (1969) and Hickman & Round (1970). The epipsammic algae are separated from any epipelic algae and detritus following the method of Moss & Round (1967). Subsamples of washed sand are then placed in incubation bottles along with filtered water and the ^{14}C source and incubated *in situ* for 3–4 h. Both light and dark incubations are performed. Afterwards the algae can be removed from the sand grains by sonication (see previous section). A suitable subsample (25 ml) is then filtered onto Millipore HA filters and the radioactivity determined. However, since this technique does not take into account extracellular production, it is preferable to add 1 ml formalin to each incubation vessel and utilize the acid bubbling technique of Schindler *et al.* (1972) to remove unincorporated inorganic ^{14}C before subsamples are removed and the incorporated activity is determined using scintillation counting.

The dissolved oxygen technique can also be used to investigate epipsammic algal productivity by adding subsamples of washed sand, and filtered water to each bottle (keeping the container and volume small) followed by exposure for several hours to light. The amount of discoloured oxygen can be estimated trimetrically by the Winkler method (Hickman, 1969).

Epipelon. Initially, workers used bell jars placed directly over the sediment in intertidal zones, and then determined productivity using either the oxygen technique or the ^{14}C technique (Pomeroy, 1960). Filtered water may be placed in the bell jars to eliminate phytoplankton primary producers—if unfiltered water is used instead a correction factor must be applied. On hard substrata a band of flexible rubber may be placed around the bottom of the jar to act as a seal (Odum & Odum, 1955). The changes in dissolved oxygen are usually estimated by the Winkler method; it is possible to obtain a continuous record of the changes in oxygen tension using electrodes and a suitable recorder (Carritt & Kanwisher, 1959; Kanwisher, 1959).

The ^{14}C method has also been used in a similar manner in a shallow lake where the littoral zone comprised small angular pebbles (Wetzel, 1963, 1964). There is no reason why this could not be adapted to marine habitats, although problems would arise in more exposed situations. Wetzel placed the plexiglass chambers on the sediments using a short rotational movement so that chambers were actually worked into the sediment. Afterwards ^{14}C was injected into the chambers underwater and a 4 h incubation period was used. At the end of this period a plate was worked under the open end and the whole sample removed (i.e. an undisturbed core of the superficial sedimentary material and overlying water). The overlying water was removed using a large syringe and the upper centimetre of sediment was then frozen. The organic material was oxidized to CO_2 by Van Slyke combustion for radio-assay in the gas phase.

Hickman (1969) developed a method of studying the productivity of the epipelic algae by separating the algae from the sediment using the tissue trapping technique of Eaton & Moss (1966). The lens tissues, along with the harvested epipelic algae, are carefully placed in filtered water in glass bottles with the ^{14}C source, and incubated for 3–4 h *in situ*. In the original method the algae had to be removed from the lens tissue. This was accomplished by shaking by inversion, and thus the percentage freed by this technique can be determined. A suitable subsample is then filtered onto Millipore HA filter membranes, for subsequent determination of incorporated activity either by a thin window Geiger-Muller counter or following Hickman & Klarer (1973) by scintillation counting. Jenkerson and Hickman (in preparation) have modified this method. After the incubation period 1 ml formalin is added and the bottles vigorously shaken to break up the lens tissue. Then the unincorporated ^{14}C is removed using the acid bubbling technique of Schindler *et al.* (1972), after which 2 ml sub-samples are placed into scintillation fluor (Aquasol, New

England Nuclear). This modification now accounts for both particulate and extracellular production.

Marshall *et al.* (1973) outlined a method for measuring the primary productivity of the 'bottom microflora' without disrupting the surficial sediments. In this method one cannot differentiate between, for example, epipelic and epipsammic primary productivity which is also the problem encountered in the 'bell jar' technique. In the method of Marshall *et al.* (1973) samples are taken with a surficial corer, kept intact and placed in separate light and dark incubation flasks containing ^{14}C and filtered water. These are incubated *in situ* with the incubation flasks attached to a weighted rack for 3–4 h, after which 1 ml neutral formalin is added. Each flask is vigorously shaken and the suspension transferred onto a Gelman type A glass-fibre filter with minimum delay. To remove any unincorporated ^{14}C about 50 ml additional filtered water is passed through the filter. A weakness of this technique is the combination of formalin with filtration, since the former will release unknown amounts of ^{14}C and, as with any filtration technique, extracellular production is not measured and the results will underestimate true production. The filters and sediment are now dried before placing them in a mortar and thoroughly grinding using a pestle. The resulting powder is transferred quantitatively to a vial and suspended in sufficient 4% Cab-O-Sil gel in toluene to bring the volume up to 30 ml. This suspension is shaken, after which three 1 ml subsamples are pipetted into separate 20 ml scintillation vials containing 15 ml 4% Cab-O-Sil gel and 1 ml fluor concentrate (8 g PPO, 0·16 g POPOP in 100 ml toluene): each vial is again shaken well. The incorporated activity is now determined using a scintillation counter. An external standard or channels ratio method is used to determine efficiency and to calculate a correction factor for quenching. Further errors could be introduced into this method by the drying process since Wallen & Geen (1971) found appreciable losses during drying and storage when working with phytoplankton. Probably the method of Hickman (1969) as modified by Jenkerson & Hickman (1983) provides the most satisfactory way of measuring epipelic algal productivity.

Macrophytes. On soft bottoms the larger seaweeds can be covered with a plexiglass cylinder into which the $^{14}CO_3$ is injected. Photosynthesis is allowed to continue for a set time, after which the cylinder is removed and the entire alga removed and counted. A dark cylinder is used as a control. The blotted samples are frozen in dry ice and portions counted after washing in filtered sea water and exposure to concentrated HCl fumes to remove any contaminant carbonate. Loss of extracellular material is common for many seaweeds and this should be taken into account.

Another method using replicate portions of large kelps is described by

Drew (1973)—it is essentially similar to the above but requires more divers and apparatus—see also p. 123 in Chapter 5.

Epiphytic algae. Two methods can be employed. In the first small pieces of the host plus the epiphytic algal population can be placed in bottles together with filtered sea water and the ^{14}C source, then incubated either *in situ* or in an illuminated water bath. After incubation the microscopic epiphytic algae can be removed from the host by ultra-sonication or large algae removed by dissection of the host. Encrusting and creeping algae, e.g. *Melobesia*, may require the development of special techniques. As mentioned earlier, sonication only removes a percentage of the algae and some are removed more easily than others; therefore, it will be necessary to use cell count data in conjunction with ^{14}C data, counting the number of cells removed by sonication and also those remaining on the host. After sonication the microscopic algae can be filtered and then counted and the radioactivity of a sub-sample determined.

Alternatively, if the host plant lends itself to scraping, the epiphytic algae can be removed in this way. This will remove a much higher percentage of cells than sonication but in some cases it is difficult to prevent contamination from pieces of the host material which themselves may absorb some ^{14}C. The scraped-off cells can then be incubated with filtered sea water and the ^{14}C source. Again an aliquot can be filtered through a membrane filter and counted.

Each host plant will present its own problems, and these methods will have to be modified accordingly. Many marine algae contain large amounts of mucilage and this will present problems owing to possible uptake of ^{14}C by the mucilage and also by the bacterial and fungal populations which will probably be associated with it.

Algae and angiosperms. As mentioned on p. 261 the most direct, and in many cases the most satisfactory method of estimating the production of the larger marine algae is by the increase or decrease in the standing crop (measured as fresh or dry weight) over a measured interval of time. Here production is measured in terms of biomass change. The ways of expressing this are given on p. 275.

The fresh weight may be defined as the weight of the plant immediately after being collected, and after superficial water has been shaken off (Baardseth & Haug, 1953). However, intertidal algae may lose water by evaporation at low tide; these should therefore be placed in sea water until the lost water is replaced and the weight is constant. The errors involved when determining fresh weight are:

1. Loss of water during exposure at low tide.
2. Water adhering to the surface of the algae.
3. Loss of water during transport and handling of the material before weighing.

Dry weight determinations should be carried out by air-drying the algae, and then, depending on the size of the algae or sample collected, all or a known proportion of the air-dried material is ground to a powder and dried at 100–150 °C for 16 h or to a constant weight (Baardseth & Haug, 1953). From the weight of the samples, the total dry weight can be calculated. The errors of dry weight determination can be attributed to:

1. Incomplete removal of water.
2. Loss of material during the drying process—this can be avoided by careful handling.
3. Respiration or other destructive processes during the time the material is kept covered before weighing. This is probably insignificant in the majority of cases.

The biomass data can then be related to the area of substratum and to a time interval, providing the samples were quantitatively taken on an area basis in the first place, i.e. taken from within quadrats of known dimensions, e.g. in the works of Hällfords *et al.* (1975).

In some members of the Laminariales, in which the blade grows from the base, the rate of growth can be estimated by punching holes in the blade (Parke, 1948; Mann, 1973). Mann advises measuring a minimum of 50 *Laminaria* plants, recording the distance of a hole punched in the frond near the base at monthly intervals. From the 50 plants a length–weight relationship is determined for each plant by plotting the total wet weight against the mean blade length of each plant. Allowance has to be made for concurrent growth of the stipe, which is lengthened by primary growth and thickened by secondary growth (Kain, 1963; Bellamy *et al.*, 1973), and also for the hapteron. See also p. 123 in Chapter 5.

In the detailed study of Knight & Parke (1950) on *Fucus vesiculosus* and *F. serratus*, areas of the shore varying in extent from 1 m^2 to broad strips 16 m wide running through the entire fucoid zone were cleared of algae and animals. The sizes of the plants in the new recolonizing populations were then measured over a time period. Several difficulties in this approach are apparent. Eggs from plants in the vicinity of the experimental area are shed continuously over a long period, so that the cleared areas become colonized by plants of differing ages. In this study the longest plants were measured at intervals on the assumption that the longest would be the oldest in such a mixed population. However, there is great variation in the vigour of the germlings, competition between the germlings, and variation due to where the germlings grow, i.e. on flat exposed surfaces or in more sheltered rock crevices. There was also a high mortality rate or de-population rate.

A second method was also used which involved marking certain plants *in situ*. Numbered chicken or hen rings were attached to the plants. Very small plants were marked by small celluloid tablets with a split into which the frond could be inserted. Very large numbers of plants had to be tagged because of the high de-population rate. The tagging of large plants was of little use because of the large amount of frond breakage.

In *Fucus*, linear growth alone is not a full measure of growth since dichotomy of the fronds and lateral frond development occurs simultaneously with linear extension. An attempt was made to estimate this development by recording the degree of 'bushiness' by counting the number of frond apices at the same time as recording the length. Here again there was great individual variation.

South & Burrows (1967) used measurements of overall and submeristem length together with maximum diameter to assess the growth of *Chorda filum*, i.e. the vegetative growth of the plant throughout the year. Again the results showed that linear increase is a poor guide to the actual amount of tissue produced. However, from such studies growth rates and cycles can be recorded and followed over a time period as long as the limitations of such methods are realised. Norton & Burrows (1969) applied similar methods to *Saccorhiza polyschides*.

It is not usually possible when using the carbon-14 and oxygen techniques to enclose the plants within containers pressed into the substrate. Therefore, the algae have to be removed from the rock and placed in incubation chambers. However, manipulation of these larger marine algae from the surface is difficult, and impossible in deeper water, since sampling from the surface damages the plants and further the samples cannot be taken from a known area. Therefore, the whole operation of sampling and placing the plants in the incubation chambers should take place at the site and underwater. The plants should be placed in the incubation chambers at their growing site since many of these plants are adapted to low light intensities, and only brief exposures to the more intense surface light could seriously alter subsequent metabolic rates and could cause irreversible damage. The water used for the incubations should be taken from the same location as the plants in order to keep conditions as close as possible to natural. Filtered sea water should be used where possible, or a correction factor applied if a large phytoplankton population is present. A further point not always recognized is that on all the larger algae there will be an epiphytic algal population; this has either to be removed (p. 270), which in many cases would be impossible without seriously damaging the host, or its productivity must be estimated separately and subtracted from the result obtained by incubating the host plus the epiphytic population.

As mentioned in an earlier section, changes in dissolved oxygen in light and dark bottles have been extensively used to determine the photosynthesis of both marine and freshwater phytoplankton, but has rarely been used for submerged marine macrophytes under field conditions. Some marine benthic algae, particularly those in deeper water, require fairly lengthy incubation periods before there is a significant change in oxygen concentration. This lengthy incubation period (>6 h) will be accompanied by increases in bacterial populations on the surfaces of the incubation containers coupled with depletion of critical nutrients and stagnation. Therefore, the incubation containers should be sufficiently large to hold a large volume of water relative to the plant volume to prevent stagnation.

An important limitation and source of error in the application of the oxygen technique to macrophytes arises from internal storage and utilization of oxygen produced by photosynthesis. The latter will probably occur in deep benthic plants, where rate of oxygen produced and rate of oxygen consumed are very nearly equal (compensation point). Also, the use of stored oxygen for respiration during periods of darkness can occur without affecting the concentration of dissolved oxygen of the incubation water under natural conditions. In all cases where oxygen techniques are employed, careful consideration must be given to possible sources of error and the results must be interpreted extremely carefully. Johnston & Cook (1968) and Johnston (1969) used the oxygen technique to study the primary productivity of *Caulerpa prolifera* in waters around the Canary Islands. Light and dark incubation bottles were used while the oxygen concentration changes were measured with a polyethylene-lead-silver electrode system.

^{14}C techniques developed for use with phytoplankton can be adapted for these macrophytic benthic associations. However, because of the very weak radiation, self-absorption by the plant tissues occurs and major changes in counting techniques are required. At the end of the incubation period with ^{14}C the plants can be placed in polyethylene bags and quick-frozen between blocks of solid carbon dioxide for transportation to the laboratory. If samples are being taken for biomass estimations at the same time, these should be taken at similar transect intervals to those of the productivity estimations. The transects for both should be perpendicular to the shore in as many areas, and including as many different types of associations as possible.

Self-absorption problems of ^{14}C radiation can be circumvented by the van Slyke wet oxidation technique for the conversion of organic carbon to carbon dioxide folowed by radio-assay in gas phase. The production rates can then be calculated in a proportional manner as in the techniques for phytoplankton. Before oxidation the plants should be subjected to fumes of concentrated hydrochloric acid for the removal of extra-cellular ^{14}C precipitated as carbonates.

Johnston & Cook (1968) also used the ^{14}C technique. They incubated fronds of *Caulerpa prolifera* with ^{14}C, washed the fronds and placed them in 80 % ethanol. Radioactivity was measured in a Packard Tricarb Scintillation Counter.

Photographic techniques can be used to estimate the production and loss of plant material of macro-algae when the plants can be tagged in the field or when permanent quadrats can be placed over areas which can then be photographed at intervals throughout the year. Seasonal increments and losses can then be determined. Techniques of this kind are applicable to deeper waters where there is less de-population than in tidal regions and where the diver is able to spend only relatively short periods in deep waters before facing serious physiological dangers. Johnston *et al.* (1969) have developed a quantitative photographic technique that requires no formal knowledge of photogrammetry, no specialized cameras and no plotting machines to plot distribution and cover of particular plants. The underwater procedure is fast and flexible to allow the technique to be used in deep waters and under difficult diving conditions.

Laboratory and ship-board measurements of primary production

Illuminated constant-temperature incubators

The incubators enable measurements of activity to be made other than by exposure in the natural body of water. The potential advantages when working aboard a ship include the saving of ship time after samples have been taken, greater ease of manipulating samples under laboratory conditions, and standardization and control of conditions of light and temperature. There are disadvantages, however, including the impossibility of being able to reproduce fully the conditions of light in the natural habitat, particularly with respect to spectral modification, angular distribution with depth etc. Also, effects, detrimental or otherwise, of high light intensities present near the water surface are not taken into account or, alternatively, they are increased unless precautions are taken.

There are two types of incubators, those using ambient light (natural sunlight), or constant light (fluorescent, incandescent light), and each type can be used in 3 ways:

1. Without light filters.
2. With spectrally neutral light filters.
3. With spectrally selective light filters.

For both types a large temperature-controlled water bath is required. Examples of suitable baths for which sunlight has been used as the light source are described by Sanders *et al.* (1962). The incubator should be exposed to ambient solar radiation, and located so as to minimize possible shading. It may be necessary (Jitts, 1963) to use a selective filter or glass plate to remove ultra-violet radiation and possibly attempt some further simulation of the spectral modification of underwater light. Examples of incubators involving the use of artificial light source are described by Steemann Nielsen & Jensen (1957), Doty & Oguri (1958, 1959), Talling (1960) and McAllister *et al.* (1964).

In measuring the light to which the samples are exposed, determinations of irradiance (energy flux: e.g. in cal cm^{-2} min^{-1}, langleys min^{-1}, kerg cm^{-2} sec^{-1}, u Einsteins m^{-2} sec^{-1} or Watts m^{-2} (cm^{-2})) are greatly preferred to determinations of illuminance (e.g. lux, foot-candles) with a photometer, as the latter will possess a selective spectral response unrelated to the action spectrum of photosynthesis. Light sensors are now available which are sensitive to photosynthetically active (available) radiation—P.A.R.—400–700 nm. Such methods, although developed for use with phytoplankton, can be adapted for work involving the benthic communities.

Expression of productivity data

In productivity studies it is often difficult to compare the primary production of different communities and habitats because of the nature of the habitats, and the different methods employed by different workers. Therefore, methods will have to be developed which will allow direct comparisons of results.

It is important to have a standardized terminology when discussing productivity to avoid misinterpretation when comparing results. Theoretical discussions of various definitions are given in Thienemann (1931), Lindeman (1942), Ivlev (1945), MacFadyen (1948), Yapp (1958), Odum (1959) and Westlake (1963). Westlake (1965) recommends definitions of biomass, primary production, primary productivity, gross and net productivity, standing crop, crop and yield, which are in basic agreement with the theoretical definitions but modified for common usage. The question of terminology has been discussed at several IBP meetings and some general agreement on symbols, terms and definitions has been reached (*IBP News*, No. 10. 1968). This is further referred to by Bagenal (1978, pp. 2–3) and is discussed in detail in relation to secondary production on pp. 286–94 of the present handbook.

The units used to express results should give an indication of either the actual values determined or the validity of the results. For example, productivity of epipelic and epipsammic algae should be expressed as mg C fixed or O_2 evolved h^{-1} m^{-2} or day m^{-2}. However, with epiphytic algae where the surface area of the host plant cannot be estimated the results will be best put on a host-weight basis. The weight of host per unit area can be estimated and from this the epiphyte production can be expressed on an area basis. Estimations from epiphytes growing on one host will not be comparable with those on another host. Those hosts whose surface area can be estimated with reasonable precision should have epiphytic productivity expressed as a rate per unit area of host.

The units for biomass, defined as weight per unit area should be in g m^{-2} or kg m^{-2} (Westlake, 1965). Such estimates are not normally made on microscopic populations; instead the weight of pigment is estimated (see p. 258).

To avoid misunderstandings, the time units for productivity should be related to the actual period considered. For example, Doty & Oguri (1957) showed that with phytoplankton there exists a daily rhythm of photosynthesis. There is a gradual drop after midday and increase from sunrise to midday. Therefore, time interval and time of day of *in situ* incubations should be stated and investigations to determine whether or not a diurnal rhythm exists should also be carried out.

References

Admiraal, W., 1977a. Influence of various concentrations of orthophosphate on the divisional rate of an estuarine benthic diatom, *Navicula arenaria*, in culture. *Marine Biology*, **42**, 1–8.

Admiraal, W., 1977b. Experiments with mixed populations of benthic estuarine diatoms in laboratory microecosystems. *Botanica Marina*, **20**, 479–485.

Admiraal, W., 1977c. Salinity tolerance of benthic estuarine diatoms as tested with a rapid polarographic measurement of photosynthesis. *Marine Biology*, **39**, 11–18.

Admiraal, W., 1980. Experiments on the ecology of benthic diatoms in the Emms–Dollard estuary. *Biol. Onderzoch Emms–Dollard Estuarium*, **3**, 125 pp.

Aleem, A.A., 1950a. Distribution and ecology of British marine littoral diatoms. *Journal of Ecology*, **38**, 75–106.

Aleem, A.A., 1950b. The diatom community inhabiting the mud-flats at Whitstable. *New Phytologist*, **49**, 174–188.

Aleem, A.A., 1951. Contribution a l'étude de la flore de Diatomées marines de la Méditerranée. I. Diatomées des eaux profonde de Banyuls-sur Mer. (Pyrénées-Orientales). *Vie et Milieu*, **2**, 44–49.

Alsterburg, G., 1925. Methoden zur Bestimmung von in Wasser gelösten elementaren Sauerstoff bei Gegenwart von salpetriges Saüre. *Biochemische Zeitschrift*, **159**, 36–47.

Alsterburg, G., 1926. Die Winklersche Bestimmungsmethode für in Wasser gelösten elementaren Sauerstoff sowie ihre Anwendung bei Anwesenheit oxydierbarer Substanzen. *Biochemische Zeitschrift*, **170**, 30–75.

Apollonio, S., 1965. Chlorophyll in Arctic sea ice. *Arctic*, **18**, 118–122.

Baardseth, E. & Haug, A., 1953. Individual variation of some constituents in brown algae, and the reliability of analytical results. *Report of the Norwegian Institute of Seaweed Research*, **2**, 23 pp.

Bagenal, T., (ed.) 1978. *Methods for Assessment of Fish Production in Fresh Waters*. International Biological Programme Handbook No. **3** (3rd edition). Blackwell Scientific Publications, Oxford, 365 pp.

Baird, I.E. & Wetzel, R.G., 1968. A method for the determination of zero thickness activity of ^{14}C labelled benthic diatoms in sand. *Limnology and Oceanography*, **13**, 379–382.

Bellamy, D.J., Wittick, A., John, D.M. & Jones, D.J., 1973. A method for the determination of seaweed production based on biomass estimates. *Monographs on Oceanographic Methodology. UNESCO, Paris*, **3**, 27–33.

Bodeanu, N., 1964. Contribution à l'étude quantitative du microphytobenthos du littoral Roumain de la Mer Noire. *Revue Roumaine de Biologie, Ser. Zool.*, **9**, 434–445.

Brockmann, C., 1935. Diatomeen und Schlick im Jade-Gebiet. *Abhandlungen hrsg. von der Senckenbergischen naturforschenden Gesellschaft*, **430**, 1–64.

Bruhns, G., 1916. Zur Sauerstoff-Bestimmung nach L.W. Winkler. III. *Chemiker-zeitung*, **40**, 985–987, 1011–1013.

Bunt, J.S., 1963. Diatoms of Antarctic sea-ice as agents of primary production. *Nature, London*, **199**, 1255–1257.

Bunt, J.S., Owens, O. van H. & Hoch, G., 1966. Exploratory studies on the physiology and ecology of a psychrophilic marine diatom. *Journal of Phycology*, **2**, 96–100.

Burkholder, P.R. & Mandelli, E.F., 1965. Productivity of micro-algae in Antarctic sea ice. *Science, New York*, **149**, 872–874.

Carpenter, E.J., Van Raalte, E.D. & Valiela, I., 1978. Nitrogen-fixation by algae in a Massachusetts salt marsh. *Limnology and Oceanography*, **23**, 318–327.

Carritt, D.E. & Kanwisher, J.W., 1959. An electrode system for measuring dissolved oxygen. *Analytical Chemistry*, **31**, 5–9.

Castenholz, R.W., 1963. An experimental study of the vertical distribution of littoral marine diatoms. *Limnology and Oceanography*, **8**, 450–462.

Creitz, G.I. & Richards, F.A., 1955. The estimation and characterization of plankton populations by pigment analysis. III. A note on the use of 'Millipore' membrane filters in the estimation of plankton pigments. *Journal of Marine Research*, **14**, 211–216.

Doty, M.S. & Oguri, M., 1957. Evidence for a photosynthetic daily periodicity. *Limnology and Oceanography*, **2**, 37–40.

Doty, M.S. & Oguri, M., 1958. Selected features of the isotopic carbon primary productivity technique. *Rapports et Procès-verbaux des Réunions, Conseil International pour l'Exploration de la Mer*, **144**, 47–55.

Doty, M.S. & Oguri, M., 1959. The carbon-fourteen technique for determining primary plankton productivity. *Pubblicazioni della Stazione Zoologica di Napoli*, **31**(suppl.), 70–94.

Drew, E.A., 1973. Primary production of large marine algae measured *in situ* using uptake of ^{14}C. *Monographs on Oceanographic Methodology, UNESCO, Paris*, **3**, 22–33.

Eaton, J.W. & Moss, B., 1966. The estimation of numbers and pigment content in epipelic algal populations. *Limnology and Oceanography*, **11**, 584–595.

Eaton, J.W. & Simpson, P., 1979. Vertical migrations of the intertidal dinoflagellate *Amphidinium herdmaniae* Kofoid & Swezy. pp. 339–345 in E. Naylor & R.G. Hartnoll (eds) *Cyclic Phenomena in Marine Plants and Animals*, Pergamon Press, Oxford, 477 pp.

Gaarder, T. & Gran, H.H., 1927. Investigations of the production of plankton in the Oslo Fjord. *Rapports et Procès-verbaux des Réunions, Conseil International pour l'Exploration de la Mer*, **42**, 1–48.

Gallagher, J.L. & Daiber, F.C., 1973. Diel rhythms in edaphic community metabolism in a Delaware salt marsh. *Ecology*, **54**, 1160–1163.

Goldman, C.R., 1963. The measurement of primary productivity and limiting factors in freshwater with carbon-14. pp. 103–115 in M.S. Doty (ed.) *Proceedings of the Conference on Primary Productivity Measurement, Marine and Freshwater, University of Hawaii, 1961.* U.S.A.E.C. TID-7633.

Grøntved, J., 1960. On the productivity of microbenthos and phytoplankton in some Danish fjords. *Meddelelser fra Danmarks, Fiskeri-og Havundersøgelser*, **N.S. 3**, 55–92.

Grøntved, J., 1962. Preliminary report on productivity of microbenthos and phytoplankton in the Danish Wadden Sea. *Meddelelser fra Danmarks, Fiskeri-og Havundersøgelser*, **N.S. 3**, 347–378.

Hällfors, G., Kangas, P. & Lappalainen, A., 1975. Littoral benthos of the Northern Baltic Sea. III. Macrobenthos of the hydrolittoral belt of filamentous algae on rocky shores in Tvärminne. *International Revue der Gesamten Hydrobiologie*, **60**, 313–330.

Harlin, M.M. & Lindbergh, J.M., 1977. Selection of substrata by seaweeds: optimal surface relief. *Marine Biology*, **40**, 33–40.

Hickman, M., 1969. Methods for determining the primary productivity of epipelic and epipsammic algal associations. *Limnology and Oceanography*, **14**, 936–941.

Hickman, M. & Klarer, D.M., 1973. Methods for measuring the primary productivity and standing crops of an epiphytic algal community to *Scirpus validus* Vahl. *International Revue der Gesamten Hydrobiologie*, **58**, 893–901.

Hickman, M. & Round, F.E., 1970. Primary production and standing crops of epipsammic and epipelic algae. *British Phycological Journal*, **5**, 247–255.

Hopkins, J.T., 1963. A study of the diatoms of the Ouse Estuary, Sussex. I. The movement of the mud-flat diatoms in response to some chemical and physical changes. *Journal of the Marine Biological Association of the United Kingdom*, **43**, 653–663.

Horner, R. & Alexander, V., 1972. Algal populations in arctic sea ice. An investigation of heterotrophy. *Limnology and Oceanography*, **17**, 454–458.

Hutchinson, G.E., 1967. *A Treatise on Limnology. Volume II. Introduction to Lake Biology and Limnoplankton.* John Wiley & Sons, New York, 1115 pp.

Ivlev, V.S., 1945. The biological productivity of waters. *Uspekhi sovremennoi biologii*, **19**, 98–120.

Jeffrey, S., 1981. An improved thin-layer chromatographic technique for marine phytoplankton pigments. *Limnology and Oceanography*, **26**, 191–197.

Jenderson, C.G. & Hickman, M., 1983. The spatial and temporal distribution of epipelic algae in a shallow eutrophic prairie-parkland lake, Alberta, Canada. *International Revue der Gesamten Hydrobiologie*, **68**, 453–471.

Jitts, H.R., 1963. The simulated *in situ* measurement of oceanic primary production. *Australian Journal of Marine and Freshwater Research*, **14**, 139–147.

Johnston, C.S., 1969. The ecological distribution and primary production of macrophytic marine algae in the Eastern Canaries. *International Revue der Gesamten Hydrobiologie*, **54**, 473–490.

Johnston, C.S. & Cook, J.P., 1968. A preliminary assessment of the techniques for measuring primary production in macrophytic marine algae. *Experientia*, **24**, 1176–1177.

Johnston, C.S., Morrison, I.A. & Maclachlan, K., 1969. A photographic method for recording the underwater distribution of marine benthic organisms. *Journal of Ecology*, **57**, 453–459.

Kain, J.M., 1963. Aspects of the biology of *Laminaria hyperborea*. II. Age, weight and length. *Journal of the Marine Biological Association of the United Kingdom*, **43**, 129–151.

Kangas, P., 1972. Quantitative sampling equipment for the littoral benthos. II. *I.B.P. in Norden*, **10**, 9–16.

Kanwisher, J., 1959. Polarographic oxygen electrode. *Limnology and Oceanography*, **4**, 210–217.

Kliffmüller, R., 1959. Zur Bestimmung des gelösten Sauerstoffs in Vorfluter und Abwasser. *Archiv für Hydrobiologie*, **56**, 113–127.

Knight, M. & Parke, M., 1950. A biological study of *Fucus vesiculosus* L. and *F. serratus* L. *Journal of the Marine Biological Association of the United Kingdom*. **29**, 439–514.

Kuznetsov, S.I., 1956. Application of radioactive isotopes to the study of processes of photosynthesis and chemosynthesis in lakes. **12**. Radioactive Isotopes and Ionizing Radiations in Agriculture Physiology and Biochemistry, *Proceedings of the International Conference on the Peaceful Uses of Atomic Energy*, pp. 363–376.

Lappalainen, A., Hällfors, G. & Kangas, P., 1977. Littoral benthos of the Northern Baltic Sea. IV. Pattern and dynamics of macrobenthos in a sandy-bottom *Zostera marina* community in Tvarminne. *International Revue der Gesamten Hydrobiologie*, **62**, 465–503.

Lindeman, R.L., 1942. The trophic–dynamic aspect of ecology.*Ecology*, **23**, 399–418.

Littler, M.M., 1973. The productivity of Hawaiian fringing-reef crustose Corallinaceae and an experimental evaluation of production methodology. *Limnology and Oceanography*, **18**, 946–952.

Littler, M.M. & Murray, S.N., 1974. The primary productivity of marine macrophytes from a rocky intertidal community. *Marine Biology*, **27**, 131–135.

Lorenzen, C.J., 1967. Determination of chlorophyll and pheo-pigments: spectrophotometric equations. *Limnology and Oceanography*, **12**, 343–346.

Luther, G., 1976. Bewuchsuntersuchungen auf Natursteinsubstraten im Gezeitenbereich des Nordsylter Wattenmeeres: Algen. *Helgoländer wissenschaftliche Meeresuntersuchungen*, **28**, 318–351.

Lund, J.W.G., Kipling, C. & Le Cren, E.D., 1958. The inverted microscope method of estimating algal numbers and the statistical basis of estimations by counting. *Hydrobiologia*, **11**, 143–170.

McAllister, C.D., Shah, N. & Strickland, J.D.H., 1964. Marine phytoplankton photosynthesis as a function of light intensity: a comparison of methods. *Journal of the Fisheries Research Board of Canada*, **21**, 159–181.

MacFadyen, A., 1948. The meaning of productivity in biological systems. *Journal of Animal Ecology*, **17**, 75–80.

Mancy, K.H. & Jaffe, T., 1966. Analysis of dissolved oxygen in natural and waste waters. *United States Public Health Service Publication*. **999–WP. 37.**

Mann, K.H., 1973. Methods for determining growth production of *Laminaria* and *Agarum*. *Monographs on Oceanographic Methodology. UNESCO, Paris*, **3**, 20–21.

Margalef, R., 1949. A new limnological method for the investigation of thin-layered epilithic communities. *Hydrobiologia*, **1**, 215–216.

Marshall, N., Skauen, D.M., Lampe, H.C. & Oviatt, C.A., 1973. Primary production of benthic microflora. *Monographs on Oceanographic Methodology, UNESCO, Paris*, **3**, 37–44.

Moss, B., 1967a. A note on the estimation of chlorophyll *a* in freshwater algal communities. *Limnology and Oceanography*, **12**, 340–342.

Moss, B., 1967b. A spectrophotometric method for the estimation of percentage degradation of chlorophylls to pheo-pigments in extracts of algae. *Limnology and Oceanography*, **12**, 335–340.

Moss, B. & Round, F.E., 1967. Observations on standing crops of epipelic and epipsammic algal communities in Shear Water, Wilts. *British Phycological Bulletin*, **3**, 241–248.

Neumann, A.C., Gebelein, C.D. & Scoffin, T.P., 1970. The composition, structure and erodability of subtidal mats, Abaco, Bahamas. *Journal of Sedimentary Petrology*, **40**, 274–297.

Neumann, A.C. & Land, L.S., 1969. Algal production of lime mud deposition in the Bight of Abaco: a budget. *Special Papers, Geological Society of America*, **121**, 219 pp.

Norton, T.A. & Burrows, E.M., 1969. Studies on marine algae of the British Isles. 7. *Saccorhiza polyschides* (Lightf.). *British Phycological Bulletin*, **4**, 19–53.

Odum, E.P., 1959. *Fundamentals of Ecology*, 2nd ed. W.B. Saunders, Philadelphia, 546 pp.

Odum, H.T., McConnell, W. & Abbott, W., 1958. The chlorophyll '*a*' of communities. *Publications of the Institute of Marine Science University of Texas*, **5**, 65–96.

Odum, H.T. & Odum, E.P., 1955. Trophic structure and productivity of a windward coral reef community on Eniwetok Atoll. *Ecological Monographs*, **25**, 291–320.

Ohle, W., 1953. Die chemische und die elektrochemische Bestimmung des molekular gelösten Sauerstoffes der Binnengewässer. *Internationale Vereinigung für Theoretische und Angewandte Limnologie, Mitt.*, **3**, 44 pp.

Palmer, J.D. & Round, F.E., (1967). Persistent vertical-migration rhythms in benthic microflora. VI. The tidal and diurnal nature of the rhythm in the diatom. *Hantzschia virgata. Biological Bulletin, Marine Biological Laboratory, Woods Hole*, **132**, 44–55.

Parke, M., 1948. Studies on British Laminariaceae. I. Growth in *Laminaria saccharina* (L.) Lamour. *Journal of the Marine Biological Association of the United Kingdom*, **27**, 651–709.

Parsons, T.R., 1963. A new method for the microdetermination of chlorophyll *c* in sea water. *Journal of Marine Research*, **21**, 164–171.

Parsons, T.R. & Strickland, J.D.H., 1963. Discussion of spectrophotometric determination of marine-plant pigments, with revised equations for ascertaining chlorophylls and carotenoids. *Journal of Marine Research*, **21**, 155–163.

Penhale, P.A., 1977. Macrophyte–epiphyte biomass and productivity in an eelgrass (*Zostera marina* L.) community. *Journal of Experimental Marine Biology and Ecology*, **26**, 211–224.

Plante, M.R., 1966. Aperçu sur les peuplements de diatomées benthiques de quelques substrats meubles du Golfe de Marseille. *Recueil des travaux de la Station marine d'Endoume*, **56**, 83–101.

Pomeroy, L.R., 1959. Algal productivity in salt marshes of Georgia. *Limnology and Oceanography*, **4**, 386–397.

Pomeroy, R., 1960. Primary productivity of Boca Ciega Bay, Florida. *Bulletin of Marine Science of the Gulf and Caribbean*, **10**, 1–10.

Pomeroy, R. & Kirschman, H.D., 1945. Determination of dissolved oxygen. Proposed modification of the Winkler method. *Industrial and Engineering Chemistry. Analytical Edition*, **17**, 715–716.

Richards, F.A. with Thompson, T.G., 1952. The estimation and characterization of plankton populations by pigment analyses. II. A spectrophotometric method for the estimation of plankton pigments. *Journal of Marine Research*, **11**, 156–172.

Rideal, S. & Stewart, C.G., 1901. Determination of dissolved oxygen in presence of nitrites and of organic matter. *Analyst*, **26**, 141–148.

Round, F.E., 1953. An investigation of two benthic algal communities in Malham Tarn, Yorkshire. *Journal of Ecology*, **41**, 174–197.

Round, F.E., 1960. The diatom flora of a salt marsh on the River Dee. *New Phytologist*, **59**, 332–348.

Round, F.E., 1965. The epipsammon; a relatively unknown freshwater algal association. *British Phycological Bulletin*, **2**, 456–462.

Round, F.E., 1966. Persistent, vertical-migration rhythms in benthic microflora. V. The effect of artificially imposed light and dark cycles. *Proceedings of the Fifth International Seaweed Symposium, Halifax, August 25–28, 1965*. Pergamon Press, Oxford, pp. 197–203.

Round, F.E., 1978. On rhythmic movement of the diatom *Amphora ovalis*. *British Phycological Journal*, **13**, 311–317.

Round, F.E., 1979a. Occurrence and rhythmic behaviour of *Tropidoneis lepidoptera* in the epipelon of Barnstable Harbor, Massachusetts, U.S.A. *Marine Biology*, **54**, 215–217.

Round, F.E., 1979b. A diatom-assemblage living below the surface of intertidal sand flats. *Marine Biology*, **54**, 219–223.

Round, F.E., 1981. *The Ecology of Algae*. Cambridge University Press, 653 pp.

Round, F.E. & Eaton, J.W., 1966. Persistent vertical-migration rhythms in benthic microflora. III. The rhythm of the epipelic algae in a freshwater pond. *Journal of Ecology*, **54**, 609–615.

Round, F.E. & Palmer, J.D., 1966. Persistent, vertical-migration rhythms in benthic microflora. II. Field and laboratory studies on diatoms from the banks of the River Avon. *Journal of the Marine Biological Association of the United Kingdom*, **46**, 191–214.

Sanders, H.L., Goudsmit, E.M., Mills, E.L. & Hampson, G.E., 1962. A study of the intertidal fauna of Barnstable Harbor, Massachusetts. *Limnology and Oceanography*, **7**, 63–79.

Schindler, D.W., Schmidt, R.V. & Reid, R.A., 1972. Acidification and bubbling as an alternative to filtration in determining phytoplankton production by the ^{14}C method. *Journal of the Fisheries Research Board of Canada*, **29**, 1627–1631.

Scoffin, T.P., 1970. The trapping and binding of subtidal carbonate sediments by marine vegetation in Bimini Lagoon, Bahama. *Journal of Sedimentary Petrology*, **40**, 249–273.

South, G.R. & Burrows, E.M., 1967. Studies on marine algae of the British Isles 5. *Chorda filum* (L.) Stackh. *British Phycological Bulletin*, **3**, 379–402.

Steele, J.H. & Baird, I.E., 1968. Production ecology of a sandy beach. *Limnology and Oceanography*, **13**, 14–25.

Steeman Nielsen, E., 1952. The use of radio-active carbon (C^{14}) for measuring organic production in the sea. *Journal du Conseil Permanent international pour l'Exploration de la Mer*, **18**, 117–140.

Steeman Nielsen, E., 1953. On organic production in the oceans. *Journal du Conseil Permanent international pour l'Exploration de la Mer*, **19**, 309–328.

Steeman Nielsen, E., 1960. Dark fixation of CO_2 and measurements of organic productivity. With remarks on chemo-synthesis. *Physiologia Plantarum*, **13**, 348–357.

Steeman Nielsen, E. & Jensen, E.A., 1957. Primary oceanic production, the autotrophic production of organic matter in the oceans. *Galathea Reports*, **1**, 49–136.

Strickland, J.D.H. & Parsons, T.R., 1968. A practical handbook of seawater analysis. *Bulletin of the Fisheries Research Board of Canada*, **167**, 311 pp.

Talling, J.F., 1960. Comparative laboratory and field studies of photosynthesis by a marine planktonic diatom. *Limnology and Oceanography*, **5**, 62–77.

Talling, J.F. & Driver, D., 1963. Some problems in the estimation of chlorophyll-A in phytoplankton. pp. 142–146 in M.S. Doty (ed.) *Proceedings of the Conference on Primary Productivity Measurement, Marine and Freshwater, University of Hawaii, 1961*. U.S. Atomic Energy Commission TID-7633, pp. 142–6.

Thayer, G.W. & Adams, M., 1975. Structural and functional aspects of a recently established *Zostera marina* community. pp. 518–540 in L.E. Cronin (ed.) *Estuarine Research*, vol. 1. Academic Press, London, 738 pp.

Theriault, E.J. & MacNamee, P.D., 1932. Dissolved oxygen in organic matter, hypochlorites and sulfite wastes. *Industrial Engineering Chemistry* (*Anal.*), **4**, 59–64.

Thienemann, A., 1931. Der Produktionsbegriff in der Biologie. *Archiv für Hydrobiologie*, **22**, 616–622.

Valiela, I. & Teal, J.M., 1974. Nutrient limitation in salt marsh vegetation. pp. 547–563 in R.J. Reimold & W.H. Queen (eds) *Ecology of Halophytes*. Academic Press, London, 605 pp.

Vollenweider, R.A. (ed.), 1969. *A Manual on Methods for Measuring Primary Production in Aquatic Environments*. International Biological Programme Handbook No. 12. Blackwell Scientific Publications, Oxford, 224 pp.

Von Stosch, H-A., 1956. Die zentrischen Grunddiatomeen. Beiträge zur Floristik und

Ökologie einer Pflanzengesellschaft der Nordsee. *Helgoländer wissenschaftliche Meeresuntersuchungen*, **5**, 273–291.

Wallen, D.G. & Geen, G.H., 1971. Light quality in relation to growth, photosynthetic rates and carbon metabolism in two species of marine plankton algae. *Marine Biology*, **10**, 34–43.

Wanders, J.B.W., 1976. The role of benthic algae in the shallow reef of Curacao (Netherlands Antilles). I. Primary productivity on the coral reef. *Aquatic Botany*, **2**, 235–270.

Westlake, D.F., 1963. Comparisons of plant productivity. *Biological Reviews of the Cambridge Philosophical Society*, **38**, 385–425.

Westlake, D.F., 1965. Theoretical aspects of the comparability of productivity data. *Memorie dell'Istituto Italiano di Idrobiologia Dott. Marco de Marchi*. **18**(suppl.), 313–322.

Wetzel, R.G., 1963. Primary productivity of periphyton. *Nature, London*, **197**, 1026–1027.

Wetzel, R.G., 1964. A comparative study of the primary productivity of higher aquatic plants, periphyton and phytoplankton in a large, shallow lake. *Internationale Revue der Gesamten Hydrobiologie*, **49**, 1–61.

Williams, R.B., 1963. Use of netting to collect mobile benthic algae. *Limnology and Oceanography*, **8**, 360–61.

Winkler, L.W., 1888. Die Bestimmung des im Wasser gelösten Sauerstoffes. *Zeitschrift für analytische Chemie*, **53**, 665–672.

Yapp, W.B., 1958. Introduction, pp. ix–xii in W.B. Yapp & D.J. Watson (eds) *The Biological Productivity of Britain*. Symposia of the Institute of Biology No. 7, London, xii + 128 pp.

Chapter 9: Energy Flow Measurements

D. J. CRISP

Introduction

Investigations of the benthos must initially be concerned with the species and communities present in various types of deposits, their feeding mechanisms, reproductive habits and rates of growth. Such an approach, though qualitative, is an essential preliminary to the investigation of a new area and will lead to the application of more quantitative methods by which estimates of the abundance of commoner organisms and ultimately of the total amount of living matter or biomass can be made. Maps have been prepared showing the variation in the total quantity of living organisms (biomass) over large areas of sea bed in the form of contours of equal biomass per unit area of bottom (isobenths). A number of examples of such surveys are given in Zenkevich (1963).

Measurement of biomass alone, important though it may be in comparing the immediately available standing crop from place to place, is quite inadequate for the purpose of predicting the predation rate or the fishery yield that a benthic community does, or could, sustain over a long period. For these purposes the rates of production of organic matter by the various members of the benthic community and their trophic dependence on one another or on sources outside the benthos must be known. The distinction between rate of production (or productivity) of organic matter by an organism or a community and the standing crop, consisting of the organism itself or the community, is fundamental. The reader is warned against some confusing uses of the term 'productivity', especially in the early literature. Zenkevich (1930) for instance defines 'isobenths' as 'contours of equal bottom productivity', but means in fact contours of equal standing crop.

The most comprehensive form that investigations of production can take is the measurement of the flow of organic matter and energy through each component of the ecosystem. The interdependence of the various parts of the marine ecosystem is made clear by dividing the flow sheet between trophic levels starting with the primary producers which utilize solar energy and fix inorganic nutrients (first trophic level), passing through the herbivores (second trophic level), the carnivores (third trophic level) and leading ultimately to the decomposers. The principles of energy flow are well described in Odum (1959) and Phillipson (1966).

The relationship between standing crop and yield, whether to the next trophic level or to man, can be considered at each level for the complete ecosystem (e.g. Teal, 1962) or more conveniently, the problem can be broken down into individual projects in which each species playing a significant part in the energy exchange at a given trophic level is studied separately (e.g. Kuenzler, 1961). The relationship between standing crop and yield for species of economic value has become a most fruitful area of study wherever the rates of predation or harvesting are important, and has led to notable advances in the theory of optimal fishing rates (Beverton & Holt, 1957; Bagenal, 1978; Pitcher & Hart, 1982).

Less effort has been put into studies of benthic production, but the same basic concepts apply and the same advantages can be reaped. For the organism or the ecosystem, food is measurable in terms of energy; hence computations of energy flow can be employed to assess actual and potential yield of exploitable benthic species, to predict means of increasing productivity, and to estimate the contributions made by benthic communities to predatory species occupying other habitats.

The zoobenthos functions mainly at the second and third trophic levels, which are among the more easily measurable components of an ecosystem. Measurement of primary production of phytobenthos is dealt with separately in Chapters 5 and 8.

Terminology

It is often difficult to compare the results of energy flow studies by different authors because of the differences in the use of terms defining the various purposes for which energy is used or conserved by the organism and because of differences in the units by which the energy or biomass is measured. Although it would be inappropriate to recommend a rigid terminology it is essential that the meanings of terms used in this handbook should be clear. Since the concepts of quantitative ecology are universal, the terms and symbols recommended generally by IBP can be used for benthos studies.

It is useful to reserve the term 'productivity' for potential rate under ideal or stated conditions, and to use the term 'production' for the actual rate of incorporation of organic matter or energy (Davis, 1963). In reporting rates of production the trophic level and whether production refers to the whole ecosystem, to a group of species, or to a single species should be stated.

Biomass (B)

Biomass is defined as the amount of living substance constituting the organisms which are being studied. Alternative terms found in the literature are 'standing crop' or 'standing stock'.

Biomass can be expressed in units of volume, mass or energy and may refer to the whole or to part of the organism. Biomass is normally determined in relation to a particular unit used to measure the environment—usually its area, but sometimes its volume. If, for example, the biomass is being related to the depth below the surface of a sediment, the amount of living matter must be expressed per unit volume of sediment at the stated depth. Usually, the whole biomass is summed throughout the depth of the deposit and is then expressed as the biomass per unit area. This approach is particularly applicable where most, or all, of the living matter lies near the surface of the sediment.

The definition of biomass in terms of living substance raises serious difficulties because of the many ways in which the living tissue can be measured. The units in which biomass can be expressed may be divided into:

1. Crude units of biomass.
2. Units measuring the mass of living tissue only.
3. Units measuring the energy content of the living tissue.

Crude units

Biomass of aquatic organisms is most simply measured by weighing the organisms whole after mopping off the external water and emptying the water from cavities external to the animal (e.g. mantle cavity of molluscs). On board ship this may be the only available method short of preservation, which has its disadvantages (see below, p. 320). Wet weight can always be converted at a later stage into a more fundamental or refined measure of living matter by conducting a suitable calibration measurement in the laboratory (p. 320). Biomass may also be measured as the live volume of the animal, measured directly by displacement, and can similarly be converted into more refined units. If any part of the animal is excluded from crude measurement of biomass (e.g. the shells of hermit crabs and molluscs, tubes of polychaetes, etc.) this should be made clear.

Units of mass of living tissue only

When expressing biomass in a more refined manner, it is usual to exclude those parts of the animal that are non-living. Such components as water and salt, the calcareous matter of the integument or skeleton, the whole of the protective houses not connected with the body (e.g. shells of hermit crabs, tubes of polychaetes and anemones) and heavy inorganic sediments present in the guts of such animals as the echinoids and polychaetes are clearly not part of the living organism. Some biologists (e.g. Thorson, 1957) also recommend excluding the organic epidermal structures such as the heavy shells of crabs and lobsters. It is clear that there is no sharp demarcation between living and non-living matter in the constitution of the animal nor is a simple technique for removing the non-living part always available. Therefore, every worker must draw the line as seems appropriate for the particular animals being studied and for the purpose of the problem in hand. The one imperative is that those parts of the animal that are excluded from its biomass should be clearly stated so that valid comparisons can be made between various investigations.

When non-living parts have been removed, the water evaporated off by drying and the ash weight of inorganic matter measured and allowed for, the remaining mass represents the dry ash-free organic matter, which is regarded by many workers as a good measure of the living substance. It can be expressed in units of biomass (dry ash-free weight) in kilograms, grams or milligrammes as appropriate.

An alternative to expressing biomass in terms of ash-free dry weight is to make the assumption that the nitrogen content is proportional to the quantity of living matter and to analyse the tissues chemically for nitrogen. Since the results of a nitrogen analysis vary slightly with the method used, the method should be stated and preferably included in the unit, e.g. biomass (mg Kjeldahl nitrogen). (See p. 354.)

Units of energy of living tissue

Measurement of biomass in terms of the energy released from the tissues when they are fully oxidized to water and carbon dioxide overcomes the philosophical problem of defining what are the 'living' and 'non-living' parts of the animal. In general, such non-living components as water and mineral matter do not yield any energy on combustion. Furthermore, energy units have precisely the same meaning for all organisms. Since energy is a conservative principle, the energy changes between parts of the ecosystem are exactly accountable by the first law of thermodynamics. A convenient and widely used unit of biomass in terms of energy is the kilogramme calorie (kcal) and the unit of biomass per unit area is conveniently expressed as kcal m^{-2}. The SNU (Standard Nutritional Unit) employed by Stamp (1958) and others is equal to 10^6 kcal and may be more suitably used for large areas. (It is unfortunate that the internationally accepted system of units (SI) recommends that the calorie be replaced by the joule. Almost all original thermochemical data and early productivity results were quoted in calories, but younger workers are employing joules. Since 1 g calorie at 15 °C ≡ 4·185 joules the transformation is not arithmetically simple. Although perhaps beneficial on theoretical grounds, the recommendation has proved counter-productive so far as measurements of biological energies are concerned, resulting in two sets of units in the literature, which make comparison of results by different workers difficult unless both units are quoted.

When biomass is to be measured in energy units it is usually necessary to measure the caloric content, or heat of combustion per g dry weight, of the biological material either directly, by burning it in a bomb calorimeter, or indirectly, by determining its approximate chemical composition as protein, carbohydrate, and lipid and applying known average values of caloric content for each component. Details of methods are given on pp. 354–63. When the average caloric content of the organic matter is known, the biomass expressed as dry ash-free weight can be converted quite simply into biomass expressed in energy units.

Energy flow (dB/dt)

Energy flow refers to rates of change of biomass. It is divisible into a number of separate processes which together constitute the whole passage of matter or energy into and out of the organism, population, or ecosystem under investigation. The appropriate IBP terms and symbols for the components of energy flow for heterotrophic organisms are as follows:

Consumption (*C*) Total intake of food or energy.

Egesta (*F*) That part of the consumption that is not absorbed but is voided as faeces.

Absorption (*Ab*) That part of the consumed energy that is not rejected as faeces.

Excreta (*U*) That part of the consumption that is absorbed and later passed out of the body as secreted material, usually in an unwanted form as, for example, in the urine. Many organisms produce a number of other exudates such as milk, mucus, shed cuticle, nematocysts etc. which are released from the organism under various circumstances. All such exudates, with the exception of gonoproducts, will be regarded as parts of the 'excreta' in this work. The combined energies $F + U$ are sometimes referred to as *rejecta*.

Assimilation (A) That part of the consumption that is retained for physiological purposes, namely, for production (including gonoproducts) and respiration, but excluding *rejecta*.

Production (P) That part of the assimilated food or energy that is retained and incorporated in the biomass of the organism, but excluding the reproductive bodies released from the organism. This may also be regarded simply as 'growth'.

Respiration (R) That part of the assimilated energy that is converted into heat, either directly or through mechanical work performed by the organism.

Gonad output (G) That part of the absorbed energy that is released as reproductive bodies. Because of its great importance in survival and recruitment this part of the energy flow is separated from excreta (U) and production (P), although some authors might regard it as being a contribution towards either of these elements of energy flow (cf. Bagenal, 1978, Chapter 10).

Yield (Y) This term is used in a narrow and varied sense as that part of the production or excreta utilized by man or by other predators. It may refer to only part of the organism (e.g. crop yields, milk yield) or it may refer to a fraction of the individuals in a population which are consumed by a predator or harvested by man (e.g. yield of fish of a particular species).

Hence, assuming conservation of energy we may write the following equations:

$$\text{Consumption, } C = P + R + G + U + F \tag{9.1}$$

$$\text{Absorption, } Ab = C - F = P + R + G + U \tag{9.2}$$

$$\text{Assimilation, } A = P + R + G \tag{9.3}$$

It should be noted that if biomass is expressed in energy units it must be conserved and the above relations hold true. On the other hand, if biomass is expressed in units of mass, the loss of weight during respiration and excretion (e.g. as water, carbon dioxide and ammonia) is usually not taken into account and the above equations, since they involve R and U, are not applicable.

All of the above components are measured in units of energy flow, dB/dt, where B is the biomass. But since changes in biomass, like biomass itself, must be referred to a defined part of the ecosystem, the energy flow will usually need to be defined either (1) as the change in biomass of a given area of the system, as in the case of a community living in a defined habitat or a population occupying a stated area, or (2) as the change in biomass as a fraction of the existing amount of standing crop present.

In definitions of the type (1) the energy flow will be expressed as energy per unit of time, per unit of area. For example, a population of *Nephthys incisa* occupying 1 m^2 of bottom was found to produce 9·34 g dry weight of biomass per year (Sanders, 1956). The units here are g m^{-2} $year^{-1}$.

In definitions of type (2) the production would have to be measured per unit biomass standing crop. Taking the same example, Sanders found that the standing crop amounted on average to 4·32 g m^{-2}. The result would then be expressed as a rate of change of biomass of 9·34 g m^{-2} $year^{-1}$ in respect of a biomass of 4·32 g m^{-2}. The units of biomass can then be cancelled and the result expressed more simply as $9{\cdot}34\,\mathrm{g\ m^{-2}\ year^{-1}} \div 4{\cdot}32\,\mathrm{g\ m^{-2}} = 2{\cdot}16\ \mathrm{year^{-1}}$. It will be seen that since the rate of increase in biomass is measured in terms of biomass, the dimensions of energy flow are simply reciprocal time. It is, in fact, desirable to express the rate of change of biomass and the amount of standing crop biomass in the same units wherever possible, so that energy flow can be expressed in a form which does not involve the units in which biomass itself is measured.

The most widely used unit of time for energy flow studies is the year. It is convenient because (a) the organisms usually return to the same nutritional and reproductive state after a year and (b) the population structure may sometimes return to a similar form and a similar level after a year has elapsed. These two factors tend to minimize changes in the biomass between its initial and final condition and thereby facilitate the calculations.

It is worth pointing out that all the components of energy flow as defined in equations 9.1, 9.2 and 9.3 must be regarded as thermodynamically equivalent forms of energy; all of them, including the term for energy spent through respiration and activity (R), refer strictly to the heat content of the system measured in relation to the fully oxidized form of the organic material (carbon dioxide, water) as the assumed zero energy state. There is no substance whatever in the claim that the division into components of production and respiration afford a basis for distinguishing between free energy and entropy changes in the living system as stated in Phillipson (1966). Slobodkin (1962) had correctly pointed out that the concept of entropy and free energy change is not yet applicable to energy flow studies.

To use heats of combustion from bomb calorimetry as a measure of the energy content of organic matter does, however, suffer from one minor source of ambiguity. Unlike the products of combustion of compounds of carbon and hydrogen, those of nitrogen, phosphorus and sulphur are imperfectly known and cannot therefore be exactly defined as they should be. Fortunately the error attributable to the formation of different oxidation products in the measurement of the caloric content of biological materials containing these minor constituents is likely to be much smaller than the other errors encountered in energy flow studies.

Coefficient of assimilation efficiency

The ratio of the food absorbed into the organism to the total amount of food ingested is defined as the 'efficiency of assimilation'. It is obvious that the IBP definition of 'assimilation' as 'physiologically useful energy' is not strictly consistent with the normal use of the term 'Assimilation efficiency'; the definitions on pp. 289–90 would logically lead one to replace the term 'Efficiency of Assimilation' by 'Efficiency of Absorption' which is what is normally implied:

$$\text{Coefficient of efficiency of Assimilation (Absorption)} = \frac{Ab}{C} = \frac{C - F}{C} \qquad (9.4)$$

It is obtained by measuring the amount of food ingested and faeces egested over an interval of time sufficiently long to be representative of the steady state. It can be defined as a ratio either of dry ash-free weights, or better, as a ratio of energies. These ratios will differ since the caloric content of food and faeces will not usually be the same. If λ_C is the caloric content and w_C the average weight of the food ingested in unit time and λ_F, w_F, corresponding values for the faeces:

$$\text{Assimilation efficiency (by wt.)} = (w_C - w_F)/w_C$$
$$\text{Assimilation efficiency (by energy)} = (\lambda_C w_C - \lambda_F w_F)/\lambda_C w_C$$

The basis of measurement of the efficiency of assimilation (absorption), whether by weight or by energy, must, therefore, be stated.

When assimilation efficiency by weight is measured it is essential that ash-free values should be used and therefore the ash content of both food and faeces must be measured. If ash is not taken into account, an ash intake at an average rate w_a will lower the apparent value of assimilation efficiency (by weight) from $(w_C - w_F)/w_C$ to $(w_C - w_F)/(w_C + w_a)$.

On the other hand, if assimilation efficiency is measured in energy units by direct weighing and calorimetry of food and faeces, it is not essential to take ash into account, provided that it is sufficiently small in amount and of such a constitution as to be calorimetrically negligible (see p. 291). The presence of ash can be neglected because the energy measured with the ash present (viz. $\lambda'_C w'_C$, $\lambda'_F w'_F$) can be shown to be the same as that measured in terms of ash-free energy ($\lambda_C w_C$, $\lambda_F w_F$) since:

$$\lambda'_C(w_C + w_a) = \lambda_C w_C \text{ (no energy from combustion of ash } w_a)$$
$$w_C + w_a = w'_C \text{ (increase in dry weight of food due to ash, } w_a)$$

Therefore, $\lambda'_C w'_C = \lambda_C w_C$; and similarly, $\lambda'_F w'_F = \lambda_F w_F$.

Coefficient of growth efficiency

Growth efficiency may be defined as the total energy of production of body tissue and gonads ΔB, $(=P+G)$, as a fraction of the energy of food ingested (C):

$$\text{Coefficient of growth efficiency} = \frac{\Delta B}{C} = \frac{P+G}{C} \tag{9.5}$$

It is also referred to as Ivlev's growth coefficient of the first order, K_1.

Growth efficiency ratios should be based on energy content rather than wet or dry weight because the storage of smaller quantities of high energy reserve, such as fat, is not necessarily a less efficient mechanism than the storage of larger quantities of low energy reserve, such as carbohydrate. If the caloric content of the tissue is not known and ΔB and C can be expressed only as wet or dry weight, the term 'conversion rate' is recommended.

The efficiency of growth is high in juveniles but usually declines rapidly with age (see Jørgensen, 1962; Carefoot, 1967b). Even in rapidly growing embryos utilizing yolk, the growth efficiency rarely exceeds 75 %. Values in excess of 70 % should, therefore, be viewed with some suspicion.

Growth efficiency varies also with the rate of food consumption because the fraction available for increase in biomass $(P+G)$ will decrease as the ration decreases, priority being given by the organism to activities connected with body maintenance and food searching (R). Thus, for example, Gerking (1955) found that the growth efficiency of the Bluegill, measured in terms of nitrogen content, increased with the ration up to a maximum of 33 %. Most of the earlier information dealing with the relationship of growth efficiency to food ration in aquatic animals refers to experiments with fish but the concepts apply equally to invertebrates, for example *Mytilus edulis* (Bayne, 1976, Fig. 5.15).

The rate of food consumption which allows maintenance activities only to be performed but no increase in biomass $(P+G=0)$ is termed the 'maintenance ration', C_m.

When the maintenance ration is known, the net or partial growth efficiency can be defined as the rate of increase in biomass for a given increase in ration above the maintenance level:

$$\text{Partial growth efficiency} = \frac{\Delta B}{C - C_m} = \frac{P+G}{C - C_m} \tag{9.6}$$

Coefficient of ecological efficiency

The term 'ecological efficiency' is employed as a measure of the efficiency of energy transfer from one trophic level to the next. The simplest definition is the fraction of the energy consumed at a given trophic level (n) that is exploited by a predator at the next trophic level ($n + 1$). Referring to the yield to predator as Y_n:

$$\text{Ecological efficiency} = \frac{Y_n}{C_n} = \frac{C_{n+1}}{C_n} \tag{9.7}$$

A different definition is given by Ivlev for the 'Dynamic ecotrophic coefficient', ε. The denominator is replaced by the production term for the first trophic level:

$$\varepsilon_{n+1} = \frac{Y_n}{P_n} = \frac{C_{n+1}}{P_n} \tag{9.8}$$

A fuller discussion of these and related coefficients is given in Bagenal (1978).

Energy budget

The equation of the total energy budget is:

$$C = P + R + G + U + F$$

and it applies to any system where all energy sources are included in the left-hand side and all energy sinks to the right-hand side. The energy budget can be applied to an individual animal and, with increasing difficulty and complexity, to a population of a single species operating at a given trophic level, or to all the organisms constituting a multi-habitat ecosystem.

Not infrequently a single species population partakes of energy exchanges at more than one trophic level—for example a filter feeding organism may consume primary producers (algae) or herbivores (microcrustacea). This complication is immaterial provided that the above equation is rigorously applied. First, the limits of the system which is being investigated must be clearly defined; for example, a single species population of stated biomass, or one occupying a stated area. Secondly, all the items of the energy budget must be entered into the equation once and once only. In particular the novice should be wary when measuring the production term P, to avoid the fallacy of counting increase in biomass twice over, first as growth of tissues and later as mortality or yield to predator. To avoid such errors the experimenter should always attempt to write an equation of the above form, to see that it includes all possible exchanges of energy in the system.

Since the energy budget must balance, should any one of the terms be particularly hard to measure, it may be omitted and found by difference. However, it is much better, whenever possible, to estimate all the terms and use the equation as a check on the accuracy of the flow sheet. Most of the energy budgets that have been prepared for large scale ecosystems, even when they could be checked in this way, have not balanced at all closely.

Nevertheless even the crude approximations hitherto obtained are preferable to no estimates at all.

Measurement of production

For some practical purposes the production term in the energy budget, P, is the only one that is essential. It is also frequently the most important quantity to be measured in fundamental studies of ecosystems and productivity.

There are two different approaches in the measurement of the total secondary production by a population of animals, which must be clearly distinguished.

The first method is to add all the growth increments of all the members of the population as they occur during the period under consideration, say one year ($t = 1$). If we write G_i as the mean instantaneous relative growth rate of an individual of weight (or biomass) w_i, then the total production is given by the equation:

$$P = \sum_{t=0}^{t=1} \sum_{0}^{N} G_i w_i \Delta t \tag{9.9}$$

where

$$G_i \Delta t = \frac{1}{w_i}\left(\frac{dw_i}{dt}\right)\Delta t$$

is the relative growth increment of individual i during the interval Δt, and w_i is the weight of this individual at this time. The sign $\sum_0^N$ implies the summation of $G_i w_i$ for all surviving individuals during the time interval Δt, which is assumed to be sufficiently short for G_i and w_i not to change appreciably. The sign $\sum_{t=0}^{t=1}$ implies the summation of the population production increments over the whole set of time intervals Δt from the start of the survey ($t = 0$) until one year later ($t = 1$).

The second approach is to ignore growth processes and to consider instead the fate of the biomass that has been produced during the period of the survey. The organic matter or energy that is produced will either remain as living matter at the end of the survey, will have died, or will have been eaten. If B_0 were the standing crop biomass at the start and B_1 the standing crop biomass after one year then clearly

$$P = (B_1 - B_0) + M = \Delta B + M$$

where M is the mortality due to all causes, including any yield Y.

If we write the instantaneous mortality rate from all causes for an individual i of weight w_i, as Z_i, then the expectation of death over the interval Δt will be $Z_i \Delta t$, and the expected loss of biomass through mortality $Z_i w_i \Delta t$.

The corresponding summation for all mortality losses during one year will then be given by an expression similar to that for production:

$$M = \sum_{t=0}^{t=1} \sum_{0}^{N} Z_i w_i \Delta t$$

$$\therefore P = \Delta B + \sum_{t=0}^{t=1} \sum_{0}^{N} Z_i w_i \Delta t \qquad (9.10)$$

In applying either of equations 9.9 or 9.10 it is necessary to have both growth and mortality data as functions of the weight of the individual. In equation 9.9, the growth rates must be known and the number of individuals surviving must be known at each time interval. In equation 9.10, the mortality rates must be known, as must the weight of each individual at each time interval. The information necessary is illustrated in Fig. 9.1, where time in years is plotted on the abscissa and the average weight of all remaining individuals N from a single recruitment, $\bar{w}$, $(=\sum_0^N w_i/N)$, is plotted on the ordinate.

It is assumed here that all individuals begin to grow at $t = 0$ and that their weight is then virtually zero. The curve can be taken to represent the mean growth rate of a single age class of small larvae of negligible weight recruited at time $t = 0$. The survivorship of recruits is also plotted, starting at N_0 and dropping steadily on account of mortality from all possible causes including

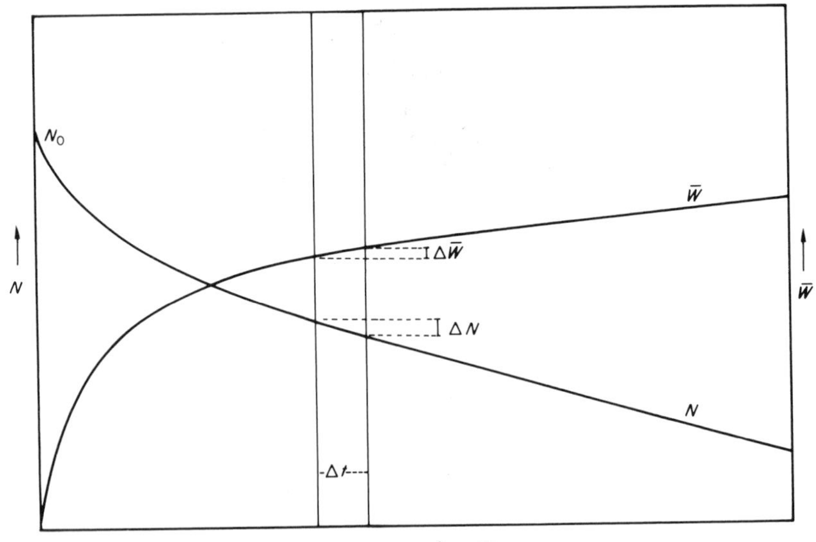

Fig. 9.1. Curves of survivorship, N, against time, t, and of mean weight, $\bar{w}$, against time for a single recruitment, N_0, of individuals of negligible weight at recruitment.

yield, predation and natural mortality. We may now apply equation 9.9 to obtain the growth increment over a small interval of time Δt during the first year of life of the recruitment. The summation of the weight increment for each individual i, $\sum_0^N G_i w_i$, has now been simplified by having values for the average weight, $\bar{w}$, since the mean growth increment for all individuals is the slope of the growth curve $(d\bar{w}/dt)$ multiplied by the time interval Δt. Hence the total increment for the population is $N(d\bar{w}/dt)\,\Delta t = N\,\Delta\bar{w}$, where N is the number of survivors at time t, and $\Delta\bar{w}$ is the growth increment as shown in Fig. 9.1. The total production by all individuals during their first year of life is therefore obtained by the summation for one year:

$$P_1 = \sum_{t=0}^{t=1} N\,\Delta\bar{w} \tag{9.11a}$$

This summation gives a valid measure of production over any period from t_1 to t_2, the more general form of equation 9.11a being:

$$P_{(t_1-t_2)} = \sum_{t=t_1}^{t=t_2} N\,\Delta\bar{w} \tag{9.11b}$$

If $t_1 = 0$ and $t_2 = \infty$ the production will be estimated for the whole life of all survivors of the initial recruitment:

$$P_\infty = \sum_{t=0}^{t=\infty} N\,\Delta\bar{w} \tag{9.11c}$$

Employing instead the second approach and applying equation 9.10 in place of 9.9, the instantaneous relative mortality rate Z_i is given by the expression

$$Z_i = \frac{1}{N}\frac{dN}{dt}$$

where $-dN/dt$ is the slope of the survivorship curve as shown in Fig. 9.1. Hence the number of individuals dying during the interval Δt will be the product of the instantaneous mortality rate and the time interval, summed over the whole population:

$$\sum_0^N \frac{1}{N}\left[\frac{dN}{dt}\right]\Delta t = \Delta N$$

and the loss of biomass during this time will therefore be $\bar{w}\,\Delta N$.

The mortality loss of the recruitment during the first year of life, M_1, is found by summing over time intervals from 0 to 1 year:

$$M_1 = \sum_{t=0}^{t=1} \bar{w}\,\Delta N$$

The biomass increment of standing stock, ΔB, created by the end of the first year will be $N_1\bar{w}_1$, where N_1 is the number surviving with average weight $\bar{w}_1$, that of an average one-year-old individual. Equation 9.10 then gives:

$$P_1 = N_1\bar{w}_1 + \sum_{t=0}^{t=1} \bar{w}\,\Delta N \qquad (9.12)$$

The estimation of production by growth increment (equation 9.11) has been associated with Allen's graphical method (Allen, 1950) described in Bagenal, 1978, pp. 203–6). The estimation of production from mortality plus residual biomass (equation 9.12) was employed by Sanders (1956, p. 393). His computation of total production falls short of the true value because the contribution by individuals more than three years old was not included, the data not extending beyond year class 2.

The results of equations 9.11 and 9.12 indicate that only the relationship between N, the number of living survivors, and $\bar{w}$, the mean weight of survivors at corresponding times, need be known to compute production. Figure 9.2 shows this relationship. The summation of production by growth increments (equation 9.11)

$$\sum_{t=0}^{t=1} N\,\Delta\bar{w}$$

is shown hatched vertically in Fig. 9.2a while summation of mortality increments

$$\sum_{t=0}^{t=1} \bar{w}\,\Delta N$$

is shown hatched horizontally and the product N_1w_1 (see equation 9.12) is shaded black (Fig. 9.2b). The unshaded areas under the curves of Fig. 9.2(a) and (b) represent the production by individuals surviving for ≥ 1 yr. It is clear that both growth and mortality approaches for estimating production (equations 9.11 and 9.12 respectively) give the same answer. Furthermore, the ultimate production by the original recruitment when all have died ($t = \infty$, $N = 0$) is the same whether growth or mortality increments are measured. It follows from equation 9.12 that when $t = \infty$ and $N = 0$, $\Delta B = 0$ and therefore

$$P_\infty = \sum_{t=0}^{t=\infty} N\,\Delta\bar{w} = \sum_{t=0}^{t=\infty} w\,\Delta N \qquad (9.13)$$

which is the total area under the curves (a) and (b) of Fig. 9.2.

Note that the curve of survivorship against mean weight is quite different from a size-frequency histogram of the population, and should not be confused with a size frequency curve. Whereas the latter includes many age groups and relates frequency to size at a given moment of time, the mean weight-survivorship curve refers to a single recruitment and covers a period of

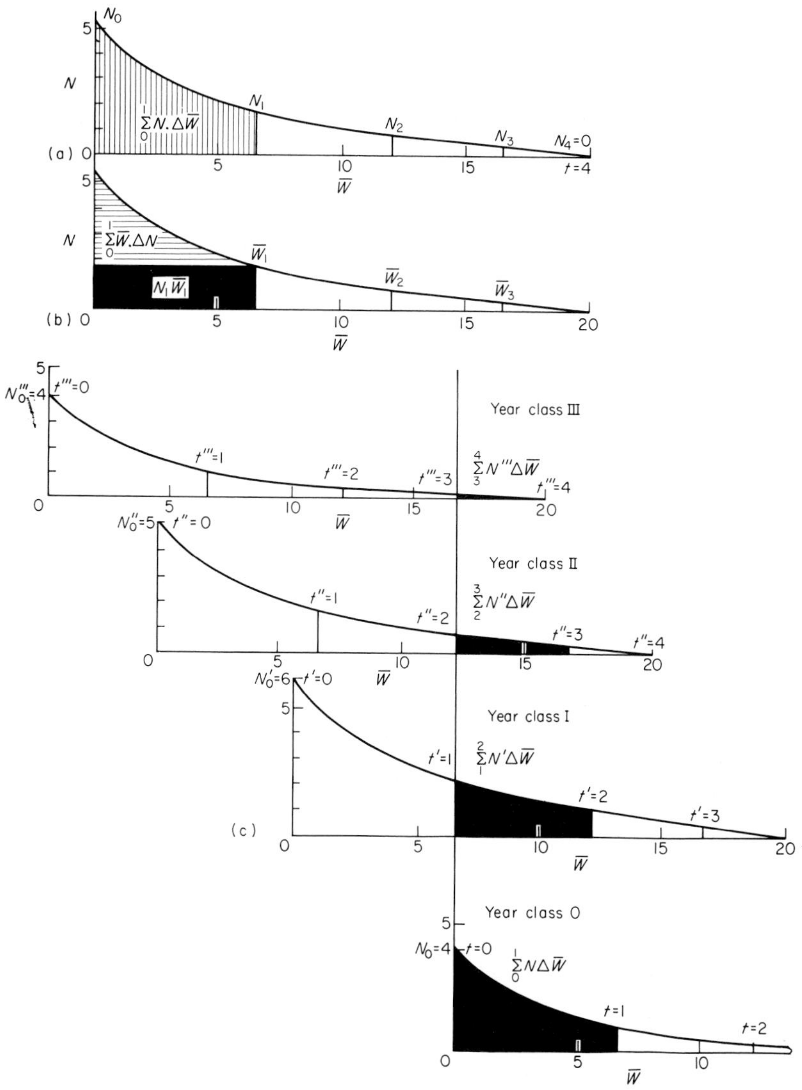

Fig. 9.2. Curves of survivorship, N, against mean weight, $\bar{w}$. $N_1 N_2 N_3 \ldots$ and $\bar{w}_1 \bar{w}_2 \bar{w}_3 \ldots$ are the values of survivorship and mean weight at the end of the first, second and third year of life respectively.

(a) Production during first year, summed as growth increments, $N \Delta \bar{w}$ (vertical shading).

(b) Production during the first year, summed as mortality losses, $\bar{w} \Delta N$ (horizontal shading), plus increase in biomass, $N_1 \bar{w}_1$ (black).

The unshaded areas represent production in subsequent years.

(c) Separate curves of survivorship—mean weight during year 0 of recruitments from year 0 and previous years.

The total production during year 0 from each consitituent recruitment is shown in black.

time equal to the whole or part of the life of the individuals. Furthermore, although size frequency histograms may have a number of peaks or modes, the mean weight-survivorship curve must show a continuous decline in numbers with increasing weight.

The precise meaning of the production term P_∞ in equations 9.11c and 9.13 is also important. It is the production by a population of N_0 recruits from the time of settlement up to the time that they have all died. The production therefore extends over a period equal to the maximum longevity of the species. An actual population will usually contain recruits settled at intervals over several years and the summation

$$P_\infty = \sum_{t=0}^{t=\infty} N \Delta \bar{w}$$

can be equated to the annual production of such a real population only under certain exceptional conditions. In a real population during any particular year, say year 0, there would be present a number of year classes contributing to production, whose original recruitment numbers 0, 1, 2, years previously might have been N_0, N_0', N_0'', the number of primes referring to the age class. Figure 9.2c shows by shading which parts of the integrated growth-survivorship curves of each of these age classes should be included to give the actual production in year 0. It will be seen from Fig. 9.2c, that if the annual recruitment is the same each year ($N_0 = N_0' = N_0''$) and the growth-mortality relationship continues along the same course in successive years, the production by all age groups in year 0 will be equal to the total production by the recruits of any one year throughout their life (P_∞). It is only when the population exists in such a steady state with constant annual recruitment, growth and mortality, that equation 9.13 can be applied to all the current age classes of a particular year to estimate production by integrating a single continuous growth frequency curve. Otherwise the annual production will vary from year to year and will be equal to

$$P_{\text{ann}} = \sum_{t=0}^{t=1} N_0 \, \mathrm{d}\bar{w} + \sum_{t=1}^{t=2} N_1' \, \mathrm{d}\bar{w} + \sum_{t=2}^{t=3} N_2'' \, \mathrm{d}\bar{w} \ldots \text{etc.} \qquad (9.14)$$

Application of production equation

Method 1. *Estimation of production by stock with no recruitment*

If there is no recruitment to the stock, the measurement of production can be carried out quite simply by counting the numbers of stock surviving (N) and measuring the mean weight ($\bar{w}$) of a sample at intervals of time close enough to allow a smooth curve of survivorship against mean weight to be drawn. The total number of stock present and the mean weight can be found by techniques described below (see p. 332). Method 1 is applicable to a non-breeding stock of organisms allowed to grow in a confined area, such as a group of fish fry in a pond (see Bagenal, 1978, pp. 203–6). It can also be applied to any identifiable recruitment of a species whose population density and mean weight can be measured continuously.

The following hypothetical example is given to illustrate the graphical method of computing production, biomass increase and mortality, based on equations 9.11, 9.12 and 9.13.

The data are given in Table 9.1. Columns I–IV represent the growth in size and decline in numbers of a fairly dense population of small organisms, such as molluscs or crustaceans, having a maximum net weight <1 g and an initial population density N_0 of 100 000 m^{-2}. The mean weight and population size are plotted against time in Fig. 9.3, which corresponds in form to the curves in Fig. 9.1. It can be seen that the numbers fall rapidly at first, then more slowly, until the end of the first year of life when the population is quickly eliminated. Growth is rapid at first; after 6 months there is a marked decrease in weight, which might be caused by reproduction; thereafter growth is slow.

Figure 9.4 shows the curve of N against $\bar{w}$, the area under which, if summed graphically, can be found to be 11·7 kg m^{-2}, which represents the total production by the 100 000 individuals initially recruited to the 1 m^2 quadrat. The computation can also be made arithmetically as demonstrated at the foot of Table 9.1. The production figures in Columns XII and XIV are based on the growth increment (equation 9.11a) and mortality (equation 9.12) approaches respectively, and agree as expected. The computation shows that, with this type of growth pattern, nearly all the production is completed before the end of the first year (see the shaded area of Fig. 9.4).

Stocks with recruitment

The factor that greatly complicates the assessment of production in a natural population is recruitment. In an isolated stock with no recruitment, survivorship, N, can be equated to the total population. This is clearly not so when recruitment takes place. In the extreme case where new individuals are recruited continuously at the same rate that older individuals die off, there would be no evident change in the population numbers, mean weight or standing stock biomass.

Method 2. For stocks with recruitment, age classes separable

The problem of production assessment in a population with steady or intermittent recruitment can be overcome if the age classes are separately recognizable. The stocks belonging to each year class can then be regarded as separate isolated populations, the survivorship and mean weight of which can be measured and integrated as in Method 1. The total production in any

Table 9.1. Growth and survivorship of a hypothetical population recruited April, 1969, to illustrate computation of production.

I Date	II Time from April 1 1969 t (years)	III Mean individual weight $\bar{w}$ (mg)	IV Population density N (thousands m^{-2})	V Standing crop $N\bar{w}$ (kg m^{-2})	VI Average value of N over period $\frac{1}{2}(N_t + N_{t+\Delta t})$	VII Average mean wt. over period $\frac{1}{2}(\bar{w}_t + \bar{w}_{t+\Delta t})$	VIII $-\Delta N$	IX $\Delta\bar{w}$	X ΔP $(= N\Delta\bar{w})$	XI ΔM $(= \bar{w}\Delta N)$	XII $\sum_0^t \Delta P$	XIII $\sum_0^t \Delta M$	XIV $\sum_0^t \Delta M + N\bar{w}$
1969													
April 1	0	0·2	100	0·02	—	—	—	—	—	—	—	—	—
May 1	0·08	5	60	0·30	80	2·6	40	4·8	0·38	0·10	0·38	0·10	0·40
June 1	0·17	30	24	0·72	42	17·5	36	25	1·05	0·63	1·43	0·73	1·45
July 1	0·25	90	20	1·80	22	60	4	60	1·32	0·24	2·75	0·97	2·77
Aug. 1	0·33	200	18	3·60	19	145	2	110	2·09	0·29	4·84	1·26	4·86
Sept. 1	0·42	350	16	5·60	17	275	2	150	2·55	0·55	7·39	1·81	7·41
Oct. 1	0·50	480	15	7·20	15·5	415	1	130	2·02	0·42	9·41	2·23	9·43
Nov. 1	0·58	600	14	8.40	14·5	540	1	120	1·73	0·54	11·14	2·77	11·17
Dec. 1	0·67	590	13	7·66	13·5	595	1	−10	−0·14	0·60	11·00	3·37	11·03
1970													
Jan. 1	0·75	580	12	6·96	12·5	585	1	−10	−0·12	0·58	10·88	3·95	10·91
March 1	0·92	610	10	6·10	11	595	2	30	0·33	1·19	11·21	5·14	11·24
May 1	1·08	630	8	5·04	9	620	2	20	0·18	1·24	11·39	6·38	11·42
July 1	1·25	660	3	1·98	5·5	645	5	30	0·17	3·23	11·56	9·61	11·59
Nov. 1	1·58	700	2	1·40	2·5	680	1	40	0·10	0·68	11·66	10·29	11·69

I	II	III	IV	V	VI	VII	VIII	IX	X	XI	XII	XIII	XIV
1971													
May 1	2·08	710	1	0·71	1·5	705	1·0	10	0·02	0·71	11·70	11·00	11·71
Nov. 1	2·58	720	0·5	0·36	0·75	715	0·5	10	negl.	0·35	11·70	11·35	11·71
1972													
May 1	3·08	730	0·2	0·15	0·35	725	0·3	10	negl.	0·22	11·70	11·57	11·72
Nov. 1	3·58	740	0·1	0·07	0·15	735	0·1	10	negl.	0·07	11·70	11·64	11·71
1973													
May 1	4·58	755	0	0	0·05	747	0·1	15	negl.	0·07	11·70	11·71	11·71
									11·70	11·74			

Method of calculation:

Mean population between April and May $= \dfrac{100+60}{2} = 80$ thousands m^{-2} (Column VI).

Increase in weight $\Delta\bar{w} = 5 - 0{\cdot}2\,mg = 4{\cdot}8 \times 10^{-6}\,kg$ (Column IX).
Production increment $\Delta P = N\Delta w = 80 \times 4{\cdot}8 \times 10^{-3}\,kg\,m^{-2} = 0{\cdot}38\,kg\,m^{-2}$ (Column X).

Mean weight between April and May $= \dfrac{(0{\cdot}2+5{\cdot}0)}{2} = 2{\cdot}6\,mg$ (Column VII) $= 2{\cdot}6 \times 10^{-6}\,kg$.

Decrease in population $\Delta N = 100 - 60$ thousands $m^{-2} = 40$ thousands m^{-2} (Column VIII).
Mortality increment $\Delta M = \bar{w}\Delta N = 2{\cdot}6 \times 40 \times 10^{-3} = 0{\cdot}104 \simeq 0{\cdot}10\,kg\,m^{-2}$ (Column XI).
Columns XII and XIII are obtained by summing the Production and Mortality increments month by month giving a total production of $11{\cdot}7\,kg\,m^{-2}$ by both methods of computation (foot of Columns X–XIV).

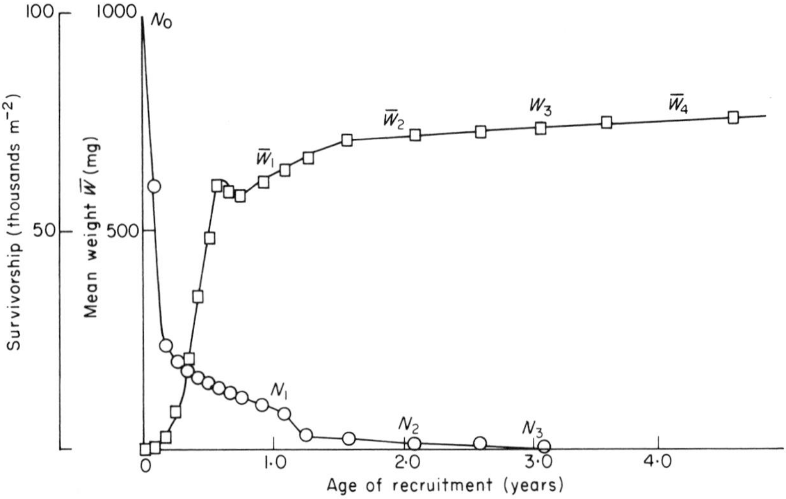

Fig. 9.3. Population and growth curves for a single recruitment, based on data from Table 9.1.

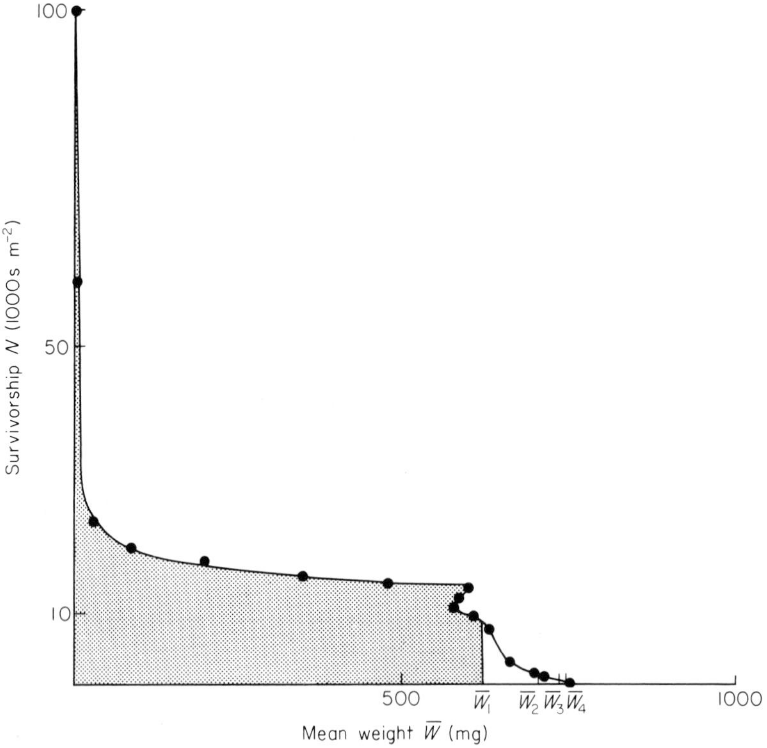

Fig. 9.4. Plot of survivorship, N, against mean weight, $\bar{w}$, for the recruitment described in Table 9.1 and Fig. 9.3. Shaded areas give the production during the first year of life of the recruitment.

particular year can then be obtained by applying equation 9.14. Methods of separating age classes are described on p. 312.

In deriving the productivity equations, and in presenting growth-survivorship curves, it was assumed for simplicity that all members of a recruitment started at zero time (Figs 9.1 and 9.2). In fact recruitment does not take place instantaneously. Whereas the numbers of all other age groups will fall continuously, the number of individuals less than a year old (by convention called 0 group individuals) will increase when recruitment starts, reaching a maximum as larval settlement on the benthic habitat begins to slow down. Such a curve is shown in Fig. 9.5a. However, the application of the summation $N\,\Delta\bar{w}$ is still valid for 0 group individuals so long as the quantity of biomass introduced by settling larvae is negligible, which is nearly always so. Since the changes in N and $\bar{w}$ are generally more rapid for 0 group individuals, more frequent sampling may be necessary for them than for the older, more slowly growing, age groups.

An example of the application of equation 9.14 to compute the production of a population in which the age groups have been separated, counted and weighed at intervals, is demonstrated in Table 9.2 and Fig. 9.5. The kind of data obtained is given in Columns I to IV of the table and the values of N and $\bar{w}$ are plotted in Fig. 9.5, the 0 group being shown inset because of the much larger numbers present. It will first be noted that the mean weight–survivorship relationship differs from that of Fig. 9.4 in being discontinuous. This results from the year classes belonging to different recruitments. It is assumed in the table that settlement begins in May, which is therefore taken as the month at which each year class begins to grow. As in Table 9.1 the calculation of growth increments is based on the equation:

$$\int N\,\mathrm{d}w = \sum \tfrac{1}{2}(N_t + N_{t+\Delta t})\,\Delta\bar{w};$$

the value of $\frac{1}{2}(N_t + N_{t+\Delta t})$ is given in Column VI, $\Delta\bar{w}$ in Column IX.

Since production by each year class must be assessed separately, a difficulty arises in applying the above approximation to obtain the mean of the last value of survivorship of one year class, (i), in this example in May, and the first value of survivorship of the next year class, ($i + 1$), in August. Assuming that the year begins and ends in May, it is necessary to extrapolate the $i + 1$ year class survivorship back from August to the beginning of growth year in May. For extrapolation of survivorship data (and, if the interval is long, for interpolation also) it is advisable to assume that survivorship follows a logarithmic law, $N = N_0e^{-Zt}$. A plot of $\log_{10}N$ against t will then give a straight line from which the value of N in the preceding May can be estimated with reasonable accuracy on the basis of the data for the rest of the year, $i + 1$. The lines for $\log_{10}N$ against time are shown in Fig. 9.5b, and the entries in Table 9.2 (Columns III & VI) based on extrapolation are shown in brackets.

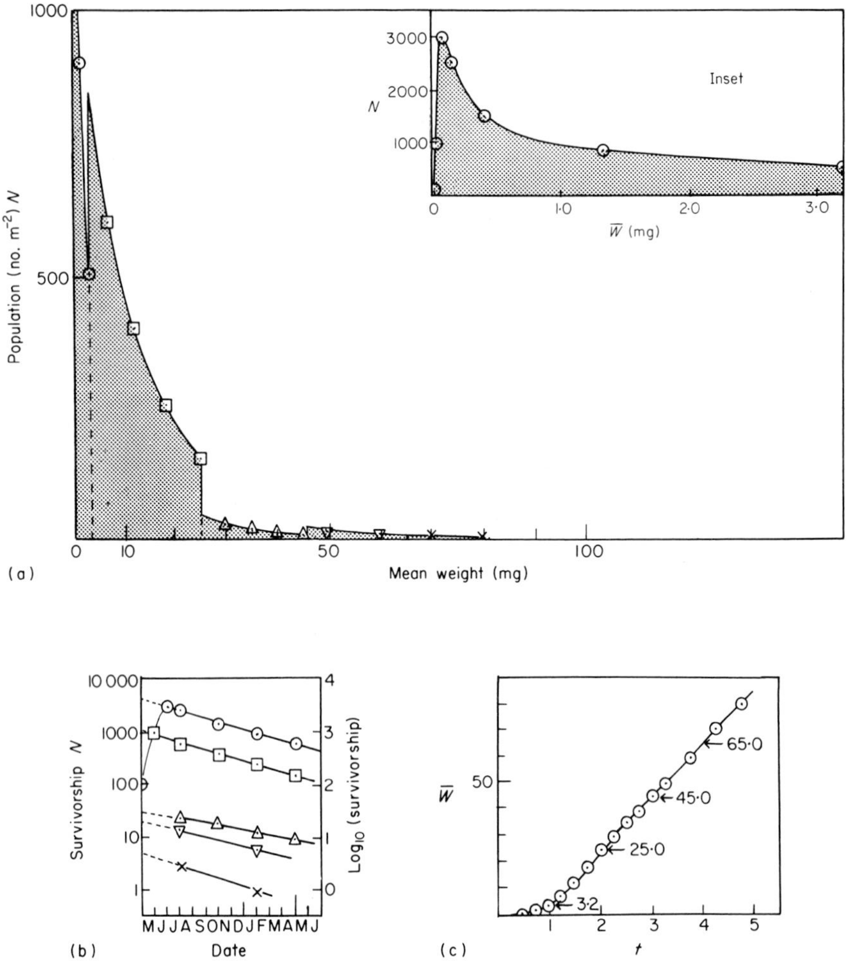

Fig. 9.5. Survivorship–mean weight curves for a population with recruitment.

(a) Plot of N against $\bar{w}$ for each annual recruitment. The discontinuous curves are: ○, year class 0 (first year individuals); □, year class I; △, year class II; ▽, year class III; ×, year class IV. Inset, plot of N against $\bar{w}$ for first year individuals. The rise in N represents the period of recruitment.

(b) Log survivorship–time curves from the slopes of which the specific mortality rate can be determined. The lines extrapolated back to the datum month of initial recruitment allow the value of N at the beginning of each year to be estimated. Extrapolated values of N are given in parenthesis against month V in column III of Table 9.2.

(c) Plot of mean weight $\bar{w}$ against time, t, using data for different age classes given in Table 9.2. The lack of any discontinuity indicates the assumption that growth rates do not vary significantly between successive annual recruitments. Arrows give mean weight at the end of each year after recruitment.

Table 9.2. Computation of production for a population with recruitment where age classes are distinguishable.

I Sampling data (month)	II Year class	III Number in year class N (m^{-2})	IV Mean wt./individual $\bar{w}$ (mg)	V Wt increment since previous sample $\Delta\bar{w}$ (mg)	VI Mean no./m^2 during period Δt N (thousands)	VII Production increment $\bar{N}\Delta\bar{w}$ ($g\,m^{-2}$)
May	0	100	0·01	—	—	
June		1000	0·04	0·03	0·505	0·015
July		3000	0·07	0·03	2·00	0·06
Aug.		2500	0·15	0·08	2·75	0·22
Nov.		1500	0·40	0·25	2·00	0·50
Feb.		900	1·32	0·92	1·20	1·10
May		500	3·20	1·88	0·70	1·31
		N'_1			$\sum_0^1 N\Delta\bar{w}$	3·205
May		(900)				
Aug.	1	600	6·80	3·6	(0·75)	2·70
Nov.		400	12·0	5·2	0·50	2·60
Mar.		250	18·0	6·0	0·325	1·95
May		150	25·0	7·0	0·20	1·40
		N''_2			$\sum_1^2 N'_1\Delta\bar{w}$	8·65
May		(32)				
Aug.	2	24	30	5·0	(0·028)	0·140
Nov.		18	35	5·0	0·021	0·105
Feb.		12	40	5·0	0·015	0·075
May		10	45	5·0	0·011	0·055
		N'''_3			$\sum_2^3 N''_2\Delta\bar{w}$	0·375
May		(22)				
Aug.	3	14	50	5·0	(0·018)	0·09
Feb.		6	60	10·0	0·010	0·10
		N''''_4			$\sum_3^4 N'''_3\Delta\bar{w}$	0·19
May		(5)				
Aug.	4	3	70	10·0	(0·004)	0·04
Feb.		1	80	10·0	0·0020	0·02
					$\sum_4^5 N''''_4\Delta\bar{w}$	0·06
Total annual production (excluding 5th and later year classes)						12·48

Since this method requires a discontinuous series of curves to be used for survivorship and growth, the break being made at a datum month corresponding to the time of recruitment, the mean weight of each year class at this time needs also to be known. It can be reasonably assumed that, unlike the rate of recruitment, the mean growth rate is not likely to differ markedly from year to year, so that a continuous curve can be constructed from the data of Table 9.2 as is shown in Fig. 9.5c. The average weight reached at the end of the year (in May) by individuals of the first, second and *n*th year classes can then be read off from this graph and the integrations of each separate year class can be carried out between the limits of these values of $\bar{w}$. Vertical lines are shown in Fig. 9.5a at the datum month where the integrations are interrupted. If the areas under each of the separate weight-survivorship curves in Fig. 9.5 are integrated, they will be found to agree with the totals given by the approximate arithmetical integration in Column VII of the table.

Method 3A and B. Assessment of production for stocks with recruitment, age classes not separable

When it is impossible to recognize age classes it becomes necessary to measure growth rates or mortality rates as a function of size and season and to carry out a summation throughout the year for each size class present. Method 3A amounts to a direct application of equation 9.9 modified by the introduction of grouping into size classes.

$$\text{Annual Production} = \sum_{t=0}^{t=1} \sum_{0}^{n} f_i G_i \bar{w}_i \, \Delta t \tag{9.15}$$

where G_i is the weight specific growth rate of size group i, $\bar{w}_i$ the mean weight of the size group and f_i the number of individuals of this size group existing in the population during the period Δt. The fraction of total standing stock biomass within the size class i is therefore $f_i\bar{w}_i$. The symbol f_i is used in place of N because it no longer represents a survivorship continuous throughout the period of the summation. The growth rate cannot necessarily be obtained from the population size distribution data because the rate of recruitment into the size class f_i is not known. Growth rates must, therefore, be obtained independently by observations on isolated individuals or groups of individuals whose identity can be established throughout the period of measurement. The methods of measuring growth rates are described below (pp. 321–3).

Details of the method of computation are best given by means of an example. A hypothetical population is divided into three size classes on the basis of length. The numbers per 100 cm^2 quadrat are assumed to have been sampled on five occasions, which include four growth periods over one completed year. The relative growth rates required are assumed to have been

Table 9.3. Computation of productivity for a population whose age classes cannot be distinguished.

Specific growth rates independently measured (see pp. 321–3):

Dates	Length class 0–5 mm G_1 (year^{-1})	5–10 mm G_2 (year^{-1})	10–15 mm G_3 (year^{-1})
April 21–July 2	50	20	5
July 2–Sept. 4	60	30	6
Sept. 4–April 21	20	10	2

Population density and mean weight data for three size groups:

	Size group 1		Size group 2		Size group 3	
Dates	Density (100 cm^{-2})	Mean wt. $\bar{w}_1$ (mg)	Density (100 cm^{-2})	Mean wt. $\bar{w}_2$ (mg)	Density (100 cm^{-2})	Mean wt. $\bar{w}_3$ (mg)
April 21	2000	0·1	300	5·1	60	17·6
Aug. 2	1200	0·5	250	5·7	60	18·1
Sept. 4	400	0·55	200	6·3	40	16·0
Dec. 21	200	0·6	160	6·0	60	17·0
April 21	1000	0·15	320	4·6	80	18·0

Computation for size group 1:

I Period of year	II Time interval Δt (years)	III Mean frequency f_1 (thousands m^{-2})	IV Mean wt. $\bar{w}_1$ (mg)	V Biomass $B_1 = f_1 w_1$ (g m^{-2})	VI Specific growth rate G_1 (year^{-1})	VII Production $G_1 B_1 \Delta t$ (kg m^{-2})
April 21–July 2	0·22	160	0·30	48	50	0·530
July 2–Sept. 4	0·20	80	0·525	42	60	0·505
Sept. 4–Dec. 21	0·31	30	0·575	17	20	0·105
Dec. 21–April 21	0·27	60	0·375	23	20	0·120
				Total production by size group 1		1·26

Summary of results of similar computations for other size groups

Period Δt	Production by size group in kg m^{-2} 1	2	3	Total during period Δt	
April 21–July 2	0·530	0·655	0·120	1·305	
July 2–Sept. 4	0·505	0·810	0·105	1·420	
Sept. 4–Dec. 21	0·105	0·205	0·050	0·360	
Dec. 21–April 21	0·120	0·205	0·065	0·390	
Total by size group	1·26	1·875	0·340	3·475	Total annual production

obtained for each period of the year and are given at the head of Table 9.3. The raw data are shown next: against each date are given the population density per quadrat and the mean weight of those individuals in the sample belonging to each size class. The computation is given in detail for size group 1. The year is divided into four periods separated by the sampling dates and the density and mean weight of the size class during each of these periods is assumed to be the average of that at the beginning and the end. These values are shown in Columns III and IV, and their product, shown in Column V, represents the average biomass during the period. The growth increment is then obtained by multiplying the standing stock biomass by the relative growth rate G_i and time interval Δt. The total production by this size group during the year will be the sum of these increments and the total annual production by the population is obtained by adding that of each size group as shown in the foot of Table 9.3.

Clearly, the larger the number of size groups and growth periods employed in the assessment, the more closely will the summation approach its true value. In order to limit the amount of calculation, the population was assumed to have been divided into only three size groups. This is, in fact, rather too coarse a division and, as a result, the relative growth rates differ too much for accurate assessment. However, the example illustrates quite characteristically the important fact that the very small numerous and fast growing young stages of size group 1, although easily overlooked in the samples, may be responsible for much more production than the more prominent, but more slowly growing, mature individuals of size group 3.

Method 3B should be mentioned for completeness. If independently assessed mortality rates are available, production by a population with indistinct year classes can be obtained by applying a modification of equation 9.10 in which i is one of the size classes into which the population has been divided:

$$\text{Production} = \sum_{0}^{1}\sum_{0}^{n} f_i Z_i \bar{w}_i \, \Delta t + \Delta B \qquad (9.16)$$

It will be seen that, unlike the corresponding equation 9.15 based on growth increments, equation 9.16 includes a term ΔB for the net change in biomass of the population. In using this method therefore, size specific mortality rates are not sufficient and must be augmented by measurements of the difference in the standing stock between the start and the finish of the annual survey.

It is usually more difficult to obtain information on mortality rates than on growth rates. The use of this method may, however, be appropriate when most of the production is accounted for by larger individuals and where mortality rates can be approximated. For example, if the rate of harvesting accounted for a large fraction of mortality and was known, mortality rates could be estimated. If the population is in a steady state condition with no appreciable change in standing stock from year to year ($\Delta B = 0$), then the production measured from the growth integral and the mortality integral of equations 9.15 and 9.16, respectively, should be the same. Kuenzler's attempt to obtain agreement between growth and mortality estimates for a population of *Modiolus demissus* (Kuenzler, 1961) was not very successful, however.

Note that the methods based on equations 9.15 and 9.16 have a wider application than to a classification only in terms of size. They can be used equally for species where age classes can be distinguished or to any division of the population into a number of subgroups, *i*, provided that each of the subgroups is reasonably homogeneous in regard to instantaneous growth rate or mortality rate.

Stratified sampling

Production by a single species population, being a function of the individual rates of growth, will vary with climate, habitat and population structure. In computing production by equations 9.15 and 9.16 above, the assumption is implicit that growth rates depend chiefly on individual size and the season of the year. In fact there will probably be significant differences in growth rate, not only from year to year but also between the various habitats that the species occupies, and over its latitudinal range. Since the investigator hopes that his estimate of production will be representative of the species as a whole in its natural environment, a large study area which includes the full range of habitat variation available to the species and therefore, different growth rates and production, should be selected. Under these circumstances the principle of stratified sampling should be applied. (See p. 5.)

The whole area should be divided into sub-areas of different habitat type on the basis of factors likely to influence growth, mortality and population structure. If these are well chosen, the sub-areas will each contain a more homogeneous population, with individuals whose size-dependent growth rates are more consistent than those in the study area treated as a whole. Separate measurements of production should be carried out in each habitat type. These estimates can then be multiplied by the fraction that the habitat type occupies as part of the whole study area, thus obtaining the value of the production of the study area as a whole.

Stratified sampling has the following advantages:

1. It provides information on differences in production between different parts of the study area.
2. If the survey of the relative areas occupied by the different habitat types is accurately carried out, stratified sampling should greatly improve the precision of the estimate of production for the study area as a whole.

3. Stratified sampling allows the possibility of predicting the productivity of other areas, where the balance of habitat types is different.

As examples of the kind of sub-division that is desirable, a benthic habitat might be separated into sub-areas on the basis of the type of deposit, since some deposits will favour certain species over others. The littoral habitat might advantageously be divided into different zones in relation to chart datum, since organisms immersed for longer periods will be likely to grow faster and to suffer greater predation than those exposed to the air only for short periods.

Multi-species studies

In the study of the production of all the species present in a selected habitat, unless an *in situ* measurement of total production can be made (see p. 350), each component species must be dealt with separately. This would be an enormous task, but for the fortunate circumstance that often arises in which only a few species account for most of the biomass and probably also for most of the production. If these species are obvious, the main effort can be concentrated on them and the data for the species whose contribution is relatively small need not be gathered with the same accuracy. In very rough estimates the minor species can be ignored altogether. It must be remembered, however, that in general the smaller the size of the individuals of a species, the more rapid their growth and metabolism is likely to be and the greater the part they will then play in the energy flow of the ecosystem. This principle applies *a fortiori* to protozoa and bacteria as well as to higher organisms.

Methods of separating age groups

Separation by means of growth marks

It has been shown that the measurement of population production is made much easier if it is possible to determine the age of the individual. Unfortunately, overt growth marks are carried by only a minority of species and these are confined to two main groups of organisms, fish and bivalve molluscs. Methods of ageing have been most fully studied in fish where growth zones or growth checks can be recognized on the hard parts, such as the scales, opercula, otoliths, spines and vertebral centra (Bagenal, 1978, Chapter 5). It is possible that methods of age determination in other vertebrate groups with hard permanent skeletons, e.g. echinoderms, may be developed.

Annual growth checks are related to fluctuating seasonal conditions and are, therefore, commonly encountered in temperate and arctic forms but are rarely found in tropical forms. Growth checks may be caused, not only by seasonal changes in the environment, but also by any event which might have a traumatic effect on growth, such as accidental disturbance of the habitat, disease, or the stress of breeding. It is essential therefore, not only to establish beyond doubt that the marks used for determining age are in fact laid down annually, but also to be able to distinguish them from all other marks caused by extraneous factors. Such adventitious marks are usually fainter and do not fit into the regular series of spacings that characterize the annual marks. The Ford–Walford technique (Walford, 1946) offers a convenient method of displaying the regularity of the intervals between annual growth marks (see p. 326).

In any collection, particularly among the older age classes, some individuals are likely to be difficult to age with certainty. They should be placed in the class to which they seem most likely to belong on the basis of apparent age and size. It is unwise to discard such doubtful specimens from production estimates because they may belong predominantly to certain age classes and their absence would then produce a systematic error in the result.

Separation by means of size-frequency histograms

The majority of benthic invertebrates are soft-bodied or regularly shed their external skeleton. In such animals the year classes can be identified only on the basis of size-frequency histograms. These are obtained by measuring the length or weight of a large number of individuals constituting a complete sample of a population and plotting the number of individuals within a stated size interval against size. A series of peaks or modes can usually be distinguished. The modes may result from irregularity in the rate of recruitment of the population. Hence, if there is an annual pattern of recruitment, the modes will probably represent separate year classes and the position of the mode along the size axis affords an approximate measure of the mean size of that year class.

Unfortunately, growth rates may vary greatly between individuals, especially if the habitat is not uniform, while the rate of growth itself usually declines with age. Consequently, only the first few modes are likely to remain distinct, those representing the older age classes becoming merged. In using this method of distinguishing between age groups, great care must therefore be exercised to ensure that the sample size is sufficiently large to authenticate the modes, especially in the older age groups. Where doubt exists simple statistical methods (e.g. χ^2 analysis) should be employed to establish that the apparent frequency fluctuations are not likely to have arisen by chance. Furthermore, should independent data exist on growth rates, it can be used as a further check by comparing the size at a given age with the position of the supposed mode.

Figure 9.6a shows a size distribution that would allow at least the first three

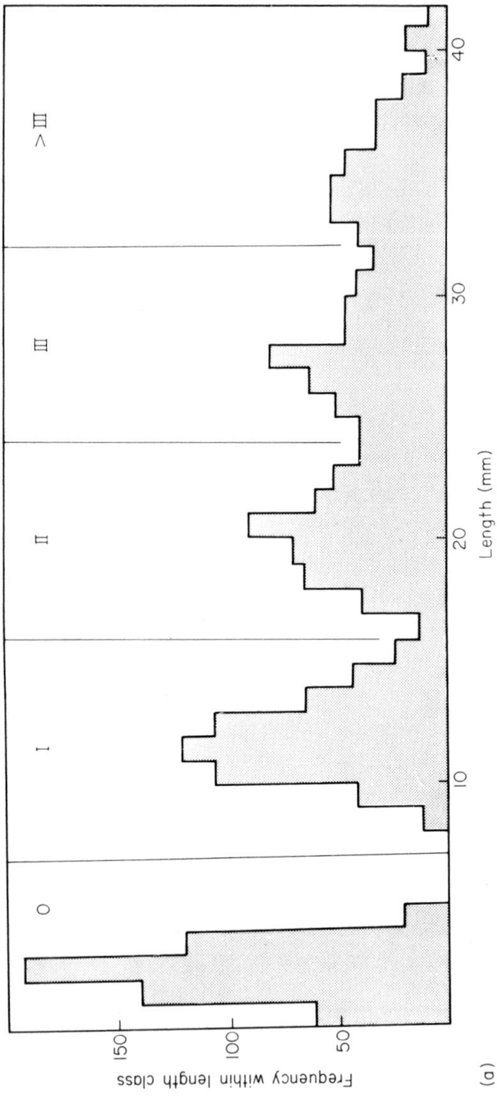
0
I
II
III
>III
150
100
50
Frequency within length class
10
20
30
40
Length (mm)
(a)

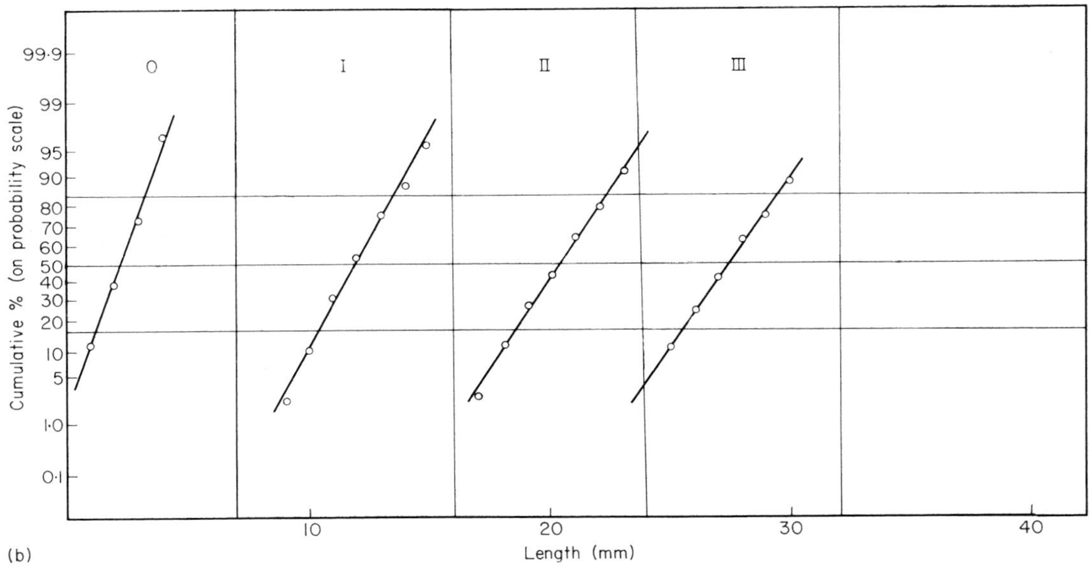

Fig. 9.6. Separation of age groups by size-frequency histograms. (a) Size-frequency histogram for a population with at least 3 distinguishable age classes. (b) The same data plotted as cumulative curves on probability paper.

modes to be distinguished. The 0 group is quite distinct and the individuals belonging to it can be classed with reasonable certainty. The remaining age groups overlap to an increasing extent and there is no way of determining the year class of those individuals whose size lies in the region of overlap. The complete separation of the 1, 2 and 3 year groups must, therefore, be arbitrary. The simplest method is to draw lines of separation perpendicular to the size axis at the points where the component distributions intersect (vertical lines, Fig. 9.6a). The individuals can then be approximately separated into their age classes by identifying the age class with the size classes falling between the vertical lines. It will be seen from Fig. 9.6a that no separation at all is possible, even by this approximate method, for individuals more than three years old. Fortunately, the contribution to production by the larger and slower growing groups is relatively small and the greater part of the total production can be measured from data on the younger age groups, whose numbers and mean weights can be obtained with reasonable accuracy. The production by the younger age groups can be measured using the survivorship-mean weight integration described under Method 2 (p. 301), but the contribution of the larger size groups can be assessed only by Methods 3A or B, which require independent measurement of size specific growth rate or mortality rate (p. 308).

An alternative to plotting histograms is to add the percentage frequencies to one another, starting with the smallest size groups, to obtain a cumulative percentage frequency curve. The cumulative percentage frequency relating to a stated size is the proportion of individuals in the total population having a size equal to or less than that stated. If the cumulative percentage frequency is plotted on the percentage scale of 'probability paper' against the size or weight to which it refers, any mode fitting the normal distribution curve will approximate to a sigmoid curve or a straight line (Harding, 1949). The successive modes can be more effectively transformed into straight lines, which are displaced from one another along the size axis, if a cumulative percentage plot is made for each, using arbitrary lines for separating the modes as in Fig. 9.6a. The modal value is the point at which these lines, extrapolated where necessary, intersect with the cumulative frequency value of 50%. The slope of the line can be used to construct the best normal curve of distribution fitting the histogram. This will be the curve with a standard deviation equal to the distance between the values read off the size axis corresponding to 50% and 16% (or 50% and 84%) on the cumulative percentage frequency curve. Figure 9.6b shows the data of Fig. 9.6a drawn on probability paper and indicates how this method linearizes sections of the cumulative frequency curve and allows the positions of the modes to be determined more objectively.

Growth rates

Productivity measurement frequently requires a knowledge of the rate of growth in terms of dry weight or energy; the latter can be obtained from the dry weight if the caloric content is known (see p. 354).

It would be extremely inconvenient to have to measure the dry weight of the large number of individuals (see p. 287) either in population production studies or in growth measurements. Moreover, to obtain individual growth rates a method of measurement must be employed that does no harm to the animal, so that its weight can be measured at intervals of time. Instead of measuring dry weight directly, it is much easier to measure the length of the animal or of some hard part of the animal, such as the length of the shell of a bivalve, the carapace length of a crustacean, or the scale, otolith or operculum length of a fish. A separate series of experiments must first be carried out to establish the relationship between the selected measure of length and the ash-free dry weight. Alternatively, volume or wet weight can be measured and converted to dry weight.

Length–weight conversion for animals with hard parts

A large number of typical animals of all sizes, extending over the full range required for later predictions, should be used to correlate length to weight. The length of the relevant part of each specimen is then measured and recorded, the specimen oven dried to constant weight, then ashed in a muffle furnace, and the weight of ash subtracted from the dry weight to obtain the ash-free weight of the individual (see p. 351). When the organisms are small it may be advantageous to dry and weigh a number of individuals of similar length and to use average weight and average length for the purposes of correlation. The relationship required is obtained by regressing $\log_{10}$ (dry weight) on $\log_{10}$ (length).

The procedure can be illustrated from a hypothetical example given in Table 9.4. The basic data are shown in Columns I and II; these are transformed to their logarithms in Columns III and IV and are plotted against each other in Fig. 9.7a. Alternatively, if log-log graph paper is available, the raw data can be plotted directly. If the part of the organism chosen for measurement provides a suitable index, the points when plotted in this form will approximately fit a straight line. The best fitting line is then either drawn by eye or by calculation using the least squares regression analysis and the appropriate formula thereby obtained. This will be in the form:

$$\log_{10} w = a + b \log_{10} l \qquad \text{or} \qquad w = al^b$$

where w is the dry weight, l the length of the selected part of the animal, a is a

Table 9.4. Length–dry weight conversion. Individuals selected of similar length, oven dried and weighed to obtain mean length, l, and dry weight, w.

I l (mm)	II w (mg)	III $\log_{10}l$	IV $\log_{10}w$
1·25	$1{\cdot}0 \times 10^{-2}$	0·10	$\bar{2}{\cdot}00$
2·2	$1{\cdot}1 \times 10^{-1}$	0·34	$\bar{1}{\cdot}04$
3·8	$6{\cdot}2 \times 10^{-1}$	0·58	$\bar{1}{\cdot}79$
6·3	1·5	0·80	0·17
10·6	8·7	1·02	0·94
15·8	28·1	1·20	1·45

The values from Columns III and IV are plotted in Fig. 9.7a, giving a line of slope 2·9 and intercept on the $\log_{10}w$ axis of $-2{\cdot}1$.
Therefore $\log_{10}w = 2{\cdot}9 \log_{10}l - 2{\cdot}1$.

constant and b an exponent having a value of the order of three. If the selected part of the animal grows in proportion to the whole of the organism, b should be exactly 3; usually, however, it is slightly but significantly different from 3. The coefficients in the above formula, particularly a, may change seasonally with the condition of the animal; it is therefore desirable to repeat the calibration experiment several times during the year.

There are now many computer programs to carry out such operations from raw data which also provide confidence limits for a and b.

Volume–dry-weight and wet-weight–dry-weight conversion for soft-bodied animals

The lengths of soft-bodied animals are usually difficult to measure precisely, and since weight is roughly proportional to the cube of the length, any errors made in measurement of length will be greatly exaggerated when transformed to weight. It is therefore better to use volume or wet weight for animals that are difficult to measure. The volume of large irregular animals can be readily obtained by displacement of water, but the results are usually less accurate than those obtained by weighing. When large numbers of animals have to be weighed, it is almost essential to have an automatic type of balance. Any external water should be removed from the animal by blotting it dry on filter paper before weighing. Volume measurement is, however, sometimes more rapid and convenient if an automatic balance is not available.

Unfortunately there is no constant relationship between the wet weight and dry tissue weight, even for a given phylum or class. Thorson (1957) provides a survey in tabular form of dry weight as a percentage of wet weight. Animals without calcareous skeletons average about 17%, those with

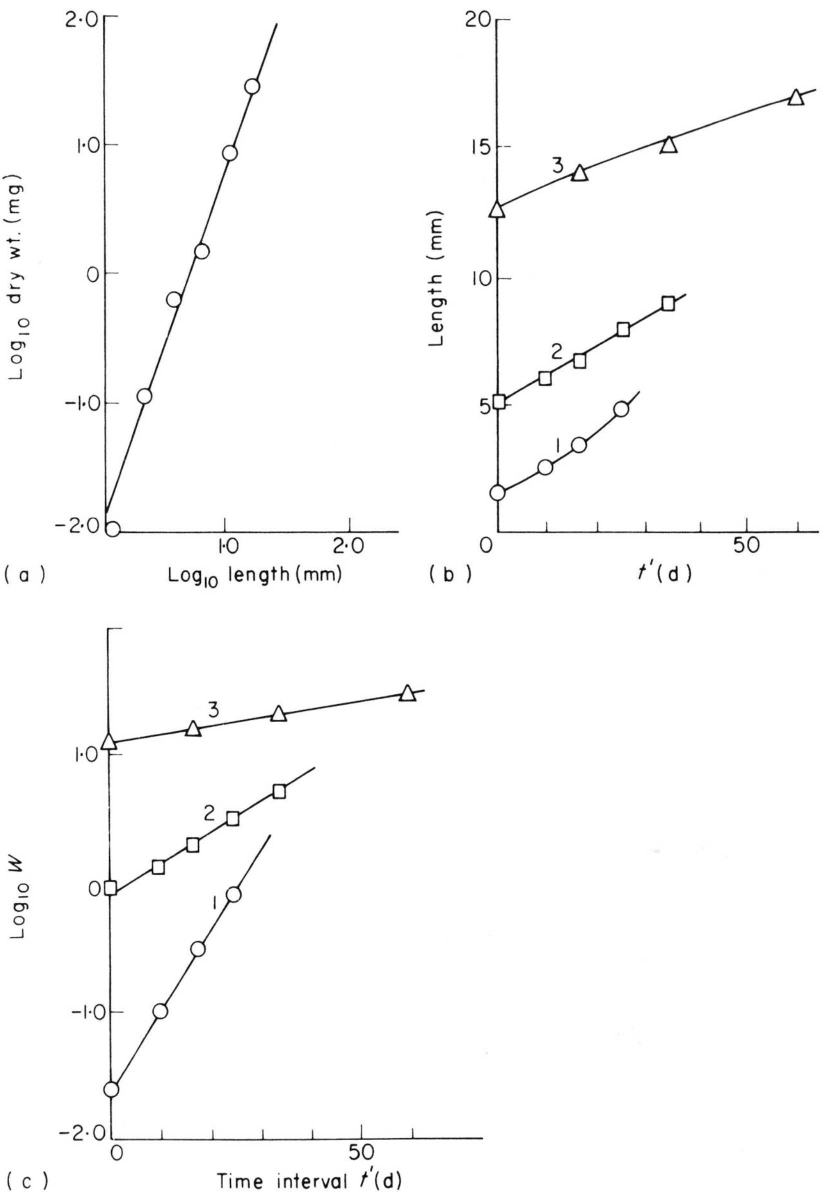

Fig. 9.7. Growth rate measurement for 3 size classes (data from Tables 9.4 and 9.5).
(a) Plot of $\log_{10}l$ against $\log_{10}w$ for length (l)–dry weight (w) conversion (see Table 9.4, applicable to all size classes).
(b) Length plotted against time in days (t') for 3 size classes 1, 2 and 3 (see Table 9.5).
(c) Growth curve plotted as $\log_{10}w$ against time (t') to obtain specific growth rate $d\log_e w/dt'$ (see Table 9.5).

skeletons (prosobranchs, lamellibranchs, echinoderms) are generally $<7\%$. For all purposes but the roughest of estimates the appropriate relationship must be determined for each separate species.

The weight of organic matter in the shells of the molluscs, crustaceans, and other animals may form a substantial fraction of the total organic matter present (e.g. Kuenzler, 1961) and must therefore be included in the dry organic weight. The hard parts of such organisms can also be included when wet weight is measured if this is found convenient, provided of course that the same procedure is adopted when correlating wet weight to ash-free dry weight. The procedure for obtaining the relationship between volume, or wet weight, and dry weight is exactly the same as for the correlation of length to weight except that the exponent b in the resulting formula will have a value close to one:

$$\log_{10} w = a + b \log_{10}[\text{wet weight or volume}]$$

Measuring animals of variable shape

Some animals with hard parts are very variable in shape; individuals of the same species of barnacle, for example, may be tall and columnar or broad and flat. In such cases, a part of the animal may provide a better index of its size than any of its major dimensions. In the case of the barnacle this is true of the opercular valves. When no single part can be found to serve as a good index of the animal's size, two or more measurements in different dimensions may be necessary. For example, bivalves such as mussels can be measured in terms of their length from the anterior (umbo) to the posterior margin. But a short mussel which is very thick laterally and high dorso-ventrally may be much heavier than one which is longer and thinner; hence length alone offers a poor criterion of biomass. In such cases it may be worthwhile to use a product of length, breadth and height to provide a 'volume index' which will correlate better with dry weight than any single measurement.

Preservation

It is most undesirable to preserve samples for the above laboratory studies in alcohol or in formalin between collection and weighing, although on a small ship in rough weather this may be unavoidable. Preservation leads to chemical changes which alter the dry organic weight and the caloric content; formalin, unless neutralized, may destroy calcareous skeletons. Deep freezing is a much preferable method of holding biological material for laboratory investigations.

Measurement of growth rates

Direct measurement of growth rate

There are considerable problems in making continuous observations of animals living freely in the sea, but slow-moving and sedentary forms present on the surface of hard substrata, which can be visited regularly by diving, can sometimes be photographed, identified and measured from time to time. Sessile forms attached to submerged frames, to piers, or suspended from an experimental raft may be recovered mechanically and measured. Live boxes containing small mobile organisms inevitably place restrictions on their movements but may provide conditions close to those in nature. In the intertidal zone it is quite easy to photograph or measure individuals of sedentary species at regular intervals of time or to mark and recapture slowly moving gastropods such as periwinkles, top shells, drills and whelks.

For the majority of soft-bodied benthic invertebrates, however, direct measurement of growth rate can only be carried out in captivity. They need suitably aerated and flushed aquaria, with the appropriate bottom deposit in the case of burrowing forms. Even if the animals can be kept healthy, it is extremely difficult to simulate natural conditions in regard to food supply. For example, it may be difficult to be sure that sufficient micro-algal food is provided for a filter feeding organism, while a predator may be given more prey and grow faster in captivity than in nature. As an example of the great divergence of growth rates when different diets are used, Carefoot's (1967a) experiments with *Aplysia punctata* may be cited. Measurements of growth rate in the laboratory should therefore be regarded strictly as a basis for productivity measurements under laboratory conditions. Application of growth rates measured in the laboratory to estimate production in field conditions must, unfortunately, always be regarded with some suspicion.

Growth rate measurement from size-frequency histograms

As described above (p. 313) the modal size in a frequency diagram may indicate the approximate mean size of a year class. The positions of the modes cannot usually be fixed with great precision, but if a time series of histograms is available in which the modes progress along the size axis, a plot of length (or weight) against time can be constructed. From this the population growth rate can be computed. Although growth rates measured from size frequency histograms are likely to be imprecise and possibly inaccurate, they do correspond to growth measurements under entirely undisturbed natural conditions.

Growth rate measurements from size distribution of animals of known age

Growth rates of certain fish, bivalves and any other animal with clear annual rings can be measured with considerable accuracy. From a large single collection of specimens of all sizes, those of normal shape and whose age can be determined with certainty should be selected, those of doubtful age or unusual shape being discarded. In ageing specimens the convention is adopted that the age group corresponds to the number of growth rings laid down. Thus the 0 group has no growth ring, age group 1 has a single ring and so on. Growth marks are usually laid down in winter. If so, 1 January can be taken arbitrarily as the date of the growth mark (1 July in the southern hemisphere) and the mean lengths measured can be related to age on the date of collection. Thus, if the collection were made on 1 July (in the northern hemisphere), the individuals with one growth ring could be classed as 1·5 years. Hence by making several measurements during the year it would be possible to interpolate on the growth curve the values for length corresponding to years and fractions of a year. In the 0 and 1 age classes, where the growth increment is large, this additional information may prove useful. In measuring the average size of each group it is desirable to use equal sample numbers, including at least 10 individuals from each age group, although, naturally, the oldest age groups become progressively less common in the sample.

In making measurements of a characteristic length of each individual (e.g. the greatest length of the shell) it is most important to define and measure precisely the same feature. When several hundreds of individuals have to be measured the operation is repetitive, tedious and sometimes tricky; the operator should therefore use calipers or a suitable mechanical jig wherever possible rather than to rely on measurements judged by eye. Where sexes can be distinguished, separate data of size against age for each sex should be collected and, if they do not differ significantly, the results can later be combined. Usually, differences in growth rates are more likely to be attributable to environmental than to sex differences. Therefore if the habitat is exposed to considerable variation in such factors as tidal current, prey availability or emersion, growth curves for individuals collected under different environmental conditions should be compared. (See stratified sampling, p. 311.)

Growth rate measurement by back calculation of lengths at a previous age

In this method a series of characteristic lengths are measured, corresponding to each of the growth rings present, so that the size of the animal itself at the time each growth ring was formed can be back calculated. This method is particularly valuable when only a few individuals are available for study or when the history of the growth of a particular individual is required. In fish, to which this method has been extensively applied, growth marks are present on relatively small parts of the animal such as the scales, otoliths etc. and the relationship between the dimensions of these parts and the length of the fish may be complex and variable; Bagenal (1978) should be consulted for details. In bivalves, the group with which this method will mainly be concerned in benthic work, the growth rings are on the shell itself and no transformation is necessary for back calculation, the shell length measured on the rings representing the actual size of the shell at an earlier age. Back measurements from growth rings can therefore be used to determine the growth increments of the younger age groups in years previous to the year of collection. If the population structure can be assumed to be similar it would then be possible to calculate variations in production in successive years from a single year's observations.

Population and individual growth rates

The results of growth rate determinations by the four methods outlined above are not equivalent. Methods 1 and 4 deal with the growth of individuals and provide data on individual growth rate. Methods 2 and 3 deal with the increase in mean length of a given age group of the population and therefore measure the population growth rate. The two concepts are different and the values will not necessarily be the same. If mortality among the larger individuals of a given age group were greater than among the smaller individuals, the mean weight of the population would not rise as fast as the weight of an individual that survived. Individual growth rates and population growth rates are therefore equal only if mortality is independent of size.

How great an error would be caused by using the wrong growth rate is not exactly known, but the error is likely to be small compared with the other inaccuracies in population sampling and the natural variations in production. Nevertheless, when production is estimated by classification into size groups (p. 308) (equation 9.15) it is the individual growth rate that should strictly be used and either the direct method of measurement or back calculation from growth rings is therefore to be recommended.

When estimating production in a population where the age classes can be identified, the integration of the survivorship-mean weight curve for each year class is free from any possible error through confusion between individual and population growth rate.

Computation of growth rate

A direct plot of length against age usually results in a sigmoid curve. The curve is not directly applicable to production estimates; it must first be transformed into a rate of increase of dry weight or energy using a formula of the type

Table 9.5. Computation of specific growth rate.
A. Lengths of identified individuals of similar initial mean length l_1 and belonging to size class 1 measured on the dates given and averaged.
Equation relating dry weight to length:

$$\log_{10} w = 2{\cdot}9 \log_{10} l - 2{\cdot}1$$

I Date	II Time interval t'(d) days	III Mean length l (mm) 0–5 mm class	IV $\log_{10} w$ calculated from above relationship
May 2	0	1·5	$\bar{2}{\cdot}40$
May 12	10	2·5	$\bar{1}{\cdot}00$
May 19	17	3·5	$\bar{1}{\cdot}48$
May 27	25	4·9	$\bar{1}{\cdot}90$

The values in Columns II and IV are plotted in Fig. 9.7c, curve 1, giving a line of slope of $d \log_{10} w/dt' = 0{\cdot}059\,d^{-1}$.
$G_1 = d \log_e w/dt$, where t is measured in years $= 0{\cdot}059 \times 2{\cdot}303 \times 365 = 50$ years^{-1}.

B. Data for larger size classes.

Date	Time interval (d)	Mean length 5–10 mm class (2) (mm)	Mean length 10–15 mm class (3) (mm)
May 2	0	5·1	12·5
May 12	10	6·0	—
May 19	17	6·7	14·0
May 27	25	8·0	—
June 5	34	9·0	15·0
July 1	60	—	17·0

These data are transformed to $\log_{10} w$ and plotted in Fig. 9.7c, curves 2 and 3, giving, respectively, slopes of 0·024 and 0·006 corresponding to estimates $G_2 = 20$ years^{-1}, $G_3 = 5$ years^{-1}.

$w = al^b$ such as that obtained by the method described on pp. 317–8 and illustrated by Table 9.4. The values of $\log_{10} w$, derived from measurements of length, wet weight, or volume, are then plotted against time, and the weight-specific instantaneous growth rate, G, can be obtained from the slope of the line at any point.

$$G = \frac{d \log_e w}{dt} = 2{\cdot}303 \frac{d \log_{10} w}{dt} \tag{9.17}$$

Table 9.5 illustrates the method as applied to consecutive measurements of

Table 9.6. Length–age data for a hypothetical bivalve with a length–weight relation $\log_{10}w$ (g) $= 2{\cdot}7 \log_{10}l$ (cm) $- 1{\cdot}6$.

I Age class (years)	II Mean shell length (cm)	III Estimated value of $\log_{10}w$ from length–weight relation
0	1·3	$\bar{2}{\cdot}70$
1	3·55	$\bar{1}{\cdot}89$
2	5·35	0·37
3	6·8	0·64
4	7·9	0·83
5	8·8	0·94

the length of individuals of the same species as that used to establish a length–weight relation in Table 9.4. Columns I to III give the raw data for the smallest size class, 0–5 mm. Column IV is the estimate of $\log_{10}w$ based on the equation derived in Fig. 9.7a:

$$\log_{10}w = 2{\cdot}9 \log_{10}l - 2{\cdot}1$$

In Fig. 9.7b, curve 1, the length l is plotted against time, t', and in Fig. 9.7c, curve 1, the same data are plotted as $\log_{10}w - t'$ to give a line from which G can be calculated by equation 9.17.

Figures 9.7b and c include also the lines for the two larger size groups computed in the same way from the data given in Table 9.5b.

A method based on measuring the slope at a particular point of the plot of $\log_{10}w$ against age is satisfactory only if the curve approximates to linearity, as do those in Fig. 9.7c. This is the situation when the growth rate of a particular size group is measured over a relatively small growth increment, but when the growth curve covers the whole life of the animal it has considerable curvature, making it very difficult to draw and measure accurately the slopes of tangents. Table 9.6, Columns I and II provide typical data for the average length of successive age groups of a bivalve. Note that the average length of individuals of year class 0, which includes all those captured before the first growth ring has been laid down, has a finite value. Consequently, zero length on the growth curve (Fig. 9.8a) starts before zero age; a result of the convention adopted for defining age classes. The length-age curve shows a considerable change of slope and when transformed to $\log_{10}w$ (Column III Table 9.6 and Fig. 9.8b) the curvature is further increased. The slopes shown at values of $\log_{10}w$ corresponding to $l = 2{\cdot}5$, 5 and $7{\cdot}5$ cm, respectively, would be very difficult to

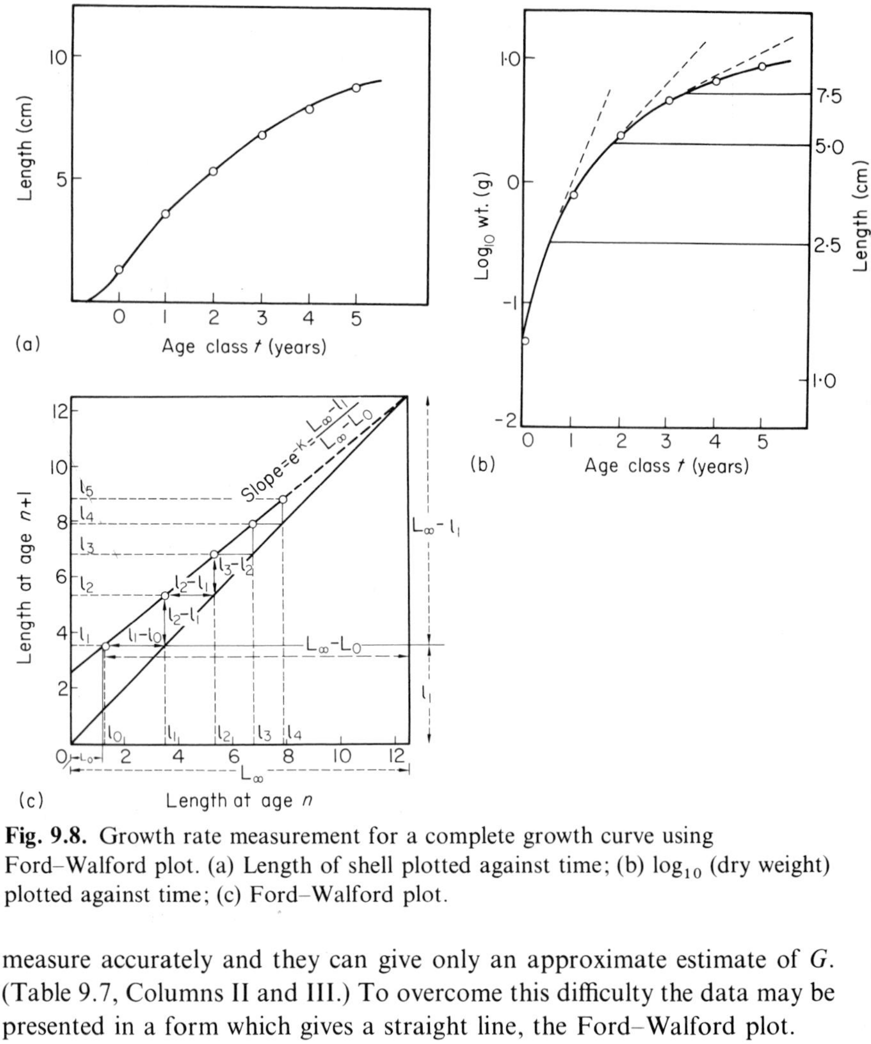

Fig. 9.8. Growth rate measurement for a complete growth curve using Ford–Walford plot. (a) Length of shell plotted against time; (b) $\log_{10}$ (dry weight) plotted against time; (c) Ford–Walford plot.

measure accurately and they can give only an approximate estimate of G. (Table 9.7, Columns II and III.) To overcome this difficulty the data may be presented in a form which gives a straight line, the Ford–Walford plot.

Ford–Walford plots

It has been found that the section of the sigmoid growth curve in which the growth rate is declining can often be fitted to the von Bertalanffy (1934) equation:

$$l_t = L_\infty - (L_\infty - L_0)e^{-Kt} \tag{9.18}$$

where L_0 and L_∞ are constants representing length at zero age, l_0, and maximum possible length respectively. Clearly the equation will not fit lengths

Table 9.7. Estimation of specific growth rate G, directly from the tangents to the growth curve of $\log_{10}$ (dry weight)–time (see Fig. 9.8b and Column III), and from the Ford–Walford plot (see Fig 9.8c and Column IV). The Ford–Walford plot gives $L_\infty = 12{\cdot}5$, $L_0 = 1{\cdot}25$, $K = 0{\cdot}23$. The weight–length relation (Table 9.6) gives exponent $b = 2{\cdot}7$.

I Length, l (cm)	II $d\log_{10}w/dt$ (approx.)	III $G = 2{\cdot}303\, d\log_{10}w/dt$ (years^{-1})	IV $G = bK[L_\infty - l]/l$ (years^{-1})
2·5	1·2	2·8	2·48
5·0	0·4	0·9	0·93
7·5	0·2	0·46	0·41

of young individuals and does not predict zero size at zero age. If $t = n$, an integral number of years, then the ratio of annual increments for successive years can be shown to be constant:

$$\frac{l_n - l_{n-1}}{l_{n+1} - l_n} = e^K$$

Hence if l_{n+1} is plotted on the ordinate against l_n on the abscissa (Fig. 9.8c) the points can be shown to fall on a straight line of slope e^{-K}, which cuts the line drawn at 45° through the common origin of l_n and l_{n+1} at the point where $l_n = l_{n+1}$, that is at the value of L_∞. The constant L_0 can be read off the plot of l_{n+1} against l_n from the value on the abscissa where the line intersects l_1 on the ordinate. This presentation of growth rate data, known as the Ford–Walford method, allows information such as that obtained by methods 3 and 4 above (and as given in Table 9.6) to be combined, enabling accurate estimates of the three growth constants L_0, L_∞ and K to be made. From these the length and relative growth rate at any age can be computed, as follows:

Differentiating equation 9.18:

$$dl/dt = (L_\infty - l)K$$

Differentiating weight-length transformation equation, $\log_e w = b \log_e l + \log_e a$:

$$d\log_e w/dl = b/l$$

hence the weight specific growth rate G is given by:

$$G = d\log_e w/dt = \frac{d\log_e w}{dl} \cdot \frac{dl}{dt} = bK[L_\infty - l]/l$$

It will be clear that b and $[L_\infty - 1]/1$ are non-dimensional but K and G have units of reciprocal time.

The value of K can be obtained from the relation:

$$K = 2{\cdot}303 \log_{10} \frac{L_\infty - L_0}{L_\infty - l_1}$$

Figure 9.8c illustrates the Ford–Walford plot of data from Table 9.6 and Table 9.7 shows how a more accurate value of G can be obtained from the Ford–Walford constants.

Methods are available for the mathematical computation of production using the constants of the fitted Bertalanffy equation and the size specific mortality rate Z in place of the empirical summation methods outlined in this chapter. They have been used for theoretical analyses of fish production (Beverton & Holt, 1957; Bagenal, 1978, p. 208). There is no reason why the same principles should not be applied to benthic production where the theoretical equations of growth and mortality fit the observational data.

Short-term growth rates

Recently two new methods have been introduced to study rates of growth over short periods of time—of the order of days rather than years. These measurements are particularly useful for bivalves when seasonal or environmental differences in growth rate may be important. Both methods strictly measure shell growth, and are, therefore, applicable only to animals with shells. Fortunately, body growth can usually be assumed to be proportional to shell growth.

The first method, originally due to Strömgren (1975), relies on measurements of increase in shell length to an accuracy of $\pm 3\,\mu$m using the diffraction pattern from a small, 1 mV output, neon helium gas laser. When a beam of coherent light is passed through a narrow slit of 100–600 μm width a very clear interference pattern of bright and dark bands is formed (Borowitz & Beiser, 1966) which can be developed on photographic film. The distance between these bands is related to the slit width (a) by the formula $a = \lambda s/d$ where λ is the wavelength of the laser light, in our case 632·8 nm, s the distance between the slit and film below, and d the distance between diffraction spots on the film. To make measurements, the animal is firmly cemented to a perspex block (Fig. 9.9) with its growing edge g very near and parallel to a straight steel edge e braced firmly to the block but capable of being moved from time to time by adjusting the set screw ss. The laser beam is shone through the slit between g and e onto a camera back with a focal plane shutter which photographs the pattern falling on it. The animal remains on all occasions cemented in exactly the same position on the block and changes in gap width are recorded daily until the shell closes against the slit. Then the reference edge, e, must be moved back. Unfortunately measurements can be made only in air or in clean water,

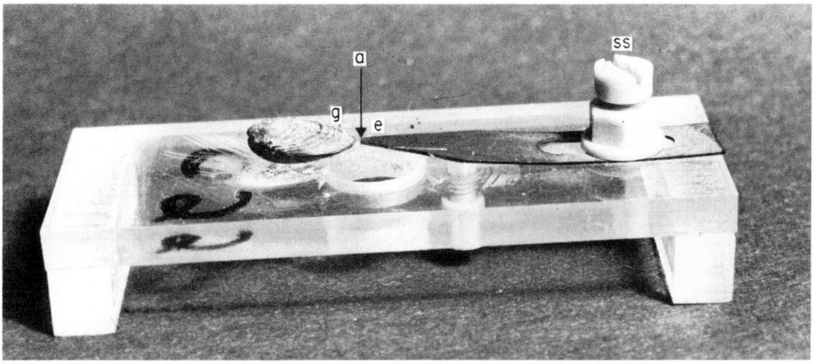

Fig. 9.9. Plastic block for measuring short term growth rate in mussels. e, steel edge; g, growing edge of shell; a, slit; ss, set screw for readjusting slit.

since the presence of suspended food particles, which also diffract the beam, obscure the required image. Table 9.8 gives an example of such results.

The second method makes use of discontinuities in shell structure formed in intertidal animals as the result of tidal emersion (Richardson *et al.*, 1979, 1980). These discontinuities consist of increments of shell growth separated by lines or 'growth bands'. They can be demonstrated by the following procedure.

The shell is embedded in plastic and is sectioned by diamond saw or hacksaw in a plane along the direction of maximum growth. For example, in a bivalve shell this plane will be normal to the surface of the shell and in a direction from the umbo along the maximum radius of the shell. A suitable slowly setting plastic is Metaserve polyester resin, type S.W. The cut surface is

Table 9.8. An example of laser measurements of mussel growth rate. $\lambda = 632{\cdot}8$ nm; $s = 1{\cdot}039$ m; $\lambda s = 657{\cdot}5 \times 10^{-9}$ m^2.

	Day				
	1	2	3	4	6
Diffraction pattern distances (d) (mm)	1·33	1·45	1·945	2·38	3·856
Gap width $= \dfrac{(657{\cdot}5)}{d}$ μm	494·4	453·4	338·0	276·3	170·1
Increment μm	—	40·9	115·4	61·7	106·2
Increment μm day^{-1}	—	40·9	115·4	61·7	53·1

Average rate of growth during period = 64·84 μm d^{-1} or 23·7 mm year^{-1}.

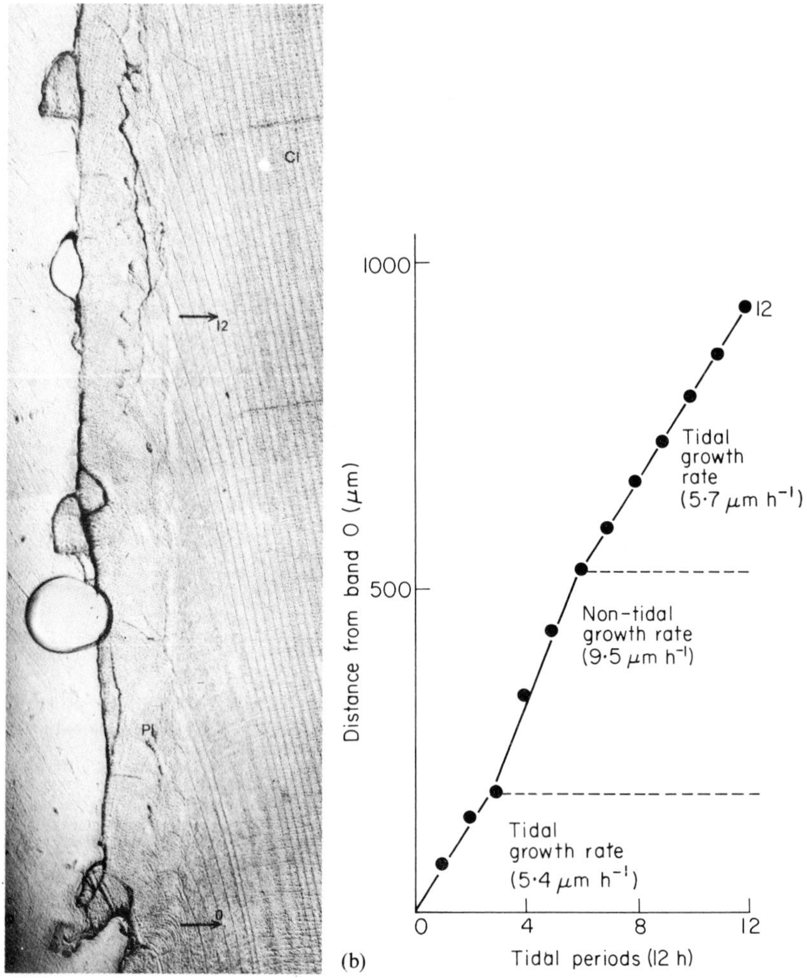

(a) (b)

Fig. 9.10. Shell increments in *Cerastoderma edule*.

(a) Peel from an experiment in which conditions in which *Cerastoderma* was growing were changed from a simulated tidal regime to a continuously immersed regime for 32 hours, and then returned to tidal conditions. Note the loss of definition of growth bands and greatly enhanced width of the bands when continuously submerged. Cl, cross-lamellar structure of shell; Pl, prismatic or reflected cross lamellar structure of shell; 0, 12, bands showing position of shell tip at start and at 12th emersion (see growth plot).

(b) Plot of growth against time using data from the peel and assuming each band represents one tidal period. Growth rates can be measured from the slope of the graph.

From Richardson *et al.* (1979).

then ground flat and polished smooth with a series of metallurgical abrasives in the following order: silicon carbide grinding paper, grit size 400; 800; 1200; and 1500. Brasso polish is then rubbed over with a napped cloth and finally the surface is dry polished with a short piled cloth such as Metaserve 'selvyt'. The shell surface should then be etched with 0·01 M hydrochloric acid for 25–40 minutes, the time varying according to the character of the shell and the contrast in detail sought. The etched shell is then briefly washed in distilled water, dried and a plastic peel applied. Blue replicating material, reference 1320 supplied by Polaron Equipment Ltd. gives good results. The peel is first floated on a recommended solvent such as ethyl acetate and, when soft, carefully placed on the etched surface so as to achieve maximum contact, avoiding air bubbles. After leaving for a few minutes to dry, it can be peeled off and examined under the light microscope in air. Phase contrast illumination is often necessary to see the growth bands sufficiently clearly for counting and measuring. Figure 9.10 illustrates how the distance between growth lines, which represent growth during a semi-diurnal tidal period averaging 12·4 h, is used to calculate growth rate. In the cockle shell illustrated, it will be seen that each growth increment is represented on the time scale by a tidal period and the position of each band, corresponding to that of the growing edges of the shell at the relevant time, is shown on the ordinate as shell length along the plane of growth. In this figure, continuous immersion resulted in more diffuse and wider sets of bands and an increase in growth rate.

Although it is best to measure the distance between bands along the direction of maximum growth, this must be converted later to the appropriate dimensions employed to measure increase in shell size. In the clam shell, the direction of maximum growth rate follows the tangent to the curved surface of the shell, so that the growth bands measure it approximately along the outer side of the curved surface, which may be referred to as the perimeter (p). A correction factor is therefore necessary to convert these growth rates into growth rate along the diameter of the shell (d), which is the dimension usually measured. The correction may be obtained pragmatically by measuring and regressing the diameter d on the length of the curved surface p for a number of shells. Richardson *et al.* (1980) obtained a ratio p/d of 1·54 for the common cockle. Gastropod shells are also marked by tidal growth lines (Ekaratne & Crisp, 1982).

The relationship between the length of the bivalve shell along the direction of maximum growth and its diameter can also be obtained using the constant, λ, of the logarithmic spiral describing its growth: $d = d_0 \exp(\lambda(\theta - \theta_0))$, where d, θ are the polar coordinates of the spiral with reference to the origin, and subscript 0 refers to time = 0. This equation will predict the value of p/d as $\sqrt{((1 + \lambda^2) \div \lambda)}$. Villmann *et al.* (1981) gives $\lambda = 0{\cdot}73$, hence $p/d = 1{\cdot}69$, which may be compared with the observed value of 1·54.

In these more complex shells, greater care is needed to cut along the direction of maximum growth. The cut should run along the outer edge of the whorl in a plane passing through the whorl axis—if the shells are not cut in the correct plane, the growth bands may not be visible. As in bivalve shells, it is necessary to relate growth rates along the direction of maximum growth to the rate of increase in the standard dimension of the shell, such as shell height as measured from the lowest point of the largest whorl to the top of the spire. The magnitude of the correction is again best found by regressing the required dimension, in this case shell height, on length measured in the direction of growth, viz. around the outer surface of the whorls, for specimens of various sizes. Ekaratne & Crisp (1983) give the formula relating shell height, H, or height increment, ΔH, to length, l, or measured tidal increment Δl along the outer side of the whorls of a turbinate shell as:

$$\frac{l}{H} = \frac{\Delta l}{\Delta H} = \frac{\sec\alpha \sec\beta}{\rho \tan\beta + 1/\lambda},$$

where α is the angle of the logarithmic spiral as given by

$$\alpha = \tan^{-1}[2\pi \sin\beta/\log_e \lambda]$$

where λ is the ratio between diameters of successive whorls (not equivalent to λ in Villmann *et al.*, 1981); β is the semi-apical angle of the spire and ρ is the ratio, in the apertural plane, of height to width of the aperture. For conical shells of limpet form they give a simple geometric relation between the increments in the longer (posterior) side of the cone and increments in maximum shell diameter across the aperture.

Shells of intertidal barnacles, and possibly of other organisms, also have tidal micro-growth lines. In barnacles, the clearest are in the sheath and at the base of the parieties (Bourget & Crisp, 1975). However, measurements of these increments would not easily be related to growth in shell diameter or other dimensions of the barnacle because of individual variation in shape.

Population density

In all methods requiring data on the survivorship of stock, N, a method of obtaining a census of the stock is necessary. There are two classical methods available: mark and recapture and direct enumeration.

Mark and recapture methods are useful for mobile animals which can be tagged and sampled on a large scale; they have been widely applied in fishery research but are rarely useful for benthic studies unless bottom dwelling, actively swimming fish and crustacea are included and can be investigated on a massive scale. The methods described in Bagenal (1978) would be applicable to such organisms. Where direct enumeration can be employed, as is possible for the sedentary or nearly sedentary organisms in the benthos, more precise information can be obtained with less effort.

Benthic species must usually be counted by means of quantitative sampling using one of the more reliable samplers to collect a quantity of deposit representing a constant area of the sea bed. When the animal being sampled shows great variation in density from place to place or between substrate types, it is important to sample the same area and habitat on all occasions in order to reduce sampling errors. If the sampler cannot be relied upon to take a constant area of the substratum but can be relied upon to collect up to a consistent depth below the surface, the volume of deposit collected can be used as a measure of the amount of habitat searched and the population density expressed in numbers per unit volume (see Chapter 6).

In habitats favourable to the investigator, such as the intertidal areas and the shallow sublittoral accessible by diving, hand sampling of quadrats may be possible and visual or photographic methods can be applied.

Mortality rates

In assessing production by Methods 1 and 2 (pp. 300–1), mortality data must be collected in the form of survivorship-time curves for single stocks (Method 1) or for age classes (Method 2). If instantaneous mortality rates are required they can be obtained by plotting $\log_{10} N$ (where N is the number of a stock or the specified age class surviving in unit area of the substratum) against t, the time of sampling. Not infrequently, a particular age group may predominate in the population—this age group should be chosen for mortality measurements in order to increase the numbers of survivors (N) in each sample and so reduce the sampling error. Also, survivorship of a year class of exceptional strength can be followed through a longer period of time than an average year class.

The type of graph obtained from survivorship data has already been shown in Fig. 9.5b; it was previously used to demonstrate extrapolation (p. 305). Plots of survivorship against time are sometimes called 'catch curves' since, if harvesting has a significant effect, the curve will descend more sharply if the animal is heavily fished. When recruitment is taking place, the catch curve may contain a rising segment, as shown by the uppermost curve of Fig. 9.5b representing the 0 year class. This segment cannot of course be used to measure either mortality or recruitment since both are taking place at once. Where recruitment can be assumed to be absent, for example in the case of first or higher year classes, or when measurements are taken outside the season of settlement, the slope of the curve of $\log_{10} N$ against time allows the instantaneous mortality rate Z to be measured:

$$Z = \frac{1}{N}\frac{dN}{dt} = d(\log_e N)/dt = 2{\cdot}303 d(\log_{10} N)/dt$$

Thus in Fig. 9.5b the slope of $\log_{10} N$ against t for the year group 1 is $-0{\cdot}8$. The age specific mortality rate, Z, is therefore $2{\cdot}303 \times 0{\cdot}8 = 1{\cdot}84$ years^{-1}.

An annual mortality rate is sometimes quoted. This may be defined as $(N_0 - N)/N_0$ and will be equal to $1 - e^{-Z}$. For the year group 1 in Fig. 9.5b the annual mortality rate will be 0·84.

Although annual recruitment may vary from year to year, when the mortality rate is measured from consecutive values of the *absolute numbers* of an identifiable year class found in unit area or unit volume of the substratum on successive occasions, the only limit to the accuracy of the method is the error associated with the method of census. The presence of later recruits will not confuse the enumeration of the survivors of a previous age class.

When it is not possible to obtain samples representing a standard area or volume of the substratum, the only method of measuring mortality may be in terms of year class frequencies, that is , the *relative fraction* of the total sample collected attributable to each year class. Since this is a *relative* and not an *absolute* measure of the strength of the year class, uneven recruitment will lead to a serious bias in the apparent mortality rate. Thus, should the rate of recruitment be high in a particular year, the whole population will increase and the proportion of all other year classes will drop. A spurious impression will be created of increased mortality rate. Similarly, over periods of poor recruitment, mortality will be under-estimated. This method of measuring mortality rate will be valid only if recruitment is regular or if the slope of the catch curve based on frequency is measured over a sufficiently long period of time to even out irregularities in recruitment.

Where there are no criteria for age determination, natural mortality rates are very difficult to determine. They may theoretically be found by observations on marked individuals which can be regarded as age classes and the change with time in their rate of recapture, R, could be used to obtain a mortality estimate from the relationship:

$$Z\,\Delta t = (R_t - R_{t+\Delta t})/R_t$$

However, for this method to be successful very large numbers of marked individuals must be added over an area that is very great in relation to the movements of individuals in the population, otherwise dissemination will be confused with mortality. The size of such an operation would usually have to be so large that it is unlikely to be applicable to most benthos investigations.

Changes in recovery *frequency* (the number of marked individuals as a fraction of the total number of individuals in the sample) can be measured more easily because, unlike measurements of recovery per unit area, they are unaffected by variations in the size of the grab samples. However, they would yield mortality rates which would suffer from bias if recruitment were irregular for the same reasons that relative year class frequencies provide an unsatisfactory basis for mortality measurements. In view of the difficulty of obtaining independent measures of mortality rates, production estimates should normally be based on measurement of growth increments.

Reproductive output

The production of eggs and sperm or of living young, together with the material associated with them (egg capsules, seminal fluid etc.), may be regarded as part of the process of growth and indeed many authors include it in the assessment of total production. The measurement of reproductive output often involves a separate series of experiments and computations and is, therefore, treated separately in this handbook.

It is virtually impossible to measure the reproductive output of most benthic species by making observations entirely in the field, but a preliminary study should be made where information is not available to establish the natural breeding season, whether the animal breeds once a year or more frequently, and the age or size at maturity.

If the species can be kept healthy and fed in the laboratory, the relation between reproductive output and the quality and amount of food intake can be examined. Examples of such studies are Marshall & Orr (1961), Paine (1965) and Carefoot (1967a, b). Laboratory observations, however, can give only an approximate idea of what occurs in the field since the conditions may be different from those in nature in some adverse respect, or they may sometimes be more conducive, for example by allowing the animal to obtain food with less expenditure of energy in searching.

For the purpose of measuring reproductive output, animals can be divided into two main categories; those which produce a single brood over a very short period of time once every year and those which breed over an extended season. The former class are characteristic of high latitudes and their reproductive output is relatively easily assessed. The latter are typical of warmer climates and the measurement of their reproductive output often presents considerable difficulty.

Annual breeders

Two basic methods are available; if the brood is shed in a form that can be handled (e.g. viviparous young, egg strings or capsules that can be collected and weighed in their entirety) the direct method can be used. If the reproductive output cannot be collected in a tangible form it is necessary to resort to the indirect method. This is necessary in the case of females producing minute planktonic eggs, and for sperm production by males.

Direct method

When there is a single annual brood it is usually possible to distinguish between gravid and spent individuals. Gravid females of widely ranging size should be brought into the laboratory just at the commencement of the spawning season. Sometimes animals are induced to spawn by the change from field to laboratory conditions, sometimes a spawning stimulus can be applied (e.g. rise of temperature, increase or decrease in illumination, the presence in the water of the reproductive products of other individuals). The eggs or young from each individual that spawns should be collected and the length or weight of the parent recorded. The reproductive products can either be weighed wet, the wet weight being converted to dry weight by means of an independently derived formula (as for whole animals see pp. 317–20), or each separate spawning can be dried and then weighed. The ash weight and caloric content of spawned products may also need to be measured, hence the dried material obtained from the spawn should be collected, homogenized, and kept for this purpose. Eggs generally contain high energy materials and their caloric content may be considerably higher than that of the parent.

A graph should be constructed from these data, showing the biomass of the reproductive products as a function of the biomass, dry weight, or length of the parent. Generally, the weight of the brood is roughly proportional to the weight of the parent for all individuals whose size exceeds that at which maturity is reached. Since, for annual breeders, this graph will represent the total annual reproductive output in relation to the size of the individuals, the only other information required to calculate reproductive output is the size frequency of the population at the breeding season, and, to allow determination of the number of actively reproducing females, the proportion of fertile females in each size class. The reproductive output by the females of the population will then be obtained from the frequency of females of each size class multiplied by the energy output for females of this size, as read off from the graph.

Unfortunately, the reproductive output of the males cannot be determined in this way, and although it is often neglected in energy flow studies, the losses can be considerable. They can sometimes be determined by the method described in the next section.

Indirect method

The energy present in the tissue of the gravid male or female before spawning must be almost equal to the energy of the spent individual plus the energy of the reproductive products. If the biomass of fully ripe males and females can be determined in relation to a characteristic length and the biomass of just spent individuals similarly determined, the difference will represent the reproductive output. When individuals in the population breed closely in synchrony, the loss of biomass can be deduced from the curve obtained by plotting against time, for each sex, the average values of individual biomass (preferably in energy units) for each age class or size class. The curves should show a discontinuity between the beginning and the end of the breeding season equal to the loss of energy at reproduction.

The weight lost during breeding is sometimes expressed in the form of changes in the ratio of weight to the cube of length and referred to as 'coefficient of condition' or 'condition factor'. As an alternative to measuring changes in the weight or energy of the whole animal, ripe and spent ovaries (or *vesiculae seminales*) can be dissected out and weighed with reference to the size of the animal from which they were obtained.

Continuously breeding species

Species which breed throughout, or for a large part of, the year will not necessarily show any discontinuity in the curve of growth or in their condition factor resulting from any sudden loss of reproductive material. Frequently, a steady state exists within the animal, the loss of weight through the shedding of reproductive products being continuously made good by assimilation. Sometimes, when assimilation cannot keep pace with breeding, the animal may gradually become spent and, in consequence, a gentle discontinuity may appear on the growth curve and a slight change be recorded in the condition factor. However, this change does not represent the whole reproductive output but only the lack of balance between assimilation and reproduction. If the indirect method, as described above, is applied in such situations it may grossly underestimate the reproductive output.

If the reproductive products are in a form that can be collected and weighed it is possible to adapt the direct method to a continuously breeding species. When the reproductive products are in the form of egg capsules, egg strings or egg masses produced at regular intervals of time, it may be possible to relate the dry weight of the egg mass produced to the size of the parent. It is then only necessary to record brood frequencies at different seasons and integrate over the year. The annual reproductive output can then be obtained from the frequency of females belonging to the post-maturity size classes, the size specific brood weights for each class of females and the total number of broods per year.

Brood frequencies may be established in various ways:

1. When the breeding cycle is synchronous and rhythmic, frequent monitoring of the state of the gonad of field samples will enable brood frequencies and the proportion of individuals breeding at each season of the year to be established. Rhythmic breeding cycles in intertidal and shallow water species are often synchronized to lunar or semi-lunar periods, and if this can be established it may assist in estimating the number of broods.

2. Where the female retains developing eggs until the next brood is due, and where the time between egg laying and hatching (brood period) can be determined by observations on ovigerous females kept in aquaria, the number of broods in the field population can be determined by sampling the population at regular intervals for the proportion of ovigerous females. The frequency of ovigerous females should then be plotted against time and integrated over the year or appropriate part of the year. Total number of broods in the population can be obtained by dividing this integral by the brood period. If there is a considerable variation in brood period at different seasons, measure the number of broods during each season using the appropriatevalue of brood period.
3. Occasionally, direct observation of brood succession in known individuals is possible in the field (e.g. Crisp & Davies, 1955). If the planktonic eggs are sufficiently large, as in some species of *Littorina*, it may be feasible to strain them from the water with fine plankton netting and count the numbers produced daily in the laboratory. However, the criticism cannot be avoided that the rate of shedding of the eggs may not be characteristic of that of the animals in the field. The laboratory population, being kept under unnatural conditions, may become out of condition, though this can be minimized by frequently renewing the animals kept in captivity. On the other hand, abnormal increases in the rate of egg laying may be induced as a result of bringing the animals into the laboratory through shock or changes in the environment.

It is usually necessary, as in the case of brooding species, to establish the approximate relation between rate of egg laying and size; to obtain the size frequency structure; and to assess the proportion of females breeding in the field, in order to compute the total output by the natural population.

Where viviparous young are produced over a long period of the year, but are sufficiently large to be collected and weighed the method outlined in this paragraph can also be applied (e.g. Heywood & Edwards, 1962).

The measurement of brood output in animals with planktonic eggs that are not readily collected and weighed and in which reproduction occurs steadily over a long period of time cannot at present be achieved. The same applies to the loss of sperm and seminal fluid from the males of such species. No information can be obtained from the size of the ovary or testes because eggs and sperm are being matured and shed continuously and the output cannot be related to the biomass of the reproductive organs. No firm recommendation can be put forward for measuring the reproductive output of organisms with this type of breeding since no satisfactory study on these lines has yet been carried out. Unfortunately, such examples are likely to be very commonly encountered among warm temperate and tropical benthic species.

Other losses of energy

Largely on account of the difficulty in their measurement, losses of energy in forms other than reproduction and respiration have largely been ignored by ecologists. Where the nature of the nitrogenous excretory products are known, it should not be difficult to measure the rate of nitrogen excretion in the laboratory, but it is generally assumed that losses in the form of low energy nitrogen compounds are unimportant.

This is certainly not true of organic secretions. Teal (1957) claimed that about 60 % of the total energy loss of the planarian *Phagocata* was in the form of mucus, while McCance & Masters (1937) found that the mollusc *Archidoris* secreted its own weight of mucus (wet weight) in 5 hours. Kuenzler (1961) determined that the amount of energy lost as byssus threads by trapped specimens of *Modiolus demissus* amounted to 2–3 % of the total energy assimilated and about half the losses through reproductive effort.

Other forms of energy loss that need to be taken into account in a full energy budget are exuviae of crustaceans, the tests of Larvacea, and the tube building materials of polychaetes. No general method can be put forward.

Ingestion and egestion

Benthic communities consist mainly of filter feeders and detritus feeders, macro feeders being in a minority. Assimilation rates are difficult to measure directly in both of the predominant groups and there are few satisfactory studies of energy flow in typically benthic species.

With few exceptions, rates of ingestion and egestion can be measured only in the laboratory. There are, in principle, two types of methods available:

1. *Direct method.* The food ingested and the faeces produced by the animal over a sufficiently long period of time to establish continuity are measured directly.
2. *Indirect method.* The concentration of an indigestible marker present in, or artificially incorporated into the food, is measured in both food and faeces, together with the rate of faecal production.

If w'_C is the dry weight of food ingested in unit time including the weight of marker and w'_F the weight of faeces egested in unit time, ϕ_C the weight fraction of the marker in the food and ϕ_F the weight fraction of the marker in the faeces, it follows that, since the marker is conserved:

$$w'_C \phi_C = w'_F \phi_F$$

if w_C is the dry weight of food alone consumed in unit time and w_F the dry weight of egestion of faeces alone in unit time then

$$w_C = w'_C(1 - \phi_C) \qquad w_F = w'_F(1 - \phi_F)\,.$$

It follows that the rate of consumption of food alone can be estimated by the relation:

$$w_C = \frac{w'_F \phi_F (1 - \phi_C)}{\phi_C}\,.$$

The rate of egestion is given by the formula:

$$w_F = w'_F(1 - \phi_F)\,.$$

The assimilation efficiency is given by the ratio:

$$\frac{\phi_F - \phi_C}{\phi_F(1 - \phi_C)}\,.$$

It will be noted that the assimilation efficiency can be obtained without measuring either ingestion or egestion rate.

This method has obvious advantages when, as is often the case, it is more difficult to collect and measure by difference the food consumption than to collect and weigh the faeces. The simplest marker is the organic ash content of the food itself, although the method is very inaccurate and perhaps unreliable if the amount of ash present is small. Conover (1966) has employed this method for measuring the ingestion rates of copepods.

Since the method assumes that all the ash absorbed in the food is present in the faeces it is essential to be sure that this is in fact the case; calcareous skeletal material ingested by carnivores might be absorbed in the gut and not appear in the faeces. Indeed, the same may apply to the ash component of the food of any organism. Soluble salts, which constitute much of the ash of some soft bodied animals, might be absorbed and later excreted through the skin, gills or kidney of the predator. Artificial markers added to the food are likely to be more reliable. Chromic oxide is used in such studies: it is not absorbed, it is not normally present in significant quantity in the food, and it can be accurately determined chemically.

Consumption by filter feeders

It is not difficult to measure the rates of filtration of filter feeding organisms by observing the rate of removal of small particles from fine suspensions. The subject was reviewed by Jørgensen (1966). Most filter feeders were believed to clear a constant volume of water in unit time independently of the concentration of food particles, provided that it was not too high. Consequently, to obtain the ingestion rate the filtration rate must be multiplied by the dry weight or energy per unit volume of the suspension. This can be done by separating the cells from a known volume of suspension by centrifugation or filtration, and obtaining the ash-free dry weight and the caloric content. The filtration rate is measured by following the concentration of food particles in the water containing the organism over a period of time. If C_0 is the original concentration, C_t the concentration after time t, V the volume of the water in millilitres, w the weight of active filter feeding organisms, the filtration rate per unit weight of the animal is given by:

$$F = \frac{0{\cdot}43(\log_{10} C_0 - \log_{10} C_t)V}{w \, . \, t} \text{ ml/unit time}$$

Care must be taken to maintain the cell suspension in a uniform state by stirring (Hildredth & Crisp, 1976), but the stirring must be sufficiently gentle to prevent faeces and pseudofaeces from being resuspended in the experimental vessel. It is often an advantage to place the animals on a gauze mesh in order to allow the faeces to drop through. Allowance must also be made for cell multiplication or cell sedimentation during the experiment. A satisfactory method of control is to use another suspension of the same concentration in a similar dish treated in the same way but without animals present. The concentration in the control suspension can be substituted in the above formula for C_0.

When a number of readings are taken at intervals of time, values of $\log_{10}(C_0/C_t)$ should be plotted against time, a straight line being fitted to the values either by eye or by the least squares method. The slope of this line, S,

$$S = \mathrm{d}\log_{10}\left[\frac{C_0}{C_t}\right] \Big/ \mathrm{d}t$$

can then be substituted in the above equation for

$$(\log_{10} C_0 - \log_{10} C_t)/t$$

The goodness-of-fit of the points to the line is an indication of the extent to which the animals have been filtering steadily and independently of cell concentration; if the line is markedly curved, the experiment should be continued until a straight line results, or be abandoned. Sometimes one or more of the individuals in the experiment has the siphons closed for the whole or part of the time. Allowance should be made for this in the calculation, by adjusting w to include active animals only. The animals must be well spaced out in an ample volume of suspension, otherwise they may filter each other's effluent, rather than the suspension.

Rates of filtration per unit weight are likely to vary with size of the animal, temperature and other environmental conditions, e.g. movement of water over the animal or the presence of food material in the water.

The concentration of particles in the suspension, usually in the form of microalgal cells from monoculture, can be measured by counting them visually in a haemocytometer cell of known area and thickness. The Coulter counter, which counts electronically the change of conductivity when each cell passes through a narrow orifice, has been employed for this purpose. In using this instrument there is always the danger that mucus secreted by the animals or by the algae may upset its operation by causing the cells to clump and block the orifice.

Instead of counting cells individually, their property of light absorption may be used to measure routinely the concentration of a suspension. It will be noted that the equation for filtration rates relies on the change in the ratio of the initial to final concentrations of the particles in the suspension. Therefore any quantity proportional to concentration can be used in place of concentration. The optical density of cell suspensions is commonly used, but a linear relation should first be shown to hold good by measuring the light absorption of a series of dilutions of the suspension. Any reliable light absorptiometer is suitable for such measurements, but an instrument which incorporates a cell with a long optical path will enable a range of more dilute suspensions to be used, corresponding more closely to cell concentrations in nature.

Somewhat different methods should be used in the case of filter feeders that utilize larger organisms such as micro-crustacea as their food. Such organisms can be counted individually using a photoelectric counter of the type described by Mitson (1963). A suitable volume of sea water containing micro-crustacea is put through the apparatus and counted. It is then poured into the aquarium containing the filter feeding organisms. Many micro-crustacea, such as the nauplius larvae of *Artemia salina* which are commonly used for feeding, react to light and the water in the aquarium must therefore be continuously stirred to ensure that the food is evenly distributed. After the time allowed for feeding, the water containing food organisms is again poured through the photoelectric counter and the number of organisms ingested can be found. The dry weight and caloric content of the micro-crustaceans can be measured by applying the usual methods to a large mass of individuals strained off on plankton silk after having been counted with this apparatus.

The pigment present in the larvae may sometimes be used to measure rates of ingestion. Ritz & Crisp (1970) applied a light absorption method for measuring the uptake of micro-crustacea by barnacles, the pigment voided in the faeces being used as a marker, taking into account that only half the pigment ingested is normally voided. The photoelectric counting apparatus was used to correlate the optical density of the pigment extracted from the faeces with the number of *Artemia* nauplii consumed during the feeding period.

In making suspension rate measurements, it should be borne in mind that many filter feeders have mechanisms for rejecting part of the food, with the result that when the suspension is too concentrated for their needs, or contains unwanted material, part of the food available is not consumed. Under natural conditions, therefore, when the animal is presented with a variety of suspended particles, some more nutritious than others, the particles actually ingested may have a much higher energy content than that of the remaining particles left in the suspension, some of the less nutritious material being removed in the pseudofaeces. Similarly some suspension feeders which utilize living plants and animals damage part of their prey in addition to that which they ingest. Such behaviour complicates the measurement of ingestion rates, and poses the question whether one should regard as 'consumption' the total loss of food organisms to the predator or only that part of the suspension which ultimately passes through the oesophagus.

In view of such difficulties in measuring the actual rate of ingestion by filter feeders some authors have used the summation equation of energy flow (equations 9.1 and 9.2, p. 290) to determine consumption or absorption by adding together all the forms of energy utilization. (For example, Kuenzler, 1961, on *Modiolus demissus* and Jørgensen, 1962, in measuring the assimilation by larval and juvenile bivalves.)

Recent studies on the filtration rates of adult and larval bivalves indicate that rates vary considerably with the density and constitution of suspended solids, thus greatly complicating estimates of consumption (see Bayne, 1976, Chapter 7).

Consumption by detritus feeders

The difficulties of measuring the rate of food intake by detritus feeders directly has led investigators to use the indirect method given above. Heywood & Edwards (1962) measured the rate of faecal production by the fresh water snail *Potamopyrgus jenkinsi* and the difference in the amount of organic carbon present in the mud and faeces, thereby allowing rates of ingestion and absorption to be calculated. The population consisted of mature animals, assumed to be non-growing, and absorption was of the same order as the sum of respiration and reproductive loss. Odum & Smalley (1959) investigating the periwinkle, *Littorina irrorata*, and Teal (1962) investigating the fiddler crab, *Uca*, adopted a similar approach.

Consumption by macro feeders

Many carnivores can be kept in the laboratory and fed quantitatively on prey organisms. Carefoot (1967a,b) fed carnivorous nudibranchs *Dendronotus frondosus* and *Archidoris britannica* on polyps of *Tubularia* and pieces of the sponge *Halichondria*, respectively. The large herbivore *Aplysia punctata* was fed on various seaweeds. The direct method was applied by providing the animals with a known weight of food and weighing the remains of the food and the faeces after each day's feeding. The feeding of *Archidoris* caused the remains of the food to be badly fragmented and therefore the indirect method was applied, using the siliceous spicules as a food marker. Silica was estimated chemically in the food and faeces. Teal (1957) fed the flat worm *Phagocata* on whole oligochaetes.

A particularly ingenious example of the measurement of food uptake in the field is described by Paine (1965). The large ophisthobranch *Navanax inermis* swallows its prey, consisting of other molluscs, whole. The remains of the shells of the prey in the gut of recently caught *Navanax* were recognized and measured. By this means the energy content of each item of prey could be calculated. The rate of ingestion was calculated from the energy derived from an examination of gut contents, divided by the rate of gut clearance observed in aquaria.

Energy losses by respiration

In the first edition of this Handbook, Crisp (1971) drew attention to the rigor of measuring the loss of metabolic energy by direct calorimetry rather than by respirometry. However, at that time, oxygen uptake measurements were recommended on account of their greater experimental simplicity. At the same time it was pointed out that the conversion of oxygen uptake into heat loss by the appropriate aerobic oxycalorific coefficient of approximately 4·8 cal (ml O_2 at n.p.t.)$^{-1}$ was based on assumptions that might not hold true.

Recent advances have fully justified this warning. Gnaiger (1981) and Hammen (1979, 1980) have shown, by direct measurements of heat output in animals such as bivalves, polychaetes and salmonid eggs, that anaerobic processes contribute to the total heat production both during hypoxia and in aerobic states. By measuring heat production and oxygen uptake simultaneously under steady state conditions, one can obtain not only thermodynamically valid data on metabolic energy loss, but also further information on the biochemical processes taking place.

Although commercially available apparatus has been used for heat output measurements (see Hammen, 1980), it is probably too early to recommend equipment for routine use. However, those studying energy flow in ecological systems should keep in close touch with these developments and interpret any results obtained by direct respirometry with caution. Respiration measurements made on bivalves, when converted into heat, may account for only two-thirds or so of the total metabolic energy loss. In the accompanying figure

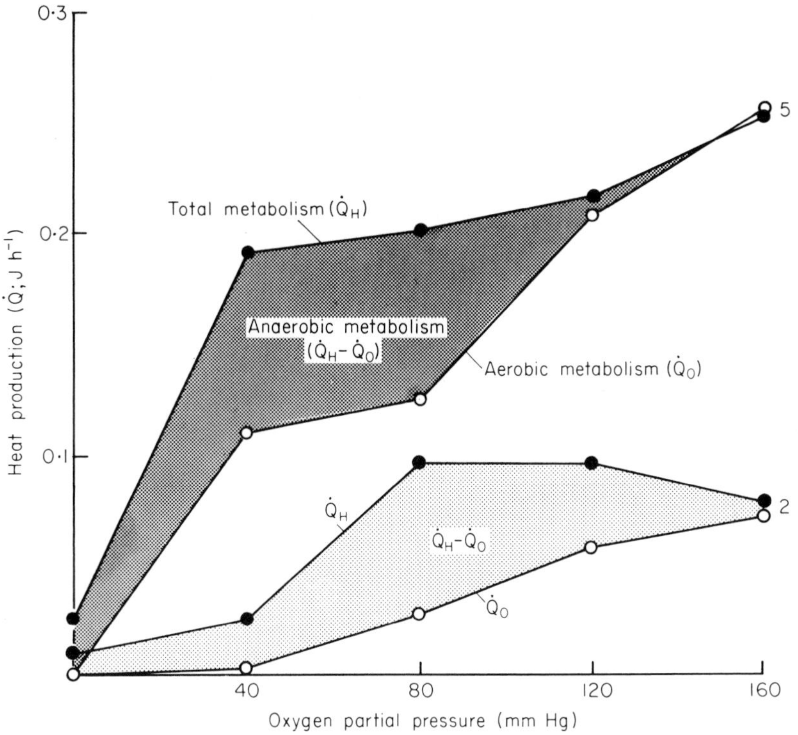

Fig. 9.11. Relationship between heat production ($\dot{Q}$; J h^{-1}) and oxygen partial pressure (P_{O_2}; mmHg) for total metabolism ($\dot{Q}_H$), aerobic metabolism ($\dot{Q}_O$) and anaerobic metabolism ($\dot{Q}_H - \dot{Q}_O$), of two mussels (2 and 5). Aerobic heat production is calculated from the oxygen consumption rate ($\dot{V}_{O_2}$; μmol/h) multiplied by the oxycalorific equivalent of glucose ($\Delta H/\Delta O = 0{\cdot}47$ J/μmol). Each point represents the integrated heat productions over approximately 20 h. (Temperature = 20–25 °C; Salinity = 18·4‰). (From Famme *et al.*, 1981.)

(9.11) taken from Famme *et al.* (1981), it can be seen that in two specimens of *Mytilus edulis* the additional heat from anaerobic metabolism starts low, increases with oxygen partial pressure up to half saturation, and then falls to a low value again at a partial pressure of oxygen corresponding to normoxia. By the same methodology, Schick (1981) noted environmental effects in sea anemones. Intertidal anemones respired aerobically but sub-tidal anemones of the same species (*Actinia equina* and *Anthopleura eligantissima*) produced a significant amount of heat anaerobically. Unless the total heat output is measured, there is no possibility of obtaining a correct energy balance.

Table 9.9 gives the oxy-calorific coefficients for converting oxygen uptake

Table 9.9. Oxycalorific equivalents. (After Elliott & Davidson, 1975.)

Compound	End products	Food		Excreta	
		cal (ml O_2 at n.t.p.)$^{-1}$	cal (mg O_2)$^{-1}$	cal (ml O_2 at n.t.p.)$^{-1}$	cal (mg O_2)$^{-1}$
Carbohydrate	CO_2, H_2O	5·04	3·53	0	0
Fat	CO_2, H_2O	4·69	3·28	0	0
Protein	NH_3	4·57	3·20	0·89	0·62
Protein	Urea	4·64	3·25	0·83	0·58
Protein	Uric acid	4·64	3·25	1·34	0·94
General		4·83	3·38	—	—

To convert cal to J, multiply by 4·185.

into energy for the three main types of nutrient. In view of the other errors involved in the measurement of energy losses through respiration, the assumption that 1 ml of oxygen at n.t.p. is equivalent to 4·8 cal, or 29 J, is sufficiently close for all practical purposes. It is equivalent to Ivlev's (1934) value expressed in alternative units of 3·34 cal, or 14 J, per mg of oxygen (1 mg of oxygen at n.t.p. occupies exactly 0·7 ml).

Some authors have converted oxygen uptake into the dry weight of food utilized, but this is not generally to be recommended. Although oxy-calorific energy equivalents are similar, the weights of nutrient corresponding to 1 ml of oxygen consumed differ greatly from one food to another. Thus if the food source contained a high proportionof fat, and had caloric content of 7·2 kcal/g dry weight (see Table 9.12) 1 ml of oxygen respired would be equivalent to only $4{\cdot}6/7{\cdot}2 \times 10^3$ g $= 0{\cdot}65$ mg dry weight of food. On the other hand, if the food were mainly of protein and carbohydrate, with a caloric content of only 4·5 kcal/g dry weight, the dry weight equivalent of 1 ml of oxygen would be approximately 1·1 mg. Therefore, unless the food source and its chemical constitution are known, errors of $\geq 50\%$ might be committed in the conversion of oxygen uptake into dry weight of food consumed.

Even the conversion of gas exchange to energy is based on two assumptions that may not always be true and are often overlooked. First, it is assumed that all the energy released when oxygen is consumed is converted into heat or work. Should some of the energy released by the oxidation of metabolites be re-utilized to form high energy chemical bonds and the products so formed stored in the animal, the energy actually released as heat at the time will be less than the quantity calculated from the product of oxygen uptake and the normal oxy-calorific equivalent.

Secondly, as described above (pp. 344–5) if part of the energy were released through anaerobic processes, the oxygen uptake would give too low a value of energy loss. Although anaerobic respiration would be inefficient and therefore less likely to occur among animals living in fully aerobic environments, it is common among micro-organisms living in deoxygenated habitats, and evidently some benthic metazoans inhabiting low oxygen environments display facultative anaerobiosis.

All laboratory methods of measuring respiration suffer from the defect that the animal is, to a greater or lesser degree, constrained in an unnatural situation during measurements. The respiration observed may in consequence be quite different from that which takes place in the natural environment, particularly in the case of active animals. In an attempt to correct this, Mann (1965) and Carefoot (1967b) found it necessary to double the observed respiration loss from laboratory experiments in order to balance the energy budget. This is an arbitrary procedure not to be generally recommended, but it indicates the present unsatisfactory state of means of measuring this large and important item in the energy flow equation. The errors are unlikely to reside in the methods of measuring oxygen uptake itself and, indeed, effort devoted to increasing the accuracy of respiration measurement in the laboratory is probably misdirected. All recognized methods suitable for the organism are likely to be sufficiently accurate. The choice of method should, therefore, rest mainly on the likelihood that the animal will be able to behave naturally in a respiration chamber.

The main effect of bringing animals into the laboratory is to alter their activity; activity changes may result in several-fold changes in metabolic rate (e.g. Newell & Northcroft, 1967). Thus Mann (1965) found that when fish are captured there is an initial burst of activity followed by a fall, especially when starved, with peaks at dawn and dusk. Many marine organisms, such as mussels and barnacles, display 'metabolic economy' when kept in the laboratory without food.

Furthermore, respiration may vary diurnally, especially in those animals that display overt rhythms, while the illumination and water movement to which the animal is exposed can change respiratory rates significantly. Whether the animal is in air or in water, and whether it is stimulated by the presence of food odours can also profoundly alter its activity (M. Crisp *et al.*, 1978).

In laboratory experiments the animals are usually unstimulated and so maintained in a state of activity that is fairly constant and approaches that of 'basal metabolism'. Under these conditions, two variables influence uptake and should be taken into account in making predictions. These are the effect of temperature, and the effect of body weight.

If measurements are made on animals of various sizes over a range of temperature and the logarithm of rate of uptake of oxygen, q, or better, the

Table 9.10. Computation of energy loss by respiration for a population.
A. Data for oxygen uptake in relation to dry weight. $\theta = 15\,°C$; $\bar{w}$ (dry wt.) = $0{\cdot}15 \times \bar{w}'$ (wet wt.); Q (cal) $= 4{\cdot}8 \times q$ (ml oxygen absorbed).

Mean wet wt., $\bar{w}'$ (g)	Mean oxygen uptake, q (ml h^{-1})	Mean dry wt., $\bar{w} = 0{\cdot}15\,\bar{w}'$ (g)	Mean energy loss, $Q = 4{\cdot}8\,q$ (cal h^{-1})	$\log_{10}\bar{w}$	$\log_{10}Q$
0·56	0·029	0·084	0·140	$\bar{2}{\cdot}92$	$\bar{1}{\cdot}15$
0·79	0·045	0·119	0·216	$\bar{1}{\cdot}08$	$\bar{1}{\cdot}33$
1·58	0·066	0·236	0·316	$\bar{1}{\cdot}37$	$\bar{1}{\cdot}50$
2·50	0·105	0·375	0·504	$\bar{1}{\cdot}57$	$\bar{1}{\cdot}70$
4·50	0·166	0·675	0·798	$\bar{1}{\cdot}76$	$\bar{1}{\cdot}90$

Equation to line (Fig. 9.12a) $\log_{10}Q = \log_{10}A_\theta + b\,\log_{10}\bar{w}$

$$\log_{10}A_\theta = 0{\cdot}02,\ b = 0{\cdot}8,\ \therefore\ Q = 1{\cdot}05\,\bar{w}^{0{\cdot}8}$$

B. Data from similar experiments at $\theta = 20°$, $10°$ and $5\,°C$ give

θ (°C)	$\log_{10}A$	b
20	0·21	0·78
15	0·02	0·80
10	$\bar{1}{\cdot}85$	0·81
5	$\bar{1}{\cdot}61$	0·79

Values of b all $= 0{\cdot}80$ within experimental error.
Figure 9.12c shows $\log_{10}A$ plotted against θ.

C. Computation of losses during the period June 21–July 31:
Time $\Delta t = 41$ days $= 984$ h.
Temperature range $= 15{\cdot}6 - 17{\cdot}8$, mean temperature $\bar{\theta} = 16{\cdot}5\,°C$.
log A for $\theta = 16{\cdot}5\,°C = 0{\cdot}07$ (Fig. 9.12c) $A = 1{\cdot}17$.
Population size distribution:

No. of size class	Size range (dry wt.) (g)	Mean dry wt., $\bar{w}_i$ (g)	$\bar{w}_i^{0{\cdot}8}$	Frequency of size class f_i (m^{-2})	$f_i\bar{w}_i^{0{\cdot}8}$
1	0·0–0·2	0·13	0·195	80	15·6
2	0·2–0·4	0·30	0·38	122	46·4
3	0·4–0·6	0·52	0·59	96	56·7
4	0·6+	0·78	0·81	12	9·7
				$\sum_0^4 f_i\bar{w}_i^{0{\cdot}8}$	= 128·4

$$A_j\Delta t_j \sum_0^4 f_i\bar{w}_i^{0{\cdot}8} = 1{\cdot}17 \times 984 \times 128{\cdot}4 = 1{\cdot}48 \times 10^5\ \mathrm{cal\,m^{-2}} = 148\ \mathrm{kcal\,m^{-2}}.$$

Table 9.10—*contd.*

D. Sum of respiration losses for the year $R = \sum_{0}^{m} A_j \Delta t_j \left[\sum_{0}^{n} f_i \bar{w}_i^{0\cdot 8} \right]$

Dates of time interval, Δt	Δt (d)	Δt (h)	Mean temperature $\bar{\theta}$(°C)	$\log_{10}A$ (Fig. 9.12c)	A	$\sum_{0}^{n} f_i \bar{w}_i^{0\cdot 8}$	$\Delta t_j A_j \sum f_i \bar{w}_i^{0\cdot 8}$
June 21–July 31	41	984	16·5	0·07	1·17	128·4	148
Aug. 1–Sept. 19	50	1200	14·2	$\bar{1}$·97	0·93	116·6	129
Sept. 20–Oct. 20	31	744	11·0	$\bar{1}$·86	0·72	115·3	62
Oct. 21–Nov. 30	41	984	7·6	$\bar{1}$·73	0·54	107·2	57
Dec. 1–Mar. 3	93	2232	5·2	$\bar{1}$·64	0·44	98·2	96
Mar. 4–April 10	38	912	7·0	$\bar{1}$·705	0·51	90·6	42
April 11–May 10	31	744	10·3	$\bar{1}$·835	0·68	152·3	77
May 11–June 20	40	960	13·4	$\bar{1}$·92	0·83	138·0	110
Total	365	8760		Total annual energy loss 721 kcal m^{-2}			

logarithm of the rate of energy loss, Q, is plotted against the logarithm of the weight, w, at each temperature, a linear relation is usually found. Table 9.10A gives an example of such data in which wet weight is converted to dry weight, and oxygen uptake to energy loss. Figure 9.12b shows the fitting of the equation $Q = A\bar{w}^b$. The value of the constant A is dependent on temperature, but the exponent b is usually constant with a value of 0·7–0·8. The assessment of respiration loss is not very sensitive to small changes in b.

If $\log_{10}$ of A is plotted against temperature, θ, a straight line can usually be drawn; this helps in evaluating A at different temperatures. An example is given in Table 9.10B and Fig. 9.12c.

To apply such results to estimate the energy loss by respiration in a field population, the population must first be divided into n weight classes, and the frequency f_i of each determined. Age classes or length classes can be substituted for weight classes if the mean weight $\bar{w}_i$ of each class can be found and if the range of weights in each class is not large. When the respiratory energy loss is required over a long period, such as a year, the summation must be divided into m separate periods during which there is no great change in temperature or population structure. For each such period of interval Δt_j the mean temperature $\bar{\theta}$ must be found and the value of A_j obtained by interpolation of $\log_{10}A$ on θ (using a graph such as Fig. 9.12c). For each size group within each period the values of f_i are each multiplied by $\bar{w}_i^b$ and summed as shown in Table 9.10C to obtain:

$$\sum_{0}^{n} f_i \bar{w}_i^b$$

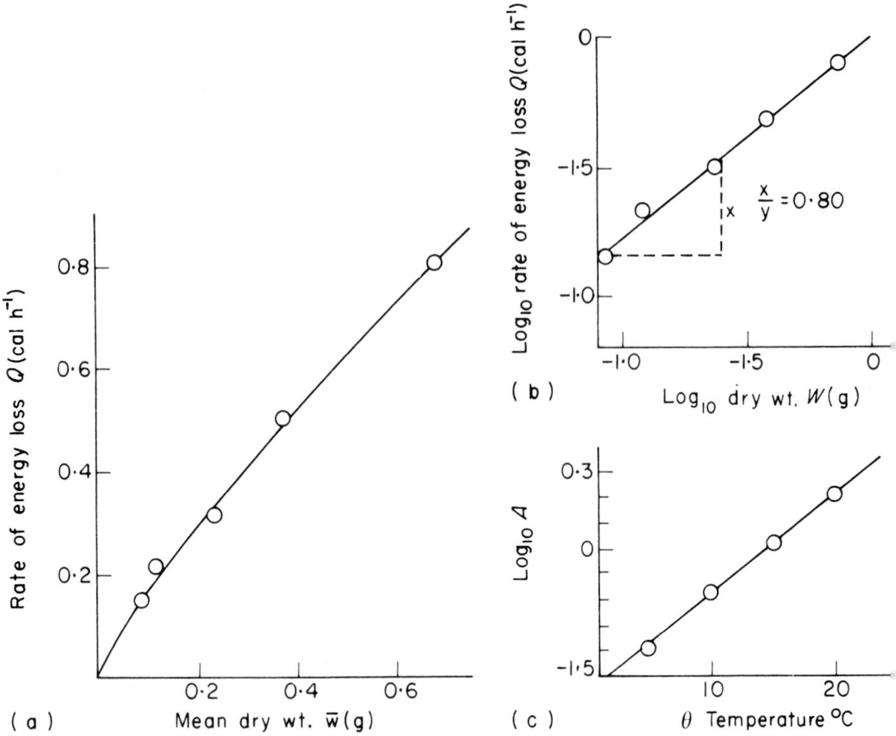

Fig. 9.12. (a) Respiratory energy loss, Q, as a function of dry weight, $\bar{w}$. (Data from Table 9.10a.) (b) Logarithmic plot of energy loss on dry weight at 15°C to obtain the equation $Q = A\bar{w}^b$ where $A = 1{\cdot}05$ and $b = 0{\cdot}80$. (c) Plot of $\log_{10}A$ against temperature θ for estimating A at any given temperature.

Then each summation must be multiplied by $A_j \Delta t_j$, appropriate to the temperature and time interval of the period, as shown in Table 9.10D. The same units of time in which A was determined (here A is expressed as cal $h^{-1}\,g^{-0{\cdot}8}$) must be used for Δt (h). The resulting summation represents the estimate of annual energy loss by respiration.

$$R = \sum_{0}^{m} A_j \Delta t_j \left(\sum_{0}^{n} f_i \bar{w}_i^b \right)$$

Methods of measuring respiration in the laboratory are given in a later section.

Measurement of oxygen uptake in the field

Bottom deposits contain large numbers of smaller metazoa, protozoa and bacteria which together form an important factor in the benthic ecosystem. To enumerate and measure the respiratory activity of each species in turn would be impossible. Nor can the deposit be studied in the laboratory without disturbing its structure, and this in turn would change the diffusion gradients and oxygen levels throughout, profoundly altering those trophic changes that are diffusion limited. Moreover, the contribution of micro-organisms to the energy flow of benthic deposits is likely to be considerably greater than that of the macro-fauna. *In situ* methods are therefore imperative.

Teal (1957) and Odum (1957a, b) have attempted such measurements by enclosing small areas of freshwater benthos in light and dark glass cylinders, funnels and bell jars and measuring changes in oxygen content of the water layers immediately above the bottom. Pamatmat (1966, 1968) inserted bell jars with plastic stirrers into marine deposits and sampled at intervals through vaccine type stoppers. The rate of oxygen uptake in the dark reported by Teal was equivalent to a loss of energy of the order of 20–50 kcal m^{-2} per month or an absorption of 4–10 mg m^{-2} h^{-1} of oxygen, Pamatmat obtained somewhat larger values of 25–30 mg oxygen m^{-2} h^{-1}. Such rates of uptake are sufficient to deplete a layer of water 2 cm deep of oxygen in about 4 h. Hence measurable changes within a single tidal cycle are possible if only a few centimetres depth of water are enclosed in the chamber.

A more elegant method which overcomes the abnormally static conditions caused by enclosing the overlying water has been described by Pamatmat (1965). A perspex channel 100 cm long, 2 cm wide and 1 cm deep was placed over the substratum and oxygen levels measured at each end while a slow flow of seawater passed through the channel. The author claimed that the flow method agreed with the bell jar method, but the results were not sufficiently precise to allow a critical comparison of respiration rates with and without water flow.

The reservation made on p. 344 concerning the use, under inappropriate circumstances, of the oxy-calorific equivalent for aerobic conditions to convert oxygen uptake into energy loss, applies with special force to *in situ* measurements of benthic deposits. These deposits contain a high proportion of micro-organisms living under conditions of depleted oxygenation. This whole subject requires much pioneer research work and is quite unsuited to any routine approach.

Laboratory methods for productivity measurements

Estimation of ash-free dry weight

Three main methods for removing water are available:

1. Heating in an oven to about 100 °C under atmospheric pressure.

2. Drying *in vacuo* at lower temperatures up to 60 °C.
3. Freeze-drying.

When the sole objective of drying is to obtain a measure of the water content, and when the dry tissue is not needed for any other purpose, all three methods would give similar results, provided that sufficient time is allowed for complete drying. Larger organisms obviously require longer to dry to constant weight, and therefore a test should always be made to ascertain the minimum time for complete drying before a routine series of measurements is undertaken.

Freeze-drying is slow, requires special apparatus, and up to 5% of the water in the tissue is very difficult to remove by this method. Oven heating at 100 °C may cause losses of the volatile tissue components, and if the temperature during the drying rises to 110 °C a loss of volatile components of soft tissues of up to 10% by weight in 48 h may result (Giese, 1967). The disadvantage of gentle drying at room temperature is that the tissues may slowly autolyse.

Drying *in vacuo* at 60 °C, a temperature which inactivates most enzymes yet does not cause serious loss through volatilization, offers a good compromise and the dry material so obtained is usually suitable for bomb calorimetry. It is very important, in carrying out a seasonal survey, to ensure that every drying operation is carried out under identical conditions. Otherwise one set of samples may experience a greater loss of weight which may be erroneously interpreted as a seasonal effect, such as a change in condition factor.

When the objective of drying is to preserve material for further biochemical study, much more careful attention must be given to the choice of method. In general the gentlest method, freeze-drying, or vacuum desiccation of initially frozen tissue over sulphuric acid or calcium chloride in the cold is to be recommended. Some workers add 0·1 ml of 10% trichloroacetic acid solution to each gram wet weight of tissue prior to drying, in order to inactivate the enzymes. If gentle drying fails to remove all the water, a small aliquot of the partially dry tissue can be sacrificed for complete removal of water by a more vigorous method to ascertain the residual proportion of water by weight.

Dried materials awaiting further treatment are best stored sealed in a deep freeze or, if kept at room temperature, under continuous drying in a desiccator.

After drying, the ash content, consisting of such substances as sea-water salts, silica, calcium phosphate and carbonate, is determined by burning off the organic matter in a muffle furnace. Four to six hours at 500–600 °C is recommended to ensure the complete destruction of all organic matter.

The main error in determining ash is caused by partial volatilization of components of the inorganic fraction, notably loss of carbon dioxide from carbonates and chemically bound water from silicates. If the total ash weight or the calcium carbonate content of the ash is known to be small, these errors may be neglected. If the carbonate content is large and the ash weight constitutes a considerable fraction of the total dry weight, as in molluscs, barnacles, echinoderms etc. the loss of weight on ignition through removal of carbon dioxide could in theory rise to 44% of the ash weight of the calcium carbonate present, resulting in a serious under-estimate of the ash. A different procedure in these cases must therefore be adopted.

Where it is practicable to remove the soft tissues from the shell without loss, as in bivalve molluscs and some gastropods, the dry organic weight of the body and of the shell can be measured separately. The former presents no difficulty. The dry ash-free weight of the shell can be found by decalcifying it in dilute hydrochloric acid, but the vigour of the effervescence may dissipate the delicate particles of shell tissue. To obviate this, calcium chelating agents in mildly acid media may be substituted for dilute acid; for example, a mixture of ethylene diamine tetracetic acid and sodium formate. The residual organic matter of the shell can be filtered off, dried and weighed on glass fibre filter paper, the tare weight of which has been previously determined.

Where the shell or skeleton cannot be mechanically separated from the tissues, for example where it is in the form of a number of small plates as in barnacles and echinoderms, it can often be separated by chemical cleaning with attendant destruction of the soft tissues. The whole animal, after being dried and weighed, is boiled in 10% by weight aqueous caustic alkali for up to one hour, a procedure that destroys all the organic tissue except for a small amount of refractory material such as slips of chitin, threads of keratin etc. which are usually of negligible weight. The cleaned calcium carbonate plates or ossicles should then be well washed free of caustic alkali and of any loosely attached organic material, dried and weighed. This procedure may not remove all the organic matter from the interior of the shell, but the organic weight compared with that of the inorganic shell is usually negligible. Similar sources of error in measuring ash weight may be encountered in animals containing skeletons of hydrated silica which lose water at high temperature. This problem is discussed by Paine (1964).

When the calcium carbonate cannot be separated by either of the above methods, for example when measuring the ash weight of a mud or sand with a high proportion of shell, the dried material should first be weighed, then treated with hydrochloric acid until effervescence has completely ceased in order to remove carbonate, washed, dried and re-weighed to obtain the weight of carbonate removed. It can then be ashed at 600 °C for 6 h to obtain the remaining ash in the usual way. A small amount of organic material may be washed out in the removal of carbonate; therefore this procedure should be adopted only if necessary because of the large amount of carbonate present.

Measurement of biomass in terms of nitrogen content

Some authors have used nitrogen content as a measure of biomass (p. 288). The classical method of measurement of nitrogen is the Kjeldahl, in which the tissue is digested in concentrated sulphuric acid containing a catalyst; copper sulphate, selenium oxide and mercuric oxides are used. After digestion, at boiling point or higher in a sealed tube, excess of strong alkali is added and the ammonia released is driven off either by passing steam through a condenser unit (Markham, 1942) or in a small Conway distillation cell at room temperature (Conway, 1947), and estimated by titration or other means. Details of methods are given in Barnes (1959, pp. 148–150). Not all the organic nitrogen is converted to ammonia, but the substances that resist conversion are not usually present in large quantities in living tissues. Nevertheless, it is advisable to report figures so obtained as 'Kjeldahl nitrogen' rather than 'total nitrogen'. If only the nitrogen contained in the protein of the tissue is required, the sample should be treated first with trichloroacetic acid to precipitate the protein while allowing most of the non-protein nitrogenous compounds to dissolve. The precipitate is then separated and used for Kjeldahl analysis. The protein: nitrogen ratio in most animal tissues is close to 6·25; hence the protein content of the sample can be approximated by multiplying the protein Kjeldahl nitrogen value by 6·25. Giese (1967) should be consulted for a critical review of methods.

Measurement of caloric content

The caloric content of a substance is defined as the energy released during complete combustion of 1 g of the dry material. The caloric content of a dry tissue homogenate made from the whole animal is an important variable in energy flow studies which enables the biomass, expressed as dry weight, to be converted into biomass expressed as energy.

Lipids have a high energy content whereas proteins and carbohydrates have only about half that of lipids (Table 9.11). Consequently, organisms with considerable reserves in the form of fat have a high average caloric content; so also do those stages in the life history which store fat, such as seeds, eggs and over-wintering adults. Seasonal changes in caloric content must be allowed for when converting dry weight into energy.

Few animals have an average caloric content outside the range 4·5–7·5 kcal g^{-1}—the majority lie between 5 and 6 kcal g^{-1} (Slobodkin & Richman, 1961).

Table 9.11. Average caloric content of biochemical components. (After Brody, 1945.)

Component	Caloric content kcal/g	J/g
Carbohydrate	4·1	$17{\cdot}16 \times 10^3$
Protein	5·65	$23{\cdot}65 \times 10^3$
Fat	9·45	$39{\cdot}55 \times 10^3$

Examples of caloric contents of a number of organisms are given in Table 9.12. These values are all on a dry weight basis.

In fisheries work, some authorities have assumed a roughly constant value for caloric content. Winberg (1956) suggests a value of 1 kcal g^{-1} wet weight (including the skeleton) for fish, a figure supported by averaging the caloric contents listed by Spector (1956) for a variety of fish flesh. However, they range from 2·2 kcal g^{-1} for salmon to 0·7 kcal g^{-1} for flounder, indicating how far the estimate may be in error for a particular case. Calculation of the above on a dry weight basis gives rather more uniform values averaging 5 kcal g^{-1} dry weight in agreement with the range stated above. Nevertheless, it is clearly advisable to measure the caloric content for any species whose productivity is being studied. A useful compendium of calorific contents has been prepared by Cummins (1967).

Table 9.12. Caloric contents of various organisms in kcal/g ash-free dry weight (=cal/mg).

Material	Caloric content kcal/g	Reference
Millipore filter membrane	3·1	Comita & Schindler (1963)
Ensis minor (bivalve)	3·5	Slobodkin (1962)
Modiolus demissus (bivalve)	4·5	Kuenzler (1961)
Sthenelais articulata (polychaete)	4·7	Slobodkin (1962)
Pandorina sp. (green alga)	4·9	Comita & Schindler (1963)
Microcystis sp. (blue-green alga)	4·8	Comita & Schindler (1963)
Eupagurus bernhardus zoea (hermit crab)	5·3	Pandian & Schumann (1967)
Artemia salina nauplius (brine shrimp)	5·9	Paffenhofer (1967)
Crangon vulgaris egg, undeveloped (shrimp)	6·4	Pandian (1967)
Calanus finmarchicus and *C. hyperboreus* (copepods)	7·4	Slobodkin (1962)
Egg yolk, birds (various)	8·0	Slobodkin (1962)

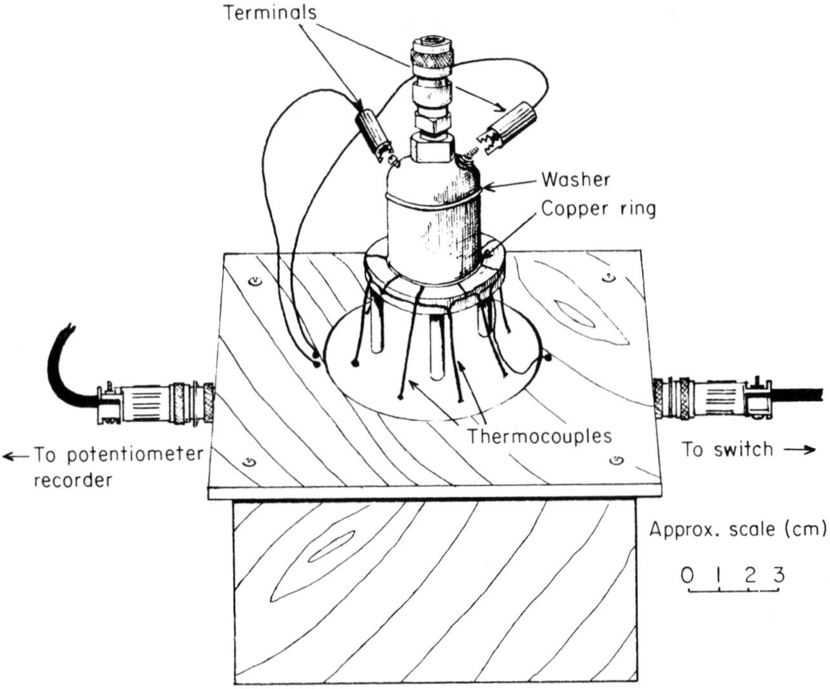

Fig. 9.13. Micro-bomb calorimeter.

Direct measurement by calorimetry

Heats of combustion are measured in bomb calorimeters, usually made of stainless steel and filled with oxygen under pressure. A bomb calorimeter of the usual size employed for biological work will burn ≤ 1 g of material; the Parr Instrument Company manufacture such an instrument for energy flow studies and its operation is described in the Parr Manual (Parr Instrument Company, 1960). For some biological work a sufficient quantity of dry tissue is not easily available and micro-bomb calorimeters must be used instead. Their principle and operation is the same. The micro-bomb calorimeter designed by Phillipson (1964) has a capacity of 8 ml and takes samples of 5–10 mg (Fig. 9.13). The stainless steel bomb rests in good thermal contact with a copper ring provided with 8 electrically insulated thermocouple junctions. The bomb stands on an aluminium base which acts as the cold junction. The temperature changes resulting from each firing are registered on a recording potentiometer (Fig. 9.14a).

Material for combustion must be thoroughly dried and ground up to form a homogeneous powder which can be compacted into a small pellet by means of a press available for this purpose. The pellet is weighed and placed on the

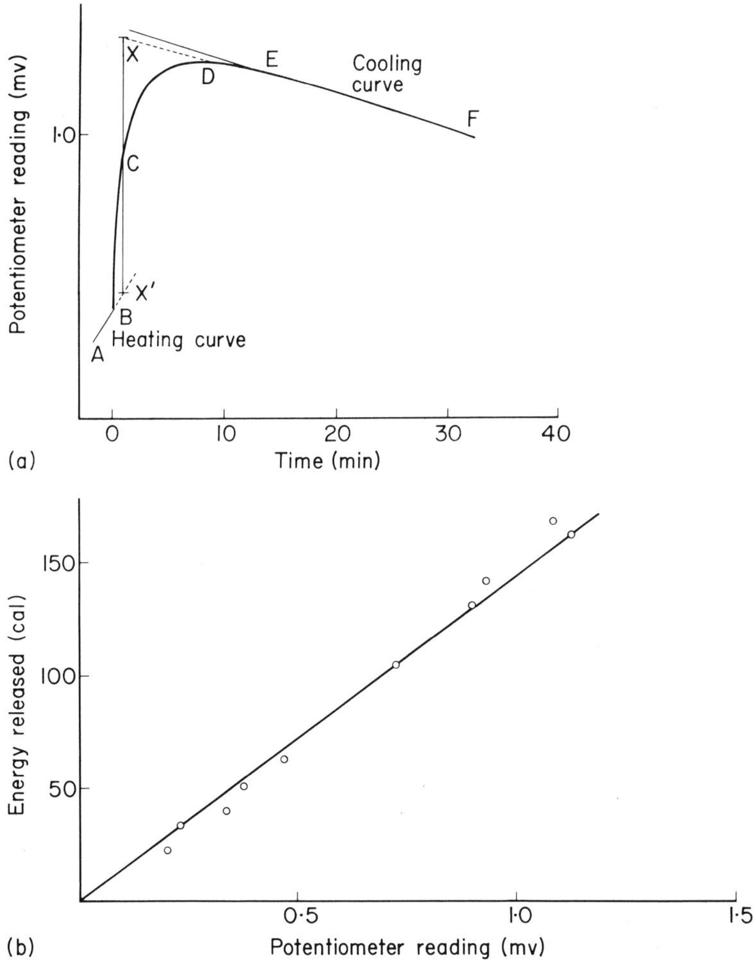

Fig. 9.14. (a) Potentiometer recording obtained from micro-bomb calorimeter. (b) Calibration using benzoic acid.

firing device, held by the fine platinum wire through which an electric charge is later passed to ignite the pellet. The platinum wire needs frequent replacement as it is liable to fuse during combustion. A small quantity of water is placed in the bomb which is flushed with oxygen from a cylinder. The bomb is then filled with oxygen under a pressure of 25–30 atm; this operation raises its temperature and it must therefore be cooled before firing.

It is most important that the specimen should burn completely and smoothly. If it burns too fast or too slowly some improvement can be achieved by compounding a known amount of benzoic acid of standard thermochemical quality into the pellet. If sputtering tends to occur, a firing chamber with a light lid may become necessary. Liquids such as oils can be burnt if contained in small inflammable capsules and small organisms can be dried and burned on Millipore filters. Obviously the capsules and filters must be calibrated beforehand for their caloric content. If the specimen is not fired by the discharge, the current passed through the platinum wire should be increased.

A successful firing results in a sudden change in the potential of the thermocouples which can be read with a sensitive potentiometer or, better, a continuous record can be made of the thermocouple potential (Fig. 9.14a). The latter allows a cooling curve correction to be applied. The potentiometer reading is calibrated by burning weighed pellets of benzoic acid, the caloric content of which is known to be of the order of 6320 kcal g^{-1}, the actual value for each batch being provided by the manufacturer (Fig. 9.14b). Addition of benzoic acid to the pellet or the calorific value of any Millipore filter or capsule must obviously be allowed for in the calculation. An example of results is given in Table 9.13.

Table 9.13. Estimation of caloric content by direct calorimetry.
A. Calibration of micro-bomb calorimeter.

I Weight of benzoic acid pellet (mg)	II Millivolts recorded on potentiometer after correcting for cooling	III Calories liberated (1 mg ≡ 6·319 cal) (cal)
3·52	0·202	22·2
5·34	0·233	33·7
6·29	0·337	39·7
8·10	0·379	51·2
9·96	0·469	62·9
16·45	0·724	104·0
20·64	0·896	130·4
22·34	0·930	141·1
25·62	1·118	161·9
26·50	1·080	167·5

Columns II and III are plotted in Fig. 9.14b, giving the relation 1 mv ≡ 144 cal.

B. Calculation of caloric content of molluscan eggs.

Dry weight of eggs 23·18 mg
Ash weight (by separate experiment) = 11·0 % of dry weight
Ash-free dry weight = 20·64 mg
Millivolts recorded = 0·88 mv (value of xx′ in Fig. 9.14a)

$$\text{Calories released per ash-free gram} = \frac{0{\cdot}88 \times 144}{20{\cdot}64} \times 1000 = 6130 \text{ cals/g}$$

Golley (1961) described small corrections for the heat generated by the acid produced and by the fuse wire, but according to Paine (1964), these corrections amount to $<1\%$ of the heat measured and are negligible. When tissues or materials containing a considerable quantity of chalk or hydrated silica are burned, carbon dioxide and water may be lost. The dissociation is endothermic hence the caloric content of such material may be seriously under-estimated.

It is possible to attempt to weigh the ash after combustion but the vigour of the reaction often disperses the ash around the walls of the bomb and leads to incorrect measurements. It is best, therefore, to ignite another sample of the material separately by the method described on p. 352 to obtain the ash weight. If the inorganic material constituting the ash loses water or carbon dioxide on ignition and this is not allowed for, the apparent dry ash-free weight of the organic material will be overestimated and the caloric content thereby further under-estimated. In some nudibranchs, errors from this cause and associated endothermic reactions can be as high as 50–60 % (Paine, 1966). Consequently, when the tissue or deposit contains large quantities of inorganic material, the caloric content can be measured more reliably by the chemical methods below.

Chemical methods

The energy content of marine sediments and of faeces deposited by sediment feeders is likely to be of particular importance in benthos work. Dry sediments do not burn well and if large amounts of benzoic acid have to be added to ensure complete combustion the calorific values obtained will be based on a small difference between the heat evolved by the benzoic acid and sediment combined and that calculated for the benzoic acid alone. The error will therefore be considerable in comparison with the quantity that is being measured. Moreover, as described above, the presence of an excess of inorganic matter may lead to endothermic changes during burning which invalidate measurement of the heat of combustion.

An alternative to calorimetry is the use of an indirect chemical method. Two approaches are possible, the first to measure the total oxidizable matter in the sample and apply an appropriate oxy-calorific coefficient to convert oxygen demand into calories. Secondly, it is possible to analyse the biochemical components and multiply each by an appropriate caloric content.

Measurement of caloric content from total oxidizable organic matter

This method consists of treating the dried material with an oxidizing agent, usually wet oxidation by dichromate, until the reaction is as complete as possible. The amount of oxidizing agent remaining is then determined by back titration or spectrophotometry. There are three weaknesses in this method.

1. Not all the organic matter, especially the protein, is oxidized. Twenty to thirty per cent of the protein present may be able to resist oxidation by dichromate (and iodate) causing an under-estimate of biologically oxidizable material. Nearly all the methods based on the original Walkley and Black dichromate wet oxidation procedure therefore embody an arbitrary factor ranging from 1·3 to 1·8 to allow for complete oxidation. Since protein is the main source of error, it is more accurate to apply a correction based on the amount of protein nitrogen, which can be measured separately, rather than to apply a universal correction (see p. 354). The assumption that 80% of this fraction is oxidized generally leads to results in good agreement with those obtained by direct calorimetry.
2. Chlorides, ferrous salts and finely divided carbon in deposits (Southward, 1952) can act as reducing agents sometimes causing a serious overestimate of the oxidizable matter. The chloride error can be eliminated by washing the material with sodium sulphate solution or by the addition of silver sulphate. Sulphide and ferrous ions can be eliminated by aerating in the presence of dilute hydrochloric acid. Nothing can be done to prevent the oxidation of fine carbon deposits since these are, in fact, particles of oxidizable carbon. They will not be digested by the animal and should therefore appear in its faeces. Consequently, if this method is used to measure the absorption of energy by the animal the effect of oxidizable carbon in both food and faeces will tend to cancel out.

   An interesting alternative approach to obtain the quality of organic matter likely to be utilized by deposit feeders has been suggested by George (1964). The percentage of organic matter in the deposit digested by enzymes is measured in place of the organic matter oxidized chemically.
3. The wet oxidation procedure estimates oxygen demand and must be converted to caloric content by an arbitrary factor. As shown on in Table 9.9 (p. 346), the energy of combustion per unit of oxygen consumed (oxy-calorific coefficient) is very similar for all the usual nutrients utilized by higher organisms. Hence the error caused by ignorance of the constitution of the organic constituents is not likely to be very serious.

Various forms of the dichromate method differing in detail have been adapted for marine work (Strickland & Parsons, 1965; Russell-Hunter *et al.*, 1968). Hughes (1969) has used an iodate oxidation method which avoids most of the errors mentioned in (1) and (2) above and has shown that the results are consistent with those obtained by direct calorimetry. The method of

Table 9.14. Computation of caloric content by wet oxidation.

A sample weighing 226 mg which by a previous ashing was known to contain 9·7 % ash, was oxidized by potassium iodate in sulphuric acid. Free iodine produced by reduction was removed by boiling and bubbling air through the digest. The remaining iodate was estimated by addition of KI and titration of liberated iodine with 0·1 N thiosulphate.

The titre difference between a blank and the sample enables the iodate reduced to be estimated:

3·567 mg iodate ≡ 1 ml of 0·1 N thiosulphate ≡ 0·6667 mg oxygen.

The observed difference in thiosulphate titre between experimental and blank for 1/20 aliquot was 20·25 ml.

Oxygen uptake by sample = 20 × 20·25 × 0·6667 mg = 270 mg.

Heat liberated by oxidation, using oxycalorific equivalent of 3·34 cal/mg oxygen = 270 × 3·34 = 902 calories.

We shall now assume that potassium iodate oxidizes only 80 % of protein present (Hughes, 1969).

Protein content (estimated as described in Table 9.15) = 45 % by weight (total wt. including ash).

Protein not oxidized = 20 % of 45 % = 9 % by wt. = 226 × 0·09 mg = 20·34 mg.

Caloric content of protein = 5·65 cal/mg.

Caloric content of unoxidized part of sample 20·34 × 5·65 = 115 cal.

Therefore, the total heat which would be liberated by complete oxidation would be 902 + 115 = 1017 cal.

Dry ash-free weight of sample = 226 × 91·3/100 mg = 0·205 g.

$$\text{Therefore caloric content per ash-free g} = \frac{1017}{0{\cdot}205} = 4961 \text{ cal/g.}$$

---

calculation of caloric content from a suitable wet oxidation method is given in Table 9.14.

Measurement of caloric content by analysis of biochemical components

It would be excessively time-consuming to analyse an organism or organic deposit into major biochemical constituents solely for the purpose of estimating caloric content. However, if the protein, carbohydrate and lipid fractions are required for some other purpose, e.g. in describing seasonal metabolic changes, the caloric content can be estimated from these results using the average values quoted by Brody (1945) and given in Table 9.11.

The estimate of the lipid present requires greater accuracy than that of the other two components because of its much higher energy content. The standard method in the past has been ether extraction by the Soxhlet procedure, but the lipid can be removed from ground homogenized dry tissue more completely by methanol-chloroform extraction in the cold. Some

Table 9.15. Examples of calculation of caloric content from biochemical analysis.

A. 10·25 g dry tissue on extraction and evaporation gave 0·82 g lipid.

$$\text{Therefore lipid content} = \frac{0{\cdot}82}{10{\cdot}25} = 8{\cdot}0\,\%.$$

1·07 g of dry tissue treated with T.C.A. and precipitate digested by Kjeldahl method, ammonia evolved titrated with 0·2 N HCl; titre = 26·1 ml.

1. Equivalent HCl ≡ 14 g protein N.

$26{\cdot}1$ ml of $0{\cdot}2$ N HCl $\equiv 14 \times 0{\cdot}2 \times 26{\cdot}2 \times 10^{-3}$ N

$= 7{\cdot}3 \times 10^{-2}$ g protein N.

This corresponds to $6{\cdot}25 \times 7{\cdot}3 \times 10^{-2}$ g protein $= 0{\cdot}456$ g protein.

Protein content = 45 %.

'Carbohydrate content' by difference = 47 %.

Caloric content $= 0{\cdot}08 \times 9{\cdot}45 + 0{\cdot}45 \times 56{\cdot}5 + 0{\cdot}47 \times 4{\cdot}1$

$= 0{\cdot}76 + 2{\cdot}54 + 1{\cdot}93 = 5{\cdot}23$ cal g^{-1} mg.

B. 10·25 g tissue extracted as above, but lipid estimated by dichromate oxidation. 20 ml of a solution of 5·15 g l^{-1} potassium dichromate added; back titration showed 3·2 ml used up in oxidizing lipid. 294 g of potassium dichromate yield 48 g oxygen.

$$\text{Therefore, oxygen equivalent of lipid} = 3{\cdot}2 \times \frac{5{\cdot}15}{1000} \times \frac{48}{294} \times 1000\,\text{mg} = 2{\cdot}69\,\text{mg}.$$

$$\text{Caloric content of lipid fraction} = 2{\cdot}69 \times \frac{3{\cdot}34}{10{\cdot}25} = 0{\cdot}82\,\text{cal}\,g^{-1}.$$

Per cent by wt. of lipid = 0·82 cal g^{-1} ÷ caloric content assumed = 0·82/9·45 – 8·7 %.

Protein fraction = 45 % by wt. (see Example 1).

'Carbohydrate' fraction = 46·3 % by wt. (by difference).

Caloric content of tissue $= 0{\cdot}82 + 2{\cdot}54 + 0{\cdot}463 \times 4{\cdot}1 = 5{\cdot}20$ cal g^{-1}.

non-lipid material may also be extracted by such polar solvents. Ansell & Trevallion (1967) followed Bligh & Dyer's (1959) wet extraction procedure which is probably the most reliable method for molluscan tissues. The lipid extraction and evaporation should strictly be carried out under nitrogen to prevent oxidation, with a trace of anti-oxidant added as a further precaution (e.g. 2.6 di tert.-butyl 4-methyl phenol). The lipid extracted can be measured gravimetrically, or estimated by direct oxidation with dichromate. In the former case (Table 9.15A) the result must be multiplied by the assumed caloric content of lipid (9·45 cal g^{-1}, Table 9.11). In the latter case the caloric content of the lipid is obtained by multiplying the oxygen equivalent determined by dichromate oxidation by the oxy-calorific coefficient (3·34 cal mg^{-1} equivalent of oxygen absorbed) (Table 9.15B).

The estimation of protein and non-protein nitrogen is dealt with on p. 354.

Direct estimation of carbohydrate is difficult (see Giese, 1967) and, in any case, the sum of the weights of all three components, fat, protein, and carbohydrate, when separately determined, is generally considerably less than the original dry ash-free weight. Therefore, in the estimation of caloric content, it is best to measure lipid, estimate protein as $6{\cdot}25 \times$ protein nitrogen, and assume that carbohydrate makes up the difference as shown in the calculations illustrated in Table 9.15. The approximation is justified because the caloric content of most organic materials including carbohydrates is similar, with the exception of fat (Table 9.11).

Measurement of oxygen uptake in the laboratory

Oxygen uptake is measured either as ml oxygen at n.t.p. or mg oxygen absorbed in unit time. The hour is the convenient unit of time for respiration experiments. Since volume measurements are made by the majority of respirometers, ml of oxygen at n.t.p. is the more widely employed unit. Volumes must of course be corrected to n.t.p. by taking the temperature, θ (°C) and barometric pressure, p (mm mercury), at the time of the experiment as well as the volume absorbed, V (ml):

$$\text{Volume at n.t.p.} = V \cdot \frac{273}{273 + \theta} \cdot \frac{p}{760} \text{ ml}$$

Results obtained by chemical methods are more usually given as mg h^{-1} oxygen. It is useful to remember the conversion factor:

$$0{\cdot}7 \text{ ml oxygen at n.t.p.} = 1 \text{ mg oxygen by weight}$$

Numerous satisfactory methods exist for the measurement of oxygen uptake, the selection of method being mainly determined by the medium in which the animal is kept during measurement. Otherwise the choice is mainly a matter of convenience, scale, availability, and suitability of the method to the animal being studied.

For measurement with animals in water

If the animal is placed in water in a sealed vessel and the oxygen concentration is measured initially and after a lapse of time, the amount taken up by the animal can be measured from the change in concentration. The concentration drop should not be excessive; for animals living in a normal aerobic environment the experimental oxygen tension should not be allowed to fall below 0·6 of the saturation level in air at the temperature of the experiment. If it should do so, respiration rate may be reduced or the animal's behaviour and consequent oxygen demand upset. On the other hand the concentration change should not be too small—it must be large enough to be measured with reasonable accuracy. The medium should therefore be fully saturated, or better, slightly supersaturated with oxygen at the start of the experiment and fall ideally to about 80 % of saturation. In order to allow for the production or uptake of oxygen by micro-organisms in the water or on the surface of the vessel, the experiment should include a control vessel of exactly similar form and origin as the experimental vessel but without the animal. If the volume of the experimental vessel is V, and the concentration of oxygen in the water x_1 units ml^{-1} initially and x_2 units ml^{-1} finally, the uptake, q, is given by:

$$q = V(x_1 - x_2)$$

Alternatively, a known constant flow of water can be maintained over the animal with a bypass of approximately equal flow through an empty vessel identical with that containing the animal, so that the differences in oxygen levels between the inflow and outflow of each vessel will give the respiration attributable to the animal alone. If x_0 is the oxygen concentration at the inflow, x_1 at the exit of the experimental vessel, and x_2 at the exit of the control vessel, the flows being $(dV/dt)_1$ and $(dV/dt)_2$ respectively, then the oxygen uptake is given by the equation:

$$q = (x_0 - x_1)(dV/dt)_1 - (x_0 - x_2)(dV/dt)_2$$

The continuous flow method has the advantage of maintaining a fairly steady oxygen tension around the animal and more closely resembles natural conditions. Furthermore, oxygen tension can be continuously recorded by an appropriate piece of apparatus such as the galvanic oxygen electrode or a dropping mercury electrode, and if the amount of dead space within the apparatus is small, recorded changes in oxygen uptake can be related to observed changes in the animal's activity. Both the sealed vessel and the continuous flow method involve the measurement of dissolved oxygen with an accuracy of the order of $\pm 1\%$ of saturation by air.

The Winkler titration is a very reliable, though tedious, method. It is the method of choice for calibrating less reliable instruments. The oxygen present in a known volume of water is absorbed by a precipitate of manganous hydroxide in a special Winkler reagent bottle which has a bevelled ground glass stopper to prevent the inclusion of air bubbles when the sample is sealed off out of contact with air. When oxygen absorption is complete, acidification in the presence of iodide liberates two iodine atoms for every oxygen molecule originally present. The iodine is then titrated with thiosulphate. Details of methods recommended for sea water can be found in Jacobsen *et al.* (1950), Barnes (1959) and Strickland & Parsons (1965).

The presence of significant amounts of oxidizing or reducing substances may invalidate the Winkler method, but in unpolluted areas and in the absence of concentrations of phytoplankton, the Winkler method determines the oxygen content of sea water satisfactorily. Modifications of the method for use in polluted water are described in Barnes (1959).

For subsampling purposes or when a large volume of water is not available, the micro-Winkler method is useful, the sample being collected out of contact with air by means of a syringe in which the reaction is carried out (Fox & Wingfield, 1938; Whitney, 1938).

When water is being transferred from an experimental vessel containing the animal to the Winkler reagent bottle, gaseous exchange with air must obviously be minimized. A clean glass siphon of narrow bore should be used to carry the water from the experimental vessel to the bottom of the Winkler bottle which should be allowed to overflow by 2 or 3 times its volume before being sealed. Siphons of plastic tubing are unsuitable, particularly if their bore is large, as they are not always fully wetted by water and may allow air bubbles to remain in the tube while the water is flowing through.

The Scholander micro-gasometric technique (Scholander *et al.*, 1955) is accurate, reliable, versatile and insensitive to pollutants in the water. The aqueous medium is drawn into an air-tight syringe and all gases dissolved in it are driven off by liberation of carbon dioxide bubbles. The liquid is then ejected, carbon dioxide absorbed in caustic alkali and the remaining bubble of oxygen and nitrogen transferred to a micro-gas analyser where the quantity of oxygen is determined by absorption in pyrogallol. The instrument blank needs to be measured using boiled gas-free water. The accuracy of oxygen determination is within 1–2 % of the oxygen present. The method is suitable for field or laboratory work and the gas analyser can be used to measure oxygen uptake by animals in air.

Polarographic methods for determining oxygen concentration are the most rapid and convenient, and allow continuous recording of oxygen tension, so that the course of oxygen uptake by an animal can be followed. The most reliable instrument of this class is the dropping mercury electrode; since the surface of the electrode is continually being renewed, it is not subject to drift and does not require frequent recalibration. However, the dropping mercury electrode is inconvenient; the diameter of the dropper and the temperature must be kept constant, the mercury must be cleaned from time to time, is toxic and is expensive, and the instrument is not suitable for use in the presence of animals. The instrument is, however, quite suitable for monitoring the oxygen level in the continuous flow type of apparatus.

More popular, and extremely convenient, are the various designs of galvanic oxygen electrodes in which the cathode consists of a noble metal separated from the liquid whose oxygen tension is to be measured by an oxygen-diffusible membrane. The cathode, in contact with a very small amount of electrolyte, is held at a potential sufficient to reduce oxygen to water and hydrogen peroxide but not sufficiently low to cause the electrolysis of the water itself. If there is no oxygen outside the membrane, the small amount of oxygen in the electrolyte is quickly reduced and the current ceases to flow. If oxygen is present in the aqueous medium outside the membrane, it will diffuse through the membrane at a rate proportional to its concentration. Each electrochemical equivalent of oxygen passing through the membrane will allow one unit of current to pass through the electrode.

Many forms of such electrodes exist and suitable electrodes for physiological purposes are marketed by Beckman and other manufacturers. Some electrodes are easily poisoned and all such instruments are very liable to drift and therefore require regular calibration of the current in oxygen-free and in air-saturated water. They must be used in a thermostat bath at constant temperature (Beckman Instrument Incorp., 1964).

Galvanic electrodes themselves consume oxygen, hence with very small organisms or those with low metabolism, electrodes with low consumption are essential and their consumption should be allowed for. Davenport (1976) describes a suitable system for small organisms.

For very small organisms and pieces of tissue the Cartesian diver respirometer can be recommended. The tissue or organism, with a small quantity of air, is enclosed in a bulb with a short neck which floats in a pressure controlled chamber. As oxygen is absorbed, the volume of gas in the diver decreases thus reducing its buoyancy. To measure the amount absorbed, the outside pressure must be reduced sufficiently to expand the bubble and return the diver to its previous position. The amount of reduction in pressure is recorded. The method is extremely sensitive and the gas volumes used excessively small: Zeuthen (1943) quotes $0{\cdot}1\ \mu$ litres.

A discussion of various forms of such apparatus, principles and computation is given in Glick (1961).

Measurement of oxygen uptake in air

Marine animals, such as mussels, barnacles, limpets and periwinkles, living intertidally, spend a significant period in air and can breathe air in a moist atmosphere. The standard constant temperature Barcroft and Warburg methods, in which the volume or pressure change is measured as oxygen is absorbed, are still in use, but calibration and computation are difficult and complicated (Dixon, 1951). Much more convenient equipment is now available incorporating a micrometer which inserts a stainless steel or plastic rod (Gilson, 1963) directly into the gas space, so maintaining constant pressure. The volume change can thus be read directly.

Suitable micrometer devices in compensating chambers are marketed by Mark Company, Randolph, Mass., U.S.A. and multi-chamber Gilson type respirometers are marketed by W.G. Flaig and Sons, Broadstairs, Kent. A conveniently assembled plastic apparatus for average sized marine organisms which can be used with this type of respirometer is described by Davies (1966). The unique magnetic susceptibility of oxygen has led to the development of instruments with high sensitivity but these are not yet widely applied in biology.

References

Allen, K.R., 1950. The computation of production in fish populations. *New Zealand Science Review*, **8**, 89.

Ansell, K.R., & Trevallion, A., 1967. Studies on *Tellina tenuis* da Costa. 1. Seasonal growth and biochemical cycle. *Journal of Experimental Marine Biology and Ecology* **1**, 220–235.

Bagenal, T., (ed.) 1978. *Methods for Assessment of Fish Production in Fresh Waters.* International Biological Programme Handbook No. **3** (3rd edition). Blackwell Scientific Publications, Oxford, 365 pp.

Barnes, H., 1959. *Apparatus and Methods of Oceanography. Part One: Chemical.* Allen & Unwin, London, 341 pp.

Bayne, B.L. (ed.) 1976. *Marine Mussels: their ecology and physiology.* International Biological Programme, **10**. Cambridge University Press, London, xvii + 506 pp.

Beckman Instruments Incorp., 1964. *Instruction manuals PG IM 2, PG.TB. 003, for oxygen macro electrode* (32 pp.) and *Physiological gas analyser* (31 pp.). Spinco Division, Palo Alto, California.

Von Bertalanffy, L., 1934. Untersuchungen über die Gesetzlichkeit des Wachstums. 1. Teil. *Archiv für Entwicklungs—Mechanik der Organismen* **131**, 613–652.

Beverton, R.J.H. & Holt, S.J., 1957. On the dynamics of exploited fish populations. *Fishery Investigations, London*, (**2**) **19**, 533 pp.

Bligh, E.G. & Dyer, W.J., 1959. A rapid method of total lipid extraction and purification. *Canadian Journal of Biochemistry and Physiology*, **37**, 911–917.

Borowitz, S. & Beiser, A., 1966. *Essentials of Physics.* Addison Wesley, London, 708 pp.

Bourget, E. & Crisp, D.J., 1975. An analysis of the growth bands and ridges of barnacle shell plates. *Journal of the Marine Biological Association of the United Kingdom*, **55**, 439–461.

Brody, S., 1945, *Bioenergetics and Growth.* Reinhold, New York, 1023 pp.

Carefoot, T.H., 1967a. Growth and nutrition of *Aplysia punctata* feeding on a variety of marine algae. *Journal of the Marine Biological Association of the United Kingdom*, **47**, 565–589.

Carefoot, T.H., 1967b. Growth and nutrition of three species of opisthobranch molluscs. *Comparative Biochemistry and Physiology*, **21**, 627–652.

Comita, W.G. & Schindler, D.W., 1963. Calorific values of microcrustacea. *Science, New York*, **140**, 1394–1395.

Conover, R.J., 1966. Assimilation of organic matter by zooplankton. *Limnology and Oceanography*, **11**, 338–345.

Conway, E.J., 1947. *Microdiffusion Analysis and Volumetric Error*, 2nd ed. Crosby Lockwood & Son Ltd., London, 357 pp.

Crisp, D.J., 1971. Energy flow measurements. Chapter 12, pp. 197–279 in N.A. Holme & A.D. McIntyre (eds) *Methods for the Study of Marine Benthos*, International Biological Programme Handbook No. 16, 1st ed. Blackwell Scientific Publications, Oxford, 334 pp.

Crisp, D.J. & Davies, P.A., 1955. Observations *in vivo* on the breeding of *Elminius modestus* grown on glass slides. *Journal of the Marine Biological Association of the United Kingdom*, **34**, 357–380.

Crisp, M., Davenport, J. & Shumway, S. E., 1978. Effects of feeding and of chemical stimulation on the oxygen uptake of *Nassarius reticulatus* (Gastropoda; Prosobranchia). *Journal of the Marine Biological Association of the United Kingdom*, **58**, 387–399.

Cummins, K.W., 1967. *Calorific Equivalents for Studies in Ecological Energetics*, 2nd ed. Pymatuning Lab. of Ecology, University of Pittsburgh, Penn., Oct. 1967, 52 pp.

Davenport, J., 1976. A technique for the measurement of oxygen consumption in small aquatic organisms. *Laboratory Practice*, **25** (10), 693–695.

Davies, P.S., 1966. A constant pressure respirometer for medium-sized animals. *Oikos*, **17**, 108–112.

Davis, C.C., 1963. On questions of production and productivity in ecology. *Archiv für Hydrogiologie*, **59**, 145–161.

Dixon, M., 1951. *Manometric Methods as Applied to the Measurement of Cell Respiration and Other Processes*, 3rd ed. Cambridge University Press, 165 pp.

Ekaratne, S.U.K. & Crisp, D.J., 1982. Tidal microgrowth bands in intertidal gastropod shells, with an evaluation of band dating techniques. *Proceedings of the Royal Society of London, B*, **214**, 305–323.

Ekaratne, S.U.K. & Crisp, D.J., 1983. A geometric analysis of growth in gastropod shells, with particular reference to turbinate forms. *Journal of the Marine Biological Association of the United Kingdom*, **63**, 777–797.

Elliott, J.M. & Davidson, W., 1975. Energy equivalents of oxygen consumption in animal energetics. *Oecologia, Berlin*, **19**, 195–201.

Famme, P., Knudsen, J. & Hansen, E.S., 1981. The effect of oxygen on the aerobic–anaerobic metabolism of the marine bivalve *Mytilus edulis* L. *Marine Biology Letters*, **2(6)**, 334–335.

Fox, H.M. & Wingfield, C.A., 1938. A portable apparatus for the determination of oxygen dissolved in a small volume of water. *Journal of Experimental Biology*, **15**, 437–445.

George, J.D., 1964. Organic matter available to the polychaete *Cirriformia tentaculata* (Montagu) living in an intertidal mudflat. *Limnology and Oceanography*, **9**, 453–455.

Gerking, S.D., 1955. Influence of rate of feeding on body composition and protein metabolism of Bluegill Sunfish. *Physiological Ecology*, **28**, 267–282.

Giese, A.C., 1967. Some methods for study of the biochemical constitution of marine invertebrates. *Oceanography and Marine Biology. Annual Review*, **5**, 159–186.

Gilson, W.E., 1963. Differential respirometer of simplified and improved design. *Science, New York*, **141**, 531–532.

Glick, D. 1961. *Quantitative Chemical Techniques in Histo and Cytochemistry. I.* John Wiley & Sons, New York, 470 pp.

Gnaiger, E., 1981. Direct calorimetry in ecological energetics. Long term monitoring of aquatic animals. pp. 155–165 in E. Marty, H.R. Oswald & H.G. Wiedemann (eds) *Angewandte Chemische Thermodynamik und Thermoanalytic*. Experientia Supplementum 37, Birkhäuser Verlag Basel, Boston, Stuttgart.

Golley, F.B., 1961. Energy values of ecological materials. *Ecology*, **42**, 581–583.

Hammen, C.S., 1979. Metabolic rates of marine bivalve molluscs determined by calorimetry. *Comparative Biochemistry and Physiology*, **62A**, 955–959.

Hammen, C.S., 1980. Total energy metabolism of marine bivalve mollusks in anaerobic and aerobic states. *Comparative Biochemistry and Physiology*, **67A**, 617–621.

Harding, J.P., 1949. The use of probability paper for the graphical analysis of polymodal frequency distributions. *Journal of the Marine Biological Association of the United Kingdom*, **28**, 141–153.

Heywood, J. & Edwards, R.W., 1962. Some aspects of the ecology of *Potamopyrgus jenkinsi* Smith. *Journal of Animal Ecology*, **31**, 239–250.

Hildredth, D.I. & Crisp, D.J., 1976. A corrected formula for calculation of filtration rate of bivalve molluscs in an experimental flowing system. *Journal of the Marine Biological Association of the United Kingdom*, **56**, 111–120.

Hughes, R.N., 1969. Appraisal of the iodate-sulphuric-acid, wet-oxidation procedure for the estimation of the caloric content of marine sediments. *Journal of the Fisheries Research Board of Canada*, **26**, 1959–1964.

Ivlev, V.S., 1934. Eine Mikromethode zur bestimming des Kaloriengehalts von Nahrstoffen. *Biochemische Zeitschrift*, **275**, pp. 49–55.

Jacobsen, J.P., Robinson, R.J. & Thompson, T.G., 1950. A review of the determination of dissolved oxygen in sea water by the Winkler method. *Publications Scientifiques. Association d'Océanographie Physique*, **11**, 22 pp.

Jørgensen, C.B., 1962. Efficiency of growth in *Mytilus edulis* and two gastropod veligers. *Nature, London*, **170**, 714.

Jørgensen, C.B., 1966. *Biology of Suspension Feeding*. Pergamon Press, London, 357 pp.

Kuenzler, E.J., 1961. Structure and energy flow of a mussel population in a Georgia salt marsh. *Limnology and Oceanography*, **6**, 191–204.

McCance, R.A. & Masters, M., 1937. The chemical composition and the acid base balance of *Archidoris britannica*. *Journal of the Marine Biological Association of the United Kingdom*, **22**, 273–279.

Mann, K.H., 1965. Energy transformations by a population of fish in the River Thames. *Journal of Animal Ecology*, **34**, 253–275.

Markham, R., 1942. A steam distillation apparatus suitable for micro-Kjeldahl analysis. *Biochemical Journal*, **36**, 790–791.

Marshall, S.M. & Orr, A.P., 1961. On the biology of *Calanus finmarchicus*. XII. The phosphorus cycle: excretion, egg production, autolysis. *Journal of the Marine Biological Association of the United Kingdom*, **41**, 463–488.

Mitson, R.B., 1963. Marine fish culture in Britain. V. An electronic device for counting the nauplii of *Artemia salina* L. *Journal du Conseil Permanent International pour l'Exploration de la Mer*, **28**, 262–269.

Newell, R.C. & Northcroft, H.R., 1967. A re-interpretation of the effect of temperature on the metabolism of certain marine invertebrates. *Journal of Zoology, London*, **151**, 277–298.

Odum, E.P., 1959. *Fundamentals of Ecology*, 2nd ed. W.B. Saunders, Philadelphia, 546 pp.

Odum, E.P. & Smalley, A.E., 1959. Comparison of population energy flow of an herbivorous and a deposit-feeding invertebrate in a saltmarsh ecosystem. *Proceedings of the National Academy of Sciences*, **45**, 617–622.

Odum, H.T., 1957a. Trophic structure and productivity of Silver Springs, Florida. *Ecological Monographs*, **27**, 55–112.

Odum, H.T., 1957b. Primary production measurements in eleven Florida springs and a marine turtle-grass community. *Limnology and Oceanography*, **2**, 85–97.

Paffenhöfer, G-A., 1967. Caloric content of the larvae of the brine shrimp *Artemia salina*. *Helgoländer wissenschaftliche Meeresuntersuchungen*, **16**, 130–135.

Paine, R.T., 1964. Ash and caloric determinations of sponge and opisthobranch tissues. *Ecology*, **45**, 384–387.

Paine, R.T., 1965. Natural history, limiting factors and energetics of the opisthobranch *Navanax inermis*. *Ecology*, **46**, 603–619.

Paine, R.T., 1966. Endothermy in bomb calorimetry. *Limnology and Oceanography*, **11**, 126–129.

Pamatmat, M.M., 1965. A continuous-flow apparatus for measuring metabolism of benthic communities. *Limnology and Oceanography*, **10**, 486–489.

Pamatmat, M.M., 1966. *The Ecology and Metabolism of a Benthic Community on an Intertidal Sandflat (False Bay, San Juan Island, Washington)* University of Washington, 250 pp.

Pamatmat, M.M., 1968. Ecology and metabolism of a benthic community on an intertidal sandflat. *Internationale Revue der Gesamten Hydrobiologie*, **53**, 211–298. (*Contribution, Department of Oceanography, University of Washington*, **No. 427**).

Pandian, T.J., 1967. Changes in chemical composition and caloric content of developing eggs of the shrimp *Crangon crangon*. *Helgoländer Wissenschaftliche Meeresuntersuchungen*, **16**, 216–224.

Pandian, T.J. & Schumann, K-H., 1967. Chemical composition and caloric content of egg and zoea of the hermit crab *Eupagurus bernhardus*. *Helgoländer Wissenschaftliche Meeresuntersuchungen*, **16**, 225–230.

Parr Instrument Company, 1960. *Oxygen Bomb Calorimetry and Oxygen Bomb Combustion Methods*. Manual No. 130, 430 Moline Illinois. Parr. Inst. Co., 56 pp.

Phillipson, J., 1964. A miniature bomb calorimeter for small biological samples. *Oikos*, **15**, 130–139.

Phillipson, J., 1966. *Ecological Energetics*. Edward Arnold, London, 57 pp.

Pitcher, T.J. & Hart, P.J.B., 1982. *Fisheries Ecology*. Croom Helm, London, 414 pp.

Richardson, C.A., Crisp, D.J. & Runham, N.W., 1979. Tidally deposited growth bands in the shell of the common cockle, *Cerastoderma edule* (L.). *Malacologia*, **18**, 277–290.

Richardson, C.A., Crisp, D.J., Runham, N.W. & Gruffydd, Ll.D., 1980. The use of tidal growth bands in the shell of *Cerastoderma edule* to measure seasonal growth rates under cool temperate and Arctic conditions. *Journal of the Marine Biological Association of the United Kingdom*, **60**, 977–989.

Ricker, W.E., 1958. Handbook of computations for biological statistics of fish populations. *Bulletin of the Fisheries Research Board of Canada*, **119**, 300 pp.

Ritz, D.A. & Crisp, D.J., 1970. Seasonal changes in feeding rate in *Balanus balanoides*. *Journal of the Marine Biological Association of the United Kingdom*, **50**, 223–240.

Russell-Hunter, W.D., Meadows, R.T., Apley, M.L. & Burky, A.J., 1968. On the use of a 'wet oxidation' method for estimates of total organic carbon in mollusc growth studies. *Proceedings of the Malacological Society of London*, **38**, 1–11.

Sanders, H.L., 1956. Oceanography of Long Island Sound, 1952–54. X. The biology of marine bottom communities. *Bulletin of the Bingham Oceanographic Collection*, **15**, 345–414.

Schick, J.M., 1981. Heat production and oxygen uptake in intertidal sea anemones from different shore heights during exposure to air. *Marine Biology Letters*, **2**, 225–236.

Scholander, P.F., Van Dam, L., Claff, C.L. & Kanwisher, J.W., 1955. Micro gasometric determination of dissolved oxygen and nitrogen. *Biological Bulletin, Marine Biological Laboratory, Woods Hole*, **109**, 328–334.

Slobodkin, L.B., 1962. Energy in animal ecology. *Advances in Ecological Research*, **1**, 69–101.

Slobodkin, L.B. & Richman, S., 1961. Calories/gm in species of animals. *Nature, London*, **191**, 299.

Southward, A.J., 1952. Organic matter in littoral deposits. *Nature, London*, **169**, 888.

Spector, W.S., 1956. *Handbook of Biological Data*. W.B. Saunders, Philadelphia, xxxvi + 584 pp.

Stamp, L.D., 1958. The land use of Britain. pp. 1–10 in W.B. Yapp & D.J. Watson (eds) *The Biological Productivity of Britain*. Symposia of the Institute of Biology, **7**, London. xii + 128 pp.

Strömgren, T., 1975. Linear measurements of growth of shells using laser diffraction. *Limnology and Oceanography*, **20**, 845–848.

Strickland, J.D.H. & Parsons, T.R., 1965. *A Manual of Sea Water Analysis*, 2nd ed. Bulletin of the Fisheries Research Board of Canada, **125**, 203 pp.

Teal, J.M., 1957. Community metabolism in a temperate cold spring. *Ecological Monographs*, **27**, 283–302.

Teal, J.M., 1962. Energy flow in the salt marsh ecosystem of Georgia. *Ecology*, **43**, 614–624.

Thorson, G., 1957. Bottom communities (sublittoral or shallow shelf). *Memoirs of the Geological Society of America*, **67**(**1**), 461–534.

Villmann, M.L.M., Skalak, R. & Villmann, O., 1981. Space-time presentations of the shell of the bivalved mollusk *Cardium edule*. *American Journal of Orthodontics*, **80**, 417–428.

Walford, L.A., 1946. A new graphic method of describing the growth of animals. *Biological Bulletin, Marine Biological Laboratory, Woods Hole*, **90**, 141–147.

Whitney, J.R., 1938. A syringe pipette method for the determination of oxygen in the field. *Journal of Experimental Biology*, **15**, 564–570.

Winberg, G.G., 1956. Rate of metabolism and food requirements of fishes. *Fisheries Research Board of Canada Translation series*, **194**, 202 pp. 1960. (Translated from the Russian.)

Zenkevich, L.A., 1930. A quantitative evaluation of the bottom fauna in the sea region about the Kanin peninsula. *Trudý Morskogo Nauchnogo Institute*, **4**(**3**), 5–23. (In Russian, with English summary.)

Zenkevich, L.A., 1963. *Biology of the Seas of the U.S.S.R.* (Translation by S. Botcharskaya.) George Allen & Unwin Ltd., London, 955 pp.

Zeuthen, E., 1943. A cartesian diver micro-respirometer with a gas volume of 0·1 μl. *Compte Rendu des Travaux du Laboratoire de Carlsberg, Ser. Chim.*, **24**, 479–518.

Index